# Fundamentals of Signal Processing

## for Sound and Vibration Engineers

# Fundamentals of Signal Processing

## for Sound and Vibration Engineers

**Kihong Shin**
*Andong National University*
*Republic of Korea*

**Joseph K. Hammond**
*University of Southampton*
*UK*

**WILEY**
John Wiley & Sons, Ltd

This publication is designed to provide accurate and authoritative information in regard to the subject matter
covered. It is sold on the understanding that the Publisher is not engaged in rendering professional services. If
professional advice or other expert assistance is required, the services of a competent professional should be sought.

*Other Wiley Editorial Offices*

John Wiley & Sons Inc., 111 River Street, Hoboken, NJ 07030, USA

Jossey-Bass, 989 Market Street, San Francisco, CA 94103-1741, USA

Wiley-VCH Verlag GmbH, Boschstr. 12, D-69469 Weinheim, Germany

John Wiley & Sons Australia Ltd, 42 McDougall Street, Milton, Queensland 4064, Australia

John Wiley & Sons (Asia) Pte Ltd, 2 Clementi Loop #02-01, Jin Xing Distripark, Singapore 129809

John Wiley & Sons Canada Ltd, 6045 Freemont Blvd, Mississauga, ONT, L5R 4J3

Wiley also publishes its books in a variety of electronic formats. Some content that appears in print may
not be available in electronic books.

*Library of Congress Cataloging-in-Publication Data*

Shin, Kihong.
  Fundamentals of signal processing for sound and vibration engineers / Kihong Shin and
Joseph Kenneth Hammond.
      p.   cm.
  Includes bibliographical references and index.
  ISBN 978-0-470-51188-6 (cloth)
 1.  Signal processing.  2. Acoustical engineering.  3. Vibration.  I. Hammond, Joseph Kenneth.
II. Title.
  TK5102.9.S5327 2007
  621.382′2—dc22                                                                    2007044557

*British Library Cataloguing in Publication Data*

A catalogue record for this book is available from the British Library

ISBN-13  978-0470-51188-6

Typeset in 10/12pt Times by Aptara, New Delhi, India.
This book is printed on acid-free paper responsibly manufactured from sustainable forestry in which at least two
trees are planted for each one used for paper production.

MATLAB® is a trademark of The MathWorks, Inc. and is used with permission. The MathWorks does not warrant
the accuracy of the text or exercises in this book. This book's use or discussion of MATLAB® software or related
products does not constitute endorsement or sponsorship by The MathWorks of a particular pedagogical approach
or particular use of the MATLAB® software.

# Contents

# Preface

This book has grown out of notes for a course that the second author has given for more years than he cares to remember – which, but for the first author who kept various versions, would never have come to this. Specifically, the Institute of Sound and Vibration Research (ISVR) at the University of Southampton has, for many years, run a Masters programme in Sound and Vibration, and more recently in Applied Digital Signal Processing. A course aimed at introducing students to signal processing has been one of the compulsory modules, and given the wide range of students' first degrees, the coverage needs to make few assumptions about prior knowledge – other than a familiarity with degree entry-level mathematics. In addition to the Masters programmes the ISVR runs undergraduate programmes in Acoustical Engineering, Acoustics with Music, and Audiology, each of which to varying levels includes signal processing modules. These taught elements underpin the wide-ranging research of the ISVR, exemplified by the four interlinked research groups in Dynamics, Fluid Dynamics and Acoustics, Human Sciences, and Signal Processing and Control. The large doctoral cohort in the research groups attend selected Masters modules and an acquaintance with signal processing is a 'required skill' (necessary evil?) in many a research project. Building on the introductory course there are a large number of specialist modules ranging from medical signal processing to sonar, and from adaptive and active control to Bayesian methods.

It was in one of the PhD cohorts that Kihong Shin and Joe Hammond made each other's acquaintance in 1994. Kihong Shin received his PhD from ISVR in 1996 and was then a postdoctoral research fellow with Professor Mike Brennan in the Dynamics Group, then joining the School of Mechanical Engineering, Andong National University, Korea, in 2002, where he is an associate professor. This marked the start of this book, when he began 'editing' Joe Hammond's notes appropriate to a postgraduate course he was lecturing – particularly appreciating the importance of including 'hands-on' exercises – using interactive MATLAB® examples. With encouragement from Professor Mike Brennan, Kihong Shin continued with this and it was not until 2004, when a manuscript landed on Joe Hammond's desk (some bits looking oddly familiar), that the second author even knew of the project – with some surprise and great pleasure.

In July 2006, with the kind support and consideration of Professor Mike Brennan, Kihong Shin managed to take a sabbatical which he spent at the ISVR where his subtle pressures – including attending Joe Hammond's very last course on signal processing at the ISVR – have distracted Joe Hammond away from his duties as Dean of the Faculty of Engineering, Science and Mathematics.

Thus the text was completed. It is indeed an introduction to the subject and therefore the essential material is not new and draws on many classic books. What we have tried to do is to bring material together, hopefully encouraging the reader to question, enquire about and explore the concepts using the MATLAB exercises or derivatives of them.

It only remains to thank all who have contributed to this. First, of course, the authors whose texts we have referred to, then the decades of students at the ISVR, and more recently in the School of Mechanical Engineering, Andong National University, who have shaped the way the course evolved, especially Sangho Pyo who spent a generous amount of time gathering experimental data. Two colleagues in the ISVR deserve particular gratitude: Professor Mike Brennan, whose positive encouragement for the whole project has been essential, together with his very constructive reading of the manuscript; and Professor Paul White, whose encyclopaedic knowledge of signal processing has been our port of call when we needed reassurance.

We would also like to express special thanks to our families, Hae-Ree Lee, Inyong Shin, Hakdoo Yu, Kyu-Shin Lee, Young-Sun Koo and Jill Hammond, for their never-ending support and understanding during the gestation and preparation of the manuscript. Kihong Shin is also grateful to Geun-Tae Yim for his continuing encouragement at the ISVR.

Finally, Joe Hammond thanks Professor Simon Braun of the Technion, Haifa, for his unceasing and inspirational leadership of signal processing in mechanical engineering. Also, and very importantly, we wish to draw attention to a new text written by Simon entitled *Discover Signal Processing: An Interactive Guide for Engineers*, also published by John Wiley & Sons, which offers a complementary and innovative learning experience.

Please note that MATLAB codes (m files) and data files can be downloaded from the Companion Website at www.wiley.com/go/shin_hammond

**Kihong Shin**
**Joseph Kenneth Hammond**

# About the Authors

**Joe Hammond**  Joseph (Joe) Hammond graduated in Aeronautical Engineering in 1966 at the University of Southampton. He completed his PhD in the Institute of Sound and Vibration Research (ISVR) in 1972 whilst a lecturer in the Mathematics Department at Portsmouth Polytechnic. He returned to Southampton in 1978 as a lecturer in the ISVR, and was later Senior lecturer, Professor, Deputy Director and then Director of the ISVR from 1992–2001. In 2001 he became Dean of the Faculty of Engineering and Applied Science, and in 2003 Dean of the Faculty of Engineering, Science and Mathematics. He retired in July 2007 and is an Emeritus Professor at Southampton.

**Kihong Shin**  Kihong Shin graduated in Precision Mechanical Engineering from Hanyang University, Korea in 1989. After spending several years as an electric motor design and NVH engineer in Samsung Electro-Mechanics Co., he started an MSc at Cranfield University in 1992, on the design of rotating machines with reference to noise and vibration. Following this, he joined the ISVR and completed his PhD on nonlinear vibration and signal processing in 1996. In 2000, he moved back to Korea as a contract Professor of Hanyang University. In Mar. 2002, he joined Andong National University as an Assistant Professor, and is currently an Associate Professor.

# 1

# Introduction to Signal Processing

Signal processing is the name given to the procedures used on measured data to reveal the information contained in the measurements. These procedures essentially rely on various transformations that are mathematically based and which are implemented using digital techniques. The wide availability of software to carry out digital signal processing (DSP) with such ease now pervades all areas of science, engineering, medicine, and beyond. This ease can sometimes result in the analyst using the wrong tools – or interpreting results incorrectly because of a lack of appreciation or understanding of the assumptions or limitations of the method employed.

This text is directed at providing a user's guide to linear system identification. In order to reach that end we need to cover the groundwork of Fourier methods, random processes, system response and optimization. Recognizing that there are many excellent texts on this,[1] why should there be yet another? The aim is to present the material from a user's viewpoint. Basic concepts are followed by examples and structured MATLAB® exercises allow the user to 'experiment'. This will not be a story with the punch-line at the end – we actually start in this chapter with the intended end point.

The aim of doing this is to provide reasons and motivation to cover some of the underlying theory. It will also offer a more rapid guide through methodology for practitioners (and others) who may wish to 'skip' some of the more 'tedious' aspects. In essence we are recognizing that it is not always necessary to be fully familiar with every aspect of the theory to be an effective practitioner. But what is important is to be aware of the limitations and scope of one's analysis.

---

[1] See for example Bendat and Piersol (2000), Brigham (1988), Hsu (1970), Jenkins and Watts (1968), Oppenheim and Schafer (1975), Otnes and Enochson (1978), Papoulis (1977), Randall (1987), etc.

*Fundamentals of Signal Processing for Sound and Vibration Engineers*
K. Shin and J. K. Hammond.    © 2008 John Wiley & Sons, Ltd

## The Aim of the Book

We are assuming that the reader wishes to understand and use a widely used approach to 'system identification'. By this we mean we wish to be able to characterize a physical process in a quantified way. The object of this quantification is that it reveals information about the process and accounts for its behaviour, and also it allows us to predict its behaviour in future environments.

The 'physical processes' could be anything, e.g. vehicles (land, sea, air), electronic devices, sensors and actuators, biomedical processes, etc., and perhaps less 'physically based' socio-economic processes, and so on. The complexity of such processes is unlimited – and being able to characterize them in a quantified way relies on the use of physical 'laws' or other 'models' usually phrased within the language of mathematics. Most science and engineering degree programmes are full of courses that are aimed at describing processes that relate to the appropriate discipline. We certainly do not want to go there in this book – life is too short! But we still want to characterize these systems – with the minimum of effort and with the maximum effect.

This is where 'system theory' comes to our aid, where we employ descriptions or models – abstractions from the 'real thing' – that nevertheless are able to capture what may be fundamentally common, to large classes of the phenomena described above. In essence what we do is simply to watch what 'a system' does. This is of course totally useless if the system is 'asleep' and so we rely on some form of activation to get it going – in which case it is logical to watch (and measure) the particular activation and measure some characteristic of the behaviour (or response) of the system.

In 'normal' operation there may be many activators and a host of responses. In most situations the activators are not separate discernible processes, but are distributed. An example of such a system might be the acoustic characteristics of a concert hall when responding to an orchestra and singers. The sources of activation in this case are the musical instruments and singers, the system is the auditorium, including the members of the audience, and the responses may be taken as the sounds heard by each member of the audience.

The complexity of such a system immediately leads one to try and conceptualize something simpler. Distributed activation might be made more manageable by 'lumping' things together, e.g. a piano is regarded as several separate activators rather than continuous strings/sounding boards all causing acoustic waves to emanate from each point on their surfaces. We might start to simplify things as in Figure 1.1.

This diagram is a model of a greatly simplified system with several actuators – and the several responses as the sounds heard by individual members of the audience. The arrows indicate a 'cause and effect' relationship – and this also has implications. For example, the figure implies that the 'activators' are unaffected by the 'responses'. This implies that there is no 'feedback' – and this may not be so.

**Figure 1.1**   Conceptual diagram of a simplified system

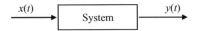

**Figure 1.2**  A single activator and a single response system

Having got this far let us simplify things even further to a single activator and a single response as shown in Figure 1.2. This may be rather 'distant' from reality but is a widely used model for many processes.

It is now convenient to think of the activator $x(t)$ and the response $y(t)$ as time histories. For example, $x(t)$ may denote a voltage, the system may be a loudspeaker and $y(t)$ the pressure at some point in a room. However, this time history model is just one possible scenario. The activator $x$ may denote the intensity of an image, the system is an optical device and $y$ may be a transformed image. Our emphasis will be on the time history model generally within a sound and vibration context.

The box marked 'System' is a convenient catch-all term for phenomena of great variety and complexity. From the outset, we shall impose major constraints on what the box represents – specifically systems that are **linear**[2] and **time invariant**.[3] Such systems are very usefully described by a particular feature, namely their response to an **ideal impulse**,[4] and their corresponding behaviour is then the **impulse response**.[5] We shall denote this by the symbol $h(t)$.

Because the system is linear this rather 'abstract' notion turns out to be very useful in predicting the response of the system to any arbitrary input. This is expressed by the **convolution**[6] of input $x(t)$ and system $h(t)$ sometimes abbreviated as

$$y(t) = h(t) * x(t) \tag{1.1}$$

where '$*$' denotes the convolution operation. Expressed in this form the system box is filled with the characterization $h(t)$ and the (mathematical) mapping or transformation from the input $x(t)$ to the response $y(t)$ is the convolution integral.

System identification now becomes the problem of measuring $x(t)$ and $y(t)$ and deducing the impulse response function $h(t)$. Since we have three quantitative terms in the relationship (1.1), but (assume that) we know two of them, then, in principle at least, we should be able to find the third. The question is: how?

Unravelling Equation (1.1) as it stands is possible but not easy. Life becomes considerably easier if we apply a transformation that maps the convolution expression to a multiplication. One such transformation is the **Fourier transform**.[7] Taking the **Fourier transform of the convolution**[8] in Equation (1.1) produces

$$Y(f) = H(f)X(f) \tag{1.2}$$

---

[*] Words in bold will be discussed or explained at greater length later.
[2] See Chapter 4, Section 4.7.
[3] See Chapter 4, Section 4.7.
[4] See Chapter 3, Section 3.2, and Chapter 4, Section 4.7.
[5] See Chapter 4, Section 4.7.
[6] See Chapter 4, Section 4.7.
[7] See Chapter 4, Sections 4.1 and 4.4.
[8] See Chapter 4, Sections 4.4 and 4.7.

where $f$ denotes frequency, and $X(f)$, $H(f)$ and $Y(f)$ are the transforms of $x(t)$, $h(t)$ and $y(t)$. This achieves the unravelling of the input–output relationship as a straightforward multiplication – in a 'domain' called the **frequency domain**.[9] In this form the system is characterized by the quantity $H(f)$ which is called the system **frequency response function (FRF)**.[10]

The problem of 'system identification' now becomes the calculation of $H(f)$, which seems easy: that is, divide $Y(f)$ by $X(f)$, i.e. divide the Fourier transform of the output by the Fourier transform of the input. As long as $X(f)$ is never zero this seems to be the end of the story – but, of course, it is not. Reality interferes in the form of 'uncertainty'. The measurements $x(t)$ and $y(t)$ are often not measured perfectly – disturbances or 'noise' contaminates them – in which case the result of dividing two transforms of contaminated signals will be of limited and dubious value.

Also, the actual excitation signal $x(t)$ may itself belong to a class of **random**[11] signals – in which case the straightforward transformation (1.2) also needs more attention. It is this 'dual randomness' of the actuating (and hence response) signal and additional contamination that is addressed in this book.

## The Effect of Uncertainty

We have referred to randomness or uncertainty with respect to both the actuation and response signal and additional noise on the measurements. So let us redraw Figure 1.2 as in Figure 1.3.

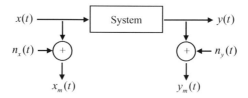

**Figure 1.3**   A single activator/response model with additive noise on measurements

In Figure 1.3, $x$ and $y$ denote the actuation and response signals as before – which may themselves be random. We also recognize that $x$ and $y$ are usually not directly measurable and we model this by including disturbances written as $n_x$ and $n_y$ which add to $x$ and $y$ – so that the actual measured signals are $x_m$ and $y_m$. Now we get to the crux of the system identification: that is, on the basis of (noisy) measurements $x_m$ and $y_m$, what is the system?

We conceptualize this problem pictorially. Imagine plotting $y_m$ against $x_m$ (ignore for now what $x_m$ and $y_m$ might be) as in Figure 1.4.

Each point in this figure is a 'representation' of the measured response $y_m$ corresponding to the measured actuation $x_m$.

System identification, in this context, becomes one of establishing a relationship between $y_m$ and $x_m$ such that it somehow relates to the relationship between $y$ and $x$. The noises are a

---

[9] See Chapter 2, Section 2.1.
[10] See Chapter 4, Section 4.7.
[11] See Chapter 7, Section 7.2.

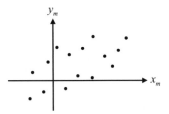

**Figure 1.4**   A plot of the measured signals $y_m$ versus $x_m$

nuisance, but we are stuck with them. This is where 'optimization' comes in. We try and find a relationship between $x_m$ and $y_m$ that seeks a 'systematic' link between the data points which suppresses the effects of the unwanted disturbances.

The simplest conceptual idea is to 'fit' a linear relationship between $x_m$ and $y_m$. Why linear? Because we are restricting our choice to the simplest relationship (we could of course be more ambitious). The procedure we use to obtain this fit is seen in Figure 1.5 where the slope of the straight line is adjusted until the match to the data seems best.

This procedure must be made systematic – so we need a measure of how well we fit the points. This leads to the need for a specific measure of fit and we can choose from an unlimited number. Let us keep it simple and settle for some obvious ones. In Figure 1.5, the closeness of the line to the data is indicated by three measures $e_y$, $e_x$ and $e_T$. These are regarded as errors which are measures of the 'failure' to fit the data. The quantity $e_y$ is an error in the $y$ direction (i.e. in the output direction). The quantity $e_x$ is an error in the $x$ direction (i.e. in the input direction). The quantity $e_T$ is orthogonal to the line and combines errors in both $x$ and $y$ directions.

We might now look at ways of adjusting the line to minimize $e_y$, $e_x$, $e_T$ or some convenient 'function' of these quantities. This is now phrased as an optimization problem. A most convenient function turns out to be an average of the squared values of these quantities ('convenience' here is used to reflect not only physical meaning but also mathematical 'niceness'). Minimizing these three different measures of closeness of fit results in three correspondingly different slopes for the straight line; let us refer to the slopes as $m_y$, $m_x$, $m_T$. So which one should we use as the best? The choice will be strongly influenced by our prior knowledge of the nature of the measured data – specifically whether we have some idea of the dominant causes of error in the departure from linearity. In other words, some knowledge of the relative magnitudes of the noise on the input and output.

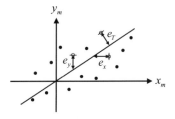

**Figure 1.5**   A linear fit to measured data

We could look to the figure for a guide:

- $m_y$ seems best when errors occur on $y$, i.e. errors on output $e_y$;
- $m_x$ seems best when errors occur on $x$, i.e. errors on input $e_x$;
- $m_T$ seems to make an attempt to recognize that errors are on both, i.e. $e_T$.

We might now ask how these rather simple concepts relate to 'identifying' the system in Figure 1.3. It turns out that they are directly relevant and lead to three different estimators for the system frequency response function $H(f)$. They have come to be referred to in the literature by the notation $\boldsymbol{H_1(f)}$, $\boldsymbol{H_2(f)}$ and $\boldsymbol{H_T(f)}$,[12] and are the analogues of the slopes $m_y$, $m_x$, $m_T$, respectively.

We have now mapped out what the book is essentially about in Chapters 1 to 10. The book ends with a chapter that looks into the implications of multi-input/output systems.

## 1.1 DESCRIPTIONS OF PHYSICAL DATA (SIGNALS)

Observed data representing a physical phenomenon will be referred to as a time history or a *signal*. Examples of signals are: temperature fluctuations in a room indicated as a function of time, voltage variations from a vibration transducer, pressure changes at a point in an acoustic field, etc. The physical phenomenon under investigation is often translated by a transducer into an electrical equivalent (voltage or current) and if displayed on an oscilloscope it might appear as shown in Figure 1.6. This is an example of a *continuous* (or *analogue*) signal.

In many cases, data are *discrete* owing to some inherent or imposed sampling procedure. In this case the data might be characterized by a sequence of numbers equally spaced in time. The sampled data of the signal in Figure 1.6 are indicated by the crosses on the graph shown in Figure 1.7.

**Figure 1.6**  A typical continuous signal from a transducer output

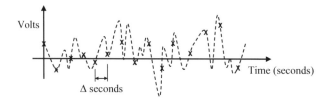

**Figure 1.7**  A discrete signal sampled at every $\Delta$ seconds (marked with $\times$)

[12] See Chapter 9, Section 9.3.

**Figure 1.8** An example of a signal where time is not the natural independent variable

For continuous data we use the notation $x(t)$, $y(t)$, etc., and for discrete data various notations are used, e.g. $x(n\Delta)$, $x(n)$, $x_n$ ($n = 0, 1, 2, \ldots$).

In certain physical situations, 'time' may not be the natural independent variable; for example, a plot of road roughness as a function of spatial position, i.e. $h(\xi)$ as shown in Figure 1.8. However, for uniformity we shall use time as the independent variable in all our discussions.

## 1.2 CLASSIFICATION OF DATA

Time histories can be broadly categorized as shown in Figure 1.9 (chaotic signals are added to the classifications given by Bendat and Piersol, 2000). A fundamental difference is whether a signal is *deterministic* or *random*, and the analysis methods are considerably different depending on the 'type' of the signal. Generally, signals are mixed, so the classifications of Figure 1.9 may not be easily applicable, and thus the choice of analysis methods may not be apparent. In many cases some prior knowledge of the system (or the signal) is very helpful for selecting an appropriate method. However, it must be remembered that this prior knowledge (or assumption) may also be a source of misleading the results. Thus it is important to remember the First Principle of Data Reduction (Ables, 1974)

> *The result of any transformation imposed on the experimental data shall incorporate and be consistent with all relevant data and be maximally non-committal with regard to unavailable data.*

It would seem that this statement summarizes what is self-evident. But how often do we contravene it – for example, by 'assuming' that a time history is zero outside the extent of a captured record?

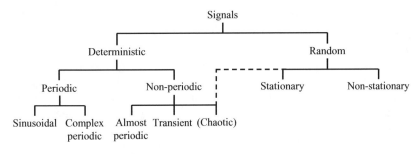

**Figure 1.9** Classification of signals

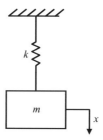

**Figure 1.10**   A simple mass–spring system

Nonetheless, we need to start somewhere and signals can be broadly classified as being either *deterministic* or *non-deterministic* (*random*). Deterministic signals are those whose behaviour can be predicted exactly. As an example, a mass–spring oscillator is considered in Figure 1.10. The equation of motion is $m\ddot{x} + kx = 0$ ($x$ is displacement and $\ddot{x}$ is acceleration). If the mass is released from rest at a position $x(t) = A$ and at time $t = 0$, then the displacement signal can be written as

$$x(t) = A \cos \left( \sqrt{k/m} \cdot t \right) \quad t \geq 0 \tag{1.3}$$

In this case, the displacement $x(t)$ is known exactly for all time. Various types of deterministic signals will be discussed later. Basic analysis methods for deterministic signals are covered in Part I of this book. Chaotic signals are not considered in this book.

Non-deterministic signals are those whose behaviour cannot be predicted exactly. Some examples are vehicle noise and vibrations on a road, acoustic pressure variations in a wind tunnel, wave heights in a rough sea, temperature records at a weather station, etc. Various terminologies are used to describe these signals, namely *random processes* (*signals*), *stochastic processes*, *time series*, and the study of these signals is called *time series analysis*. Approaches to describe and analyse random signals require probabilistic and statistical methods. These are discussed in Part II of this book.

The classification of data as being deterministic or random might be debatable in many cases and the choice must be made on the basis of knowledge of the physical situation. Often signals may be modelled as being a mixture of both, e.g. a deterministic signal 'embedded' in unwanted random disturbances (noise).

In general, the purpose of signal processing is the extraction of information from a signal, especially when it is difficult to obtain from direct observation. The methodology of extracting information from a signal has three key stages: (i) acquisition, (ii) processing, (iii) interpretation. To a large extent, signal acquisition is concerned with *instrumentation*, and we shall treat some aspects of this, e.g. **analogue-to-digital conversion**.[13] However, in the main, we shall assume that the signal is already acquired, and concentrate on stages (ii) and (iii).

---

[13] See Chapter 5, Section 5.3.

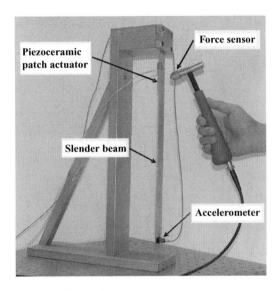

**Figure 1.11**   A laboratory setup

## Some 'Real' Data

Let us now look at some signals measured experimentally. We shall attempt to fit the observed time histories to the classifications of Figure 1.9.

(a) Figure 1.11 shows a laboratory setup in which a slender beam is suspended vertically from a rigid clamp. Two forms of excitation are shown. A small piezoceramic PZT (Piezoelectric Zirconate Titanate) patch is used as an actuator which is bonded on near the clamped end. The instrumented hammer (impact hammer) is also used to excite the structure. An accelerometer is attached to the beam tip to measure the response. We shall assume here that digitization effects (**ADC quantization, aliasing**)[14] have been adequately taken care of and can be ignored. A sharp tap from the hammer to the structure results in Figures 1.12(a) and (b). Relating these to the classification scheme, we could reasonably refer to these as deterministic transients. Why might we use the deterministic classification? Because we expect replication of the result for 'identical' impacts. Further, from the figures the signals appear to be essentially noise free. From a systems points of view, Figure 1.12(a) is $x(t)$ and 1.12(b) is $y(t)$ and from these two signals we would aim to deduce the characteristics of the beam.

(b) We now use the PZT actuator, and Figures 1.13(a) and (b) now relate to a random excitation. The source is a **band-limited**,[15] **stationary**,[16] **Gaussian process**,[17] and in the steady state (i.e. after starting transients have died down) the response should also be stationary. However, on the basis of the visual evidence the response is not evidently stationary (or is it?), i.e. it seems modulated in some way. This demonstrates the difficulty in classification. As it

[14] See Chapter 5, Sections 5.1–5.3.
[15] See Chapter 5, Section 5.2, and Chapter 8, Section 8.7.
[16] See Chapter 8, Section 8.3.
[17] See Chapter 7, Section 7.3.

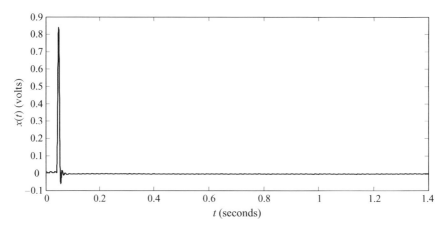

(a) Impact signal measured from the force sensor (impact hammer)

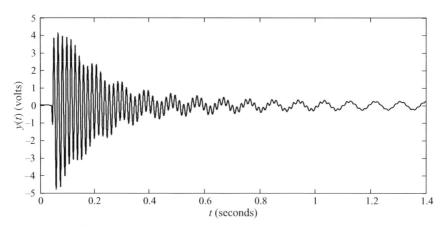

(b) Response signal to the impact measured from the accelerometer

**Figure 1.12**   Example of deterministic transient signals

happens, the response is a narrow-band stationary random process (due to the filtering action of the beam) which is characterized by an amplitude-modulated appearance.

(c) Let us look at a signal from a machine rotating at a constant rate. A tachometer signal is taken from this. As in Figure 1.14(a), this is one that could reasonably be classified as periodic, although there are some discernible differences from period to period – one might ask whether this is simply an additive low-level noise.

(d) Another repetitive signal arises from a telephone tone shown in Figure 1.14(b). The tonality is 'evident' from listening to it and its appearance is 'roughly' periodic; it is tempting to classify these signals as 'almost periodic'!

(e) Figure 1.15(a) represents the signal for a transformer 'hum', which again perceptually has a repetitive but complex structure and visually appears as possibly periodic with additive noise – or (perhaps) narrow-band random.

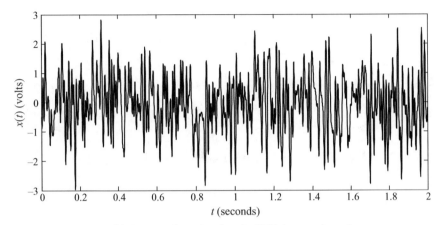

(a) Input random signal to the PZT (actuator) patch

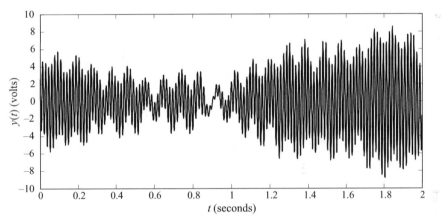

(b) Response signal to the random excitation measured from the accelerometer

**Figure 1.13**   Example of stationary random signals

Figure 1.15(b) is a signal created by adding noise (broadband) to the telephone tone signal in Figure 1.14(b). It is not readily apparent that Figure 1.15(b) and Figure 1.15(a) are 'structurally' very different.

(f) Figure 1.16(a) is an acoustic recording of a helicopter flyover. The non-stationary structure is apparent – specifically, the increase in amplitude with reduction in range. What is not apparent are any other more complex aspects such as frequency modulation due to movement of the source.

(g) The next group of signals relate to practicalities that occur during acquisition that render the data of limited value (in some cases useless!).

The jagged stepwise appearance in Figure 1.17 is due to quantization effects in the ADC – apparent because the signal being measured is very small compared with the voltage range of the ADC.

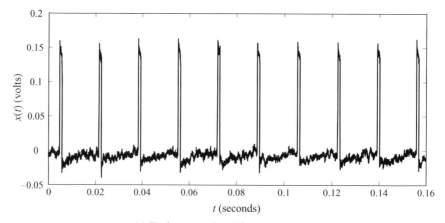

(a) Tachometer signal from a rotating machine

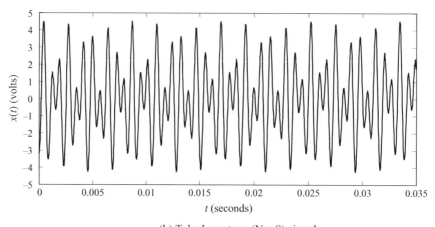

(b) Telephone tone (No. 8) signal

**Figure 1.14**   Example of periodic (and almost periodic) signals

(h) Figures 1.18(a), (b) and (c) all display flats at the top and bottom (positive and negative) of their ranges. This is characteristic of 'clipping' or saturation. These have been synthesized by clipping the telephone signal in Figure 1.14(b), the band-limited random signal in Figure 1.13(a) and the accelerometer signal in Figure 1.12(b). Clipping is a nonlinear effect which 'creates' spurious frequencies and essentially destroys the credibility of any Fourier transformation results.

(i) Lastly Figures 1.19(a) and (b) show what happens when 'control' of an experiment is not as tight as it might be. Both signals are the free responses of the cantilever beam shown in Figure 1.11. Figure 1.19(a) shows the results of the experiment performed on a vibration-isolated optical table. The signal is virtually noise free. Figure 1.19(b) shows the results of the same experiment, but performed on a normal bench-top table. The signal is now contaminated with noise that may come from various external sources. Note that we may not be able to control our experiments as carefully as in Figure 1.19(a), but, in fact, it is a signal as in

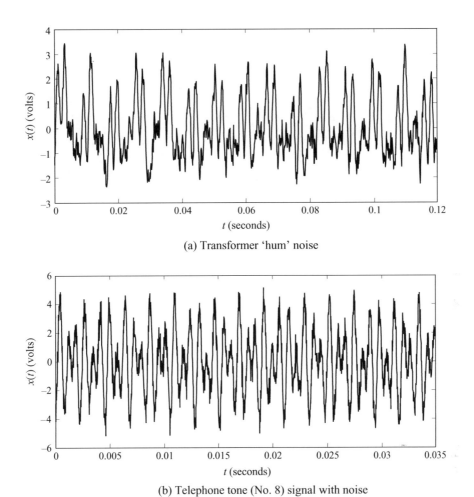

(a) Transformer 'hum' noise

(b) Telephone tone (No. 8) signal with noise

**Figure 1.15**  Example of periodic signals with additive noise

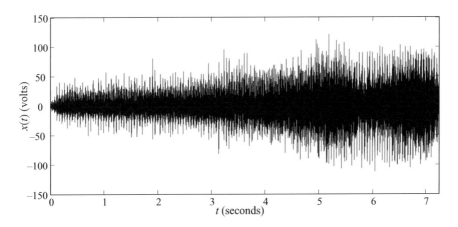

**Figure 1.16**  Example of a non-stationary signal (helicopter flyover noise)

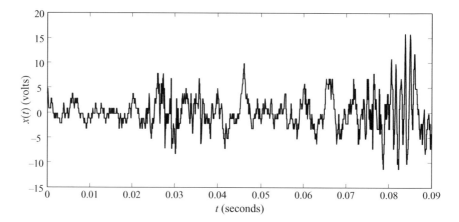

**Figure 1.17**   Example of low dynamic range

Figure 1.19(b) which we often deal with. Thus, the nature of uncertainty in the measurement process is again emphasized (see Figure 1.3).

## The Next Stage

Having introduced various classes of signals we can now turn to the principles and details of how we can model and analyse the signals. We shall use Fourier-based methods – that is, we essentially model the signal as being composed of sine and cosine waves and tailor the processing around this idea. We might argue that we are imposing/assuming some prior information about the signal – namely, that sines and cosines are appropriate descriptors. Whilst this may seem constraining, such a 'prior model' is very effective and covers a wide range of phenomena. This is sometimes referred to as a *non-parametric* approach to signal processing.

So, what might be a 'parametric' approach? This can again be related to modelling. We may have additional 'prior information' as to how the signal has been generated, e.g. a result of filtering another signal. This notion may be extended from the knowledge that this generation process is indeed 'physical' to that of its being 'notional', i.e. another model. Specifically Figure 1.20 depicts this when $s(t)$ is the 'measured' signal, which is conceived to have arisen from the action of a system being driven by a very fundamental signal – in this case so-called **white noise**[18] $w(t)$.

Phrased in this way the analysis of the signal $s(t)$ can now be transformed into a problem of determining the details of the system. The system could be characterized by a set of parameters, e.g. it might be mathematically represented by differential equations and the parameters are the coefficients. Set up like this, the analysis of $s(t)$ becomes one of system parameter estimation – hence this is a parametric approach.

The system could be linear, time varying or nonlinear depending on one's prior knowledge, and could therefore offer advantages over Fourier-based methods. However, we shall not be pursuing this approach in this book and will get on with the Fourier-based methods instead.

---

[18] See Chapter 8, Section 8.6.

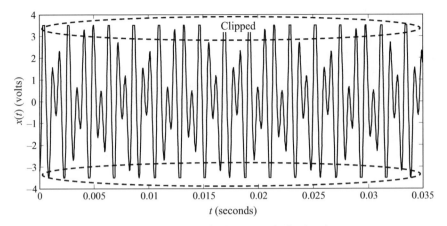

(a) Clipped (almost) periodic signal

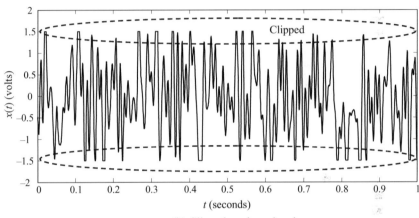

(b) Clipped random signal

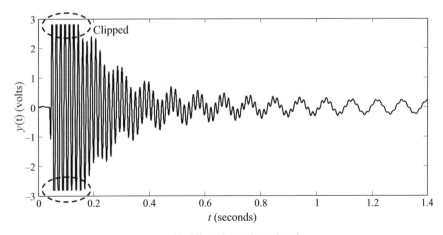

(c) Clipped transient signal

**Figure 1.18** Examples of clipped signals

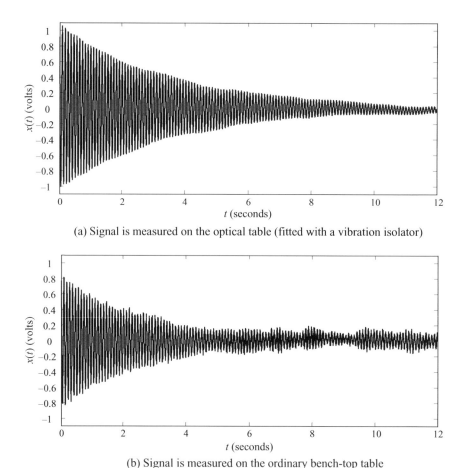

(a) Signal is measured on the optical table (fitted with a vibration isolator)

(b) Signal is measured on the ordinary bench-top table

**Figure 1.19**   Examples of experimental noise

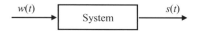

**Figure 1.20**   A white-noise-excited system

We have emphasized that this is a book for practitioners and users of signal processing, but note also that there should be sufficient detail for completeness. Accordingly we have chosen to highlight some main points using a light grey background. From Chapter 3 onwards there is a reasonable amount of mathematical content; however, a reader may wish to get to the main points quickly, which can be done by using the highlighted sections. The details supporting these points are in the remainder of the chapter adjacent to these sections and in the appendices. Examples and MATLAB exercises illustrate the concepts. A superscript notation is used to denote the relevant MATLAB example given in the last section of the chapter, e.g. see the superscript ($^{M2.1}$) in page 21 for MATLAB Example 2.1 given in page 26.

# Part I
## Deterministic Signals

# 2

# Classification of Deterministic Data

## Introduction

As described in Chapter 1, deterministic signals can be classified as shown in Figure 2.1. In this figure, chaotic signals are not considered and the sinusoidal signal and more general periodic signals are dealt with together. So deterministic signals are now classified as periodic, almost periodic and transient, and some basic characteristics are explained below.

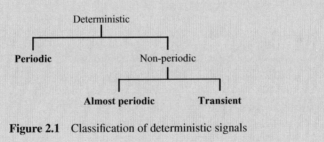

**Figure 2.1**   Classification of deterministic signals

## 2.1 PERIODIC SIGNALS

Periodic signals are defined as those whose waveform repeats exactly at regular time intervals. The simplest example is a sinusoidal signal as shown in Figure 2.2(a), where the time interval for one full cycle is called the period $T_P$ (in seconds) and its reciprocal $1/T_P$ is called the frequency (in hertz). Another example is a triangular signal (or sawtooth wave), as shown in Figure 2.2(b). This signal has an abrupt change (or discontinuity) every $T_P$ seconds. A more

*Fundamentals of Signal Processing for Sound and Vibration Engineers*
K. Shin and J. K. Hammond.   © 2008 John Wiley & Sons, Ltd

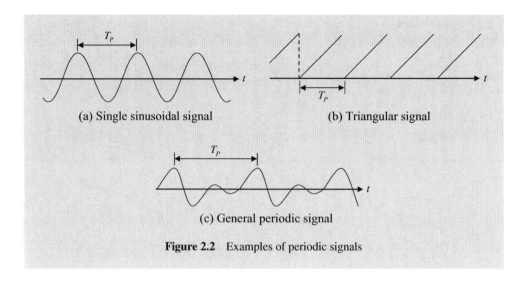

(a) Single sinusoidal signal                                    (b) Triangular signal

(c) General periodic signal

**Figure 2.2**   Examples of periodic signals

general periodic signal is shown in Figure 2.2(c) where an arbitrarily shaped waveform repeats with period $T_P$.

In each case the mathematical definition of periodicity implies that the behaviour of the wave is unchanged for all time. This is expressed as

$$x(t) = x(t + nT_P) \quad n = \pm 1, \ \pm 2, \ \pm 3, \ldots \tag{2.1}$$

For cases (a) and (b) in Figure 2.2, explicit mathematical descriptions of the wave are easy to write, but the mathematical expression for the case (c) is not obvious. The signal (c) may be obtained by measuring some physical phenomenon, such as the output of an accelerometer placed near the cylinder head of a constant speed car engine. In this case, it may be more useful to consider the signal as being made up of simpler components. One approach to this is to 'transform' the signal into the 'frequency domain' where the details of periodicities of the signal are clearly revealed. In the frequency domain, the signal is decomposed into an infinite (or a finite) number of frequency components. The periodic signals appear as discrete components in this frequency domain, and are described by a Fourier series which is discussed in Chapter 3. As an example, the frequency domain representation of the amplitudes of the triangular wave (Figure 2.2(b)) with a period of $T_P = 2$ seconds is shown in Figure 2.3. The components in the frequency domain consist of the fundamental frequency $1/T_P$ and its harmonics $2/T_P, 3/T_P, \ldots$, i.e. all frequency components are 'harmonically related'.

However, there is hardly ever a perfect periodic signal in reality even if the signal is carefully controlled. For example, almost all so-called periodic signals produced by a signal generator used in sound and vibration engineering are not *perfectly* periodic owing to the limited precision of the hardware and noise. An example of this may be a telephone keypad tone that usually consists of two frequency components (assume the ratio of the two frequencies is a rational number — see Section 2.2). The measured time data of the telephone tone of keypad '8' are shown in Figure 2.4(a), where it seems to be a periodic signal. However, when it is

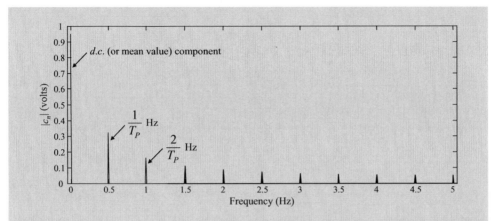

**Figure 2.3** Frequency domain representation of the amplitudes of a triangular wave with a period of
$$T_p = 2$$

transformed into the frequency domain, we may find something different. The telephone tone of keypad '8' is designed to have frequency components at 852 Hz and 1336 Hz only. This measured telephone tone is transformed into the frequency domain as shown in Figures 2.4(b) (linear scale) and (c) (log scale). On a linear scale, it seems to be composed of the two frequencies. However, there are in fact, many other frequency components that may result if the signal is not perfectly periodic, and this can be seen by plotting the transform on a log scale as in Figure 2.4(c).

Another practical example of a signal that may be considered to be periodic is transformer hum noise (Figure 2.5(a)) whose dominant frequency components are about 122 Hz, 366 Hz and 488 Hz, as shown in Figure 2.5(b). From Figure 2.5(a), it is apparent that the signal is not periodic. However, from Figure 2.5(b) it is seen to have a periodic structure contaminated with noise.

From the above two practical examples, we note that most periodic signals in practical situations are not 'truly' periodic, but are 'almost' periodic. The term 'almost periodic' is discussed in the next section.

## 2.2 ALMOST PERIODIC SIGNALS[M2.1] (This superscript is short for MATLAB Example 2.1)

The name 'almost periodic' seems self-explanatory and is sometimes called quasi-periodic, i.e. it looks periodic but in fact it is not if observed closely. We shall see in Chapter 3 that suitably selected sine and cosine waves may be added together to represent cases (b) and (c) in Figure 2.2. Also, even for apparently simple situations the sum of sines and cosines results in a wave which never repeats itself *exactly*. As an example, consider a wave consisting of two sine components as below

$$x(t) = A_1 \sin(2\pi p_1 t + \theta_1) + A_2 \sin(2\pi p_2 t + \theta_2) \tag{2.2}$$

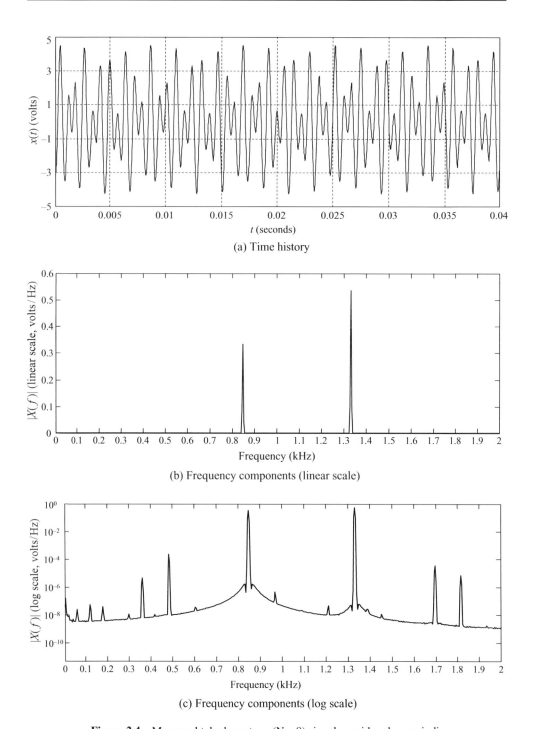

(a) Time history

(b) Frequency components (linear scale)

(c) Frequency components (log scale)

**Figure 2.4**   Measured telephone tone (No. 8) signal considered as periodic

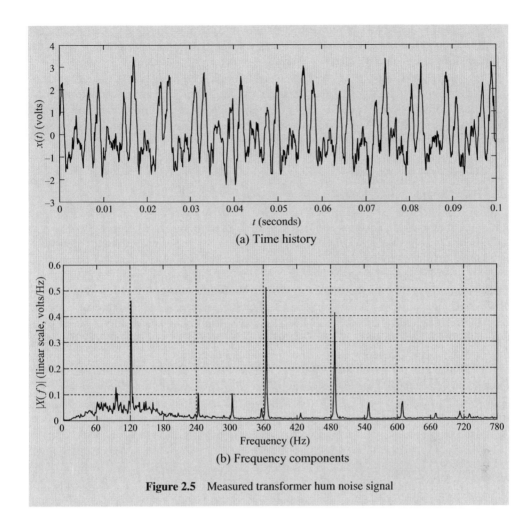

**Figure 2.5** Measured transformer hum noise signal

where $A_1$ and $A_2$ are amplitudes, $p_1$ and $p_2$ are the frequencies of each sine component, and $\theta_1$ and $\theta_2$ are called the phases. If the frequency ratio $p_1/p_2$ is a rational number, the signal $x(t)$ is periodic and repeats at every time interval of the smallest common period of both $1/p_1$ and $1/p_2$. However, if the ratio $p_1/p_2$ is irrational (as an example, the ratio $p_1/p_2 = 2/\sqrt{2}$ is irrational), the signal $x(t)$ never repeats. It can be argued that the sum of two or more sinusoidal components is periodic only if the ratios of all pairs of frequencies are found to be rational numbers (i.e. ratio of integers). A possible example of an almost periodic signal may be an acoustic signal created by tapping a slightly asymmetric wine glass.

However, the representation (model) of a signal as the addition of simpler (sinusoidal) components is very attractive – whether the signal is truly periodic or not. In fact a method which predated the birth of Fourier analysis uses this idea. This is the so-called Prony series (de Prony, 1795; Spitznogle and Quazi, 1970; Kay and Marple, 1981; Davies, 1983). The basic components here have the form $Ae^{-\sigma t}\sin(\omega t + \phi)$ in which there are four parameters for each component – namely, amplitude $A$, frequency $\omega$, phase $\phi$ and an additional feature $\sigma$ which controls the decay of the component.

Prony analysis fits a sum of such components to the data using an optimization procedure. The parameters are found from a (nonlinear) algorithm. The nonlinear nature of the optimization arises because (even if $\sigma = 0$) the frequency $\omega$ is calculated for each component. This is in contrast to Fourier methods where the frequencies are fixed once the period $T_P$ is known, i.e. only amplitudes and phases are calculated.

## 2.3 TRANSIENT SIGNALS

The word 'transient' implies some limitation on the duration of the signal. Generally speaking, a transient signal has the property that $x(t) = 0$ when $t \to \pm\infty$; some examples are shown in Figure 2.6. In vibration engineering, a common practical example is impact testing (with a hammer) to estimate the frequency response function (FRF, see Equation (1.2)) of a structure. The measured input force signal and output acceleration signal from a simple cantilever beam experiment are shown in Figure 2.7. The frequency characteristic of this type of signal is very different from the Fourier series. The discrete frequency components are replaced by the concept of the signal containing a continuum of frequencies. The mathematical details and interpretation in the frequency domain are presented in Chapter 4.

Note also that the modal characteristics of the beam allow the transient response to be modelled as the sum of decaying oscillations, i.e. ideally matched to the Prony series. This allows the Prony model to be 'fitted to' the data (see Davies, 1983) to estimate the amplitudes, frequencies, damping and phases, i.e. a parametric approach.

## 2.4 BRIEF SUMMARY AND CONCLUDING REMARKS

1. Deterministic signals are largely classified as periodic, almost periodic and transient signals.
2. Periodic and almost periodic signals have discrete components in the frequency domain.
3. Almost periodic signals may be considered as periodic signals having an infinitely long period.
4. Transient signals are analysed using the Fourier integral (see Chapter 4).

Chapters 1 and 2 have been introductory and qualitative. We now add detail to these descriptions and note again that a quick 'skip-through' can be made by following the highlighted sections. MATLAB examples are also presented with enough detail to allow the reader to try them and to understand important features (MATLAB version 7.1 is used, and Signal Processing Toolbox is required for some MATLAB examples).

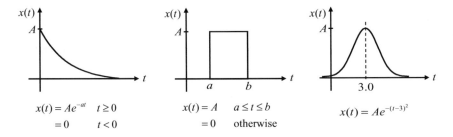

**Figure 2.6**   Examples of transient signals

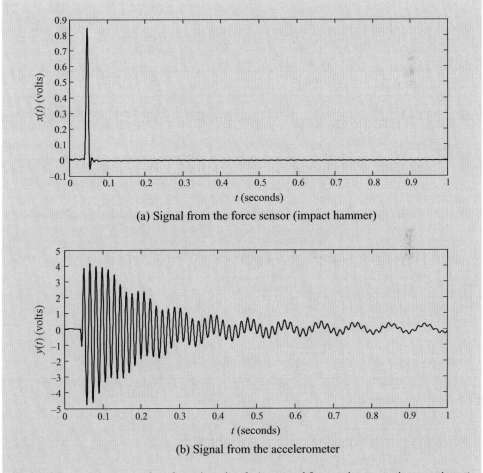

(a) Signal from the force sensor (impact hammer)

(b) Signal from the accelerometer

**Figure 2.7**   Practical examples of transient signals (measured from an impact testing experiment)

## 2.5 MATLAB EXAMPLES[1]

---

***Example 2.1:*** **Synthesis of periodic signals and almost periodic signals**
(see Section 2.2)

Consider Equation (2.2) for this example, i.e.

$$x(t) = A_1 \sin (2\pi p_1 t + \theta_1) + A_2 \sin (2\pi p_2 t + \theta_2)$$

Let the amplitudes $A_1 = A_2 = 1$ and phases $\theta_1 = \theta_2 = 0$ for convenience.

---

Case 1: Periodic signal with frequencies $p_1 = 1.4$ Hz and $p_2 = 1.5$ Hz.
Note that the ratio $p_1/p_2$ is rational, and the smallest common period of both $1/p_1$ and $1/p_2$ is '10', thus the period is 10 seconds in this case.

---

| Line | MATLAB code | Comments |
|------|-------------|----------|
| 1 | clear all | Removes all local and global variables (this is a good way to start a new MATLAB script). |
| 2 | A1=1; A2=1; Theta1=0; Theta2=0; p1=1.4; p2=1.5; | Define the parameters for Equation (2.2). Semicolon (;) separates statements and prevents displaying the results on the screen. |
| 3 | t=[0:0.01:30]; | The time variable $t$ is defined as a row vector from zero to 30 seconds with a step size 0.01. |
| 4 | x=A1*sin(2*pi*p1*t+Theta1) +A2*sin(2*pi*p2*t+Theta2); | MATLAB expression of Equation (2.2). |
| 5 | plot(t, x) | Plot the results of $t$ versus $x$ ($t$ on abscissa and $x$ on ordinate). |
| 6 | xlabel('\itt\rm (seconds)'); ylabel('\itx\rm(\itt\rm)') | Add text on the horizontal (xlabel) and on the vertical (ylabel) axes. '\it' is for italic font, and '\rm' is for normal font. Readers may find more ways of dealing with graphics in the section 'Handle Graphics Objects' in the MATLAB Help window. |
| 7 | grid on | Add grid lines on the current figure. |

---

[1] MATLAB codes (m files) and data files can be downloaded from the Companion Website (www.wiley.com/go/shin_hammond).

**Results**

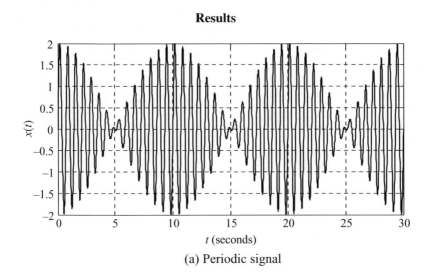

(a) Periodic signal

**Comments:** It is clear that this signal is periodic and repeats every 10 seconds, i.e. $T_P = 10$ seconds, thus the fundamental frequency is 0.1 Hz. The frequency domain representation of the above signal is shown in Figure (b). Note that the amplitude of the fundamental frequency is zero and thus does not appear in the figure. This applies to subsequent harmonics until 1.4 Hz and 1.5 Hz. Note also that the frequency components 1.4 Hz and 1.5 Hz are 'harmonically' related, i.e. both are multiples of 0.1 Hz.

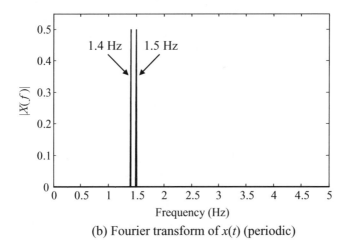

(b) Fourier transform of $x(t)$ (periodic)

Case 2:  Almost periodic signal with frequencies $p_1 = \sqrt{2}$ Hz and $p_2 = 1.5$ Hz.
Note that the ratio $p_1/p_2$ is now irrational, so there is no common period of both $1/p_1$ and $1/p_2$.

| Line | MATLAB code | Comments |
|------|-------------|----------|
| 1 | clear all | |
| 2 | A1=1; A2=1; Theta1=0; | |
| | Theta2=0; p1=sqrt(2); p2=1.5; | Exactly the same script as in the previous |
| 3 | t=[0:0.01:30]; | case except 'p1=1.4' is replaced with |
| 4 | x=A1*sin(2*pi*p1*t+Theta1) | 'p1=sqrt(2)'. |
| | +A2*sin(2*pi*p2*t+Theta2); | |
| 5 | plot(t, x) | |
| 6 | xlabel('\itt\rm (seconds)'); | |
| | ylabel('\itx\rm(\itt\rm)') | |
| 7 | grid on | |

**Results**

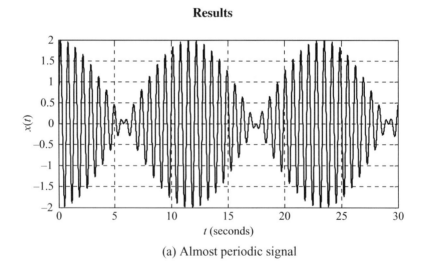

(a) Almost periodic signal

**Comments:** One can find that this signal is *not* periodic if it is observed carefully by closely investigating or magnifying appropriate regions. The frequency domain representation of the above signal is shown in Figure (b). Since the signal is not exactly periodic, the usual concept of the fundamental frequency does not hold. However, it may be considered that the periodicity of this signal is infinite, i.e. the fundamental frequency is '0 Hz' (this concept leads us to the Fourier integral which is discussed in Chapter 4). The spread of the frequency components in the figure is not representative of the true frequency components in the signal, but results from the truncation of the signal, i.e. it is a windowing effect (see Sections 3.6 and 4.11 for details).

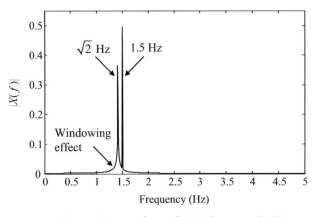

(b) Fourier transform of $x(t)$ (almost periodic)

# 3

# Fourier Series

## Introduction

This chapter describes the simplest of the signal types – periodic signals. It begins with the ideal situation and the basis of Fourier decomposition, and then, through illustrative examples, discusses some of the practical issues that arise. The delta function is introduced, which is very useful in signal processing. The chapter concludes with some examples based on the MATLAB software environment.

The presentation is reasonably detailed, but to assist the reader in skipping through to find the main points being made, some equations and text are highlighted.

## 3.1 PERIODIC SIGNALS AND FOURIER SERIES

Periodic signals are analysed using Fourier series. The basis of Fourier analysis of a periodic signal is the representation of such a signal by adding together sine and cosine functions of appropriate frequencies, amplitudes and relative phases. For a single sine wave

$$x(t) = X \sin(\omega t + \phi) = X \sin(2\pi f t + \phi) \tag{3.1}$$

where    $X$    is amplitude,
         $\omega$    is circular (angular) frequency in radians per unit time (rad/s),
         $f$    is (cyclical) frequency in cycles per unit time (Hz),
         $\phi$    is phase angle with respect to the *time origin* in radians.

The period of this sine wave is $T_P = 1/f = 2\pi/\omega$ seconds. A *positive* phase angle $\phi$ shifts the waveform to the *left* (a lead or advance) and a *negative* phase angle to the *right* (a lag or delay), where the time shift is $\phi/\omega$ seconds. When $\phi = \pi/2$ the wave becomes a

---

*Fundamentals of Signal Processing for Sound and Vibration Engineers*
K. Shin and J. K. Hammond.    © 2008 John Wiley & Sons, Ltd

cosine wave. The Fourier series (for periodic signal) is now described. A periodic signal, $x(t)$, is shown in Figure 3.1 and satisfies

$$x(t) = x(t + nT_P) \quad n = \pm 1, \pm 2, \pm 3, \dots \tag{3.2}$$

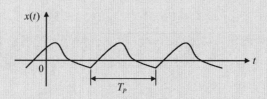

**Figure 3.1**   A period signal with a period $T_P$

With a few exceptions such periodic functions may be represented by

$$x(t) = \frac{a_0}{2} + \sum_{n=1}^{\infty} \left[ a_n \cos\left( \frac{2\pi nt}{T_P} \right) + b_n \sin\left( \frac{2\pi nt}{T_P} \right) \right] \tag{3.3}$$

The fundamental frequency is $f_1 = 1/T_P$ and all other frequencies are multiples of this. $a_0/2$ is the d.c. level or *mean value* of the signal. The reason for wanting to use a representation of the form (3.3) is because it is useful to decompose a 'complicated' signal into a sum of 'simpler' signals – in this case, sine and cosine waves. The amplitude and phase of each component can be obtained from the coefficients $a_n$, $b_n$, as we shall see later in Equation (3.12). These coefficients are calculated from the following expressions:

$$\frac{a_0}{2} = \frac{1}{T_P} \int_0^{T_P} x(t)dt = \frac{1}{T_P} \int_{-T_P/2}^{T_P/2} x(t)dt : \quad \text{mean value}$$

$$a_n = \frac{2}{T_P} \int_0^{T_P} x(t)\cos\left( \frac{2\pi nt}{T_P} \right) dt = \frac{2}{T_P} \int_{-T_P/2}^{T_P/2} x(t)\cos\left( \frac{2\pi nt}{T_P} \right) dt \quad n = 1, 2, \dots \tag{3.4}$$

$$b_n = \frac{2}{T_P} \int_0^{T_P} x(t)\sin\left( \frac{2\pi nt}{T_P} \right) dt = \frac{2}{T_P} \int_{-T_P/2}^{T_P/2} x(t)\sin\left( \frac{2\pi nt}{T_P} \right) dt \quad n = 1, 2, \dots$$

We justify the expressions (3.4) for the coefficients $a_n$, $b_n$ as follows. Suppose we wish to add up a set of 'elementary' functions $u_n(t)$, $n = 1, 2, \dots$, so as to represent a function $x(t)$, i.e. we want $\sum_n c_n u_n(t)$ to be a 'good' representation of $x(t)$. We may write $x(t) \approx \sum_n c_n u_n(t)$, where $c_n$ are coefficients to be found. Note that we cannot assume *equality* in this expression. To find the coefficients $c_n$, we form an error function $e(t) = x(t) - \sum_n c_n u_n(t)$ and select the $c_n$ so as to minimize some function of $e(t)$, e.g.

$$J = \int_0^{T_P} e^2(t)dt \tag{3.5}$$

Since the $u_n(t)$ are chosen functions, $J$ is a function of $c_1, c_2, \ldots$ only, so in order to minimize $J$ we need necessary conditions as below:

$$\frac{\partial J}{\partial c_m} = 0 \quad \text{for} \quad m = 1, 2, \ldots \tag{3.6}$$

The function $J$ is

$$J(c_1, c_2, \ldots) = \int_0^{T_P} \left( x(t) - \sum_n c_n u_n(t) \right)^2 dt$$

and so Equation (3.6) becomes

$$\frac{\partial J}{\partial c_m} = \int_0^{T_P} 2 \left( x(t) - \sum_n c_n u_n(t) \right) (-u_m(t)) dt = 0 \tag{3.7}$$

Thus the following result is obtained:

$$\int_0^{T_P} x(t) u_m(t) dt = \sum_n c_n \int_0^{T_P} u_n(t) u_m(t) dt \tag{3.8}$$

At this point we can see that a very desirable property of the 'basis set' $u_n$ is that

$$\int_0^{T_P} u_n(t) u_m(t) dt = 0 \quad \text{for } n \neq m \tag{3.9}$$

i.e. they should be 'orthogonal'.

Assuming this is so, then using the orthogonal property of Equation (3.9) gives the required coefficients as

$$c_m = \frac{\int_0^{T_P} x(t) u_m(t) dt}{\int_0^{T_P} u_m^2(t) dt} \tag{3.10}$$

Equation (3.10) is the equivalent of Equation (3.4) for the particular case of selecting sines and cosines as the basic elements. Specifically Equation (3.4) utilizes the following results of orthogonality:

$$\int_0^{T_P} \cos\left(\frac{2\pi mt}{T_P}\right) \sin\left(\frac{2\pi nt}{T_P}\right) dt = 0 \quad \text{for all } m, n$$

$$\left.\begin{array}{l} \int_0^{T_P} \cos\left(\frac{2\pi mt}{T_P}\right) \cos\left(\frac{2\pi nt}{T_P}\right) dt = 0 \\[2mm] \int_0^{T_P} \sin\left(\frac{2\pi mt}{T_P}\right) \sin\left(\frac{2\pi nt}{T_P}\right) dt = 0 \end{array}\right\} \quad \text{for } m \neq n \tag{3.11}$$

$$\int_0^{T_P} \cos^2\left(\frac{2\pi nt}{T_P}\right) dt = \int_0^{T_P} \sin^2\left(\frac{2\pi nt}{T_P}\right) dt = \frac{T_P}{2}$$

We began by referring to the amplitude and phase of the components of a Fourier series. This is made explicit now by rewriting Equation (3.3) in the form

$$x(t) = \frac{a_0}{2} + \sum_{n=1}^{\infty} M_n \cos(2\pi n f_1 t + \phi_n) \tag{3.12}$$

where   $f_1 = 1/T_P$ is the fundamental frequency,

$M_n = \sqrt{a_n^2 + b_n^2}$ are the amplitudes of frequencies at $nf_1$,

$\phi_n = \tan^{-1}(-b_n/a_n)$ are the phases of the frequency components at $nf_1$.

Note that we have assumed that the summation of the components does indeed accurately represent the signal $x(t)$, i.e. we have tacitly assumed the sum converges, and furthermore converges to the signal. This is discussed further in what follows.

### An Example (A Square Wave)

As an example, let us find the Fourier series of the function defined by

$$
\begin{aligned}
x(t) &= -1 & -\frac{T}{2} < t < 0 \\
&&&\text{and}\quad x(t + nT) = x(t) \quad n = \pm 1, \pm 2, \ldots \tag{3.13} \\
&= 1 & 0 < t < \frac{T}{2}
\end{aligned}
$$

where the function can be drawn as in Figure 3.2.

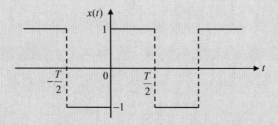

**Figure 3.2**   A periodic square wave signal

From Figure 3.2, it is apparent that the mean value is zero, so

$$\frac{a_0}{2} = \frac{1}{T} \int_{-T/2}^{T/2} x(t)dt = 0 \quad \text{(mean value)} \tag{3.14}$$

and the coefficients $a_n$ and $b_n$ are

$$a_n = \frac{2}{T} \int_{-T/2}^{T/2} x(t) \cos\left(\frac{2\pi nt}{T}\right) dt$$

$$= \frac{2}{T} \left[ \int_{-T/2}^{0} -\cos\left(\frac{2\pi nt}{T}\right) dt + \int_{0}^{T/2} \cos\left(\frac{2\pi nt}{T}\right) dt \right] = 0$$

(3.15)

$$b_n = \frac{2}{T} \int_{-T/2}^{T/2} x(t) \sin\left(\frac{2\pi nt}{T}\right) dt$$

$$= \frac{2}{T} \left[ \int_{-T/2}^{0} -\sin\left(\frac{2\pi nt}{T}\right) dt + \int_{0}^{T/2} \sin\left(\frac{2\pi nt}{T}\right) dt \right] = \frac{2}{n\pi}(1 - \cos n\pi)$$

So Equation (3.13) can be written as

$$x(t) = \frac{4}{\pi} \left[ \sin\left(\frac{2\pi t}{T}\right) + \frac{1}{3}\sin\left(\frac{2\pi 3t}{T}\right) + \frac{1}{5}\sin\left(\frac{2\pi 5t}{T}\right) + \cdots \right]$$

(3.16)

We should have anticipated that only a sine wave series is necessary. This follows from the fact that the square wave is an 'odd' function and so does not require the cosine terms which are 'even' (even and odd functions will be commented upon later in this section).

Let us look at the way the successive terms on the right hand side of Equation (3.16) affect the representation. Let $\omega_1 = 2\pi f_1 = 2\pi/T$, so that

$$x(t) = \frac{4}{\pi} \left[ \sin \omega_1 t + \frac{1}{3} \sin 3\omega_1 t + \frac{1}{5} \sin 5\omega_1 t + \cdots \right]$$

(3.17)

Consider 'partial sums' of the series above and their approximation to $x(t)$, i.e. denoted by $S_n(t)$, the sum of $n$ terms, as in Figure 3.3:

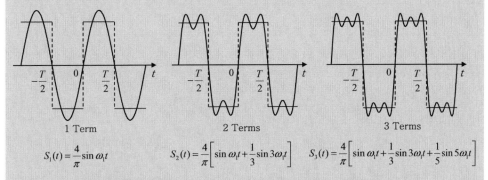

$$S_1(t) = \frac{4}{\pi} \sin \omega_t t$$

1 Term

$$S_2(t) = \frac{4}{\pi} \left[ \sin \omega_t t + \frac{1}{3} \sin 3\omega_t t \right]$$

2 Terms

$$S_3(t) = \frac{4}{\pi} \left[ \sin \omega_t t + \frac{1}{3} \sin 3\omega_t t + \frac{1}{5} \sin 5\omega_t t \right]$$

3 Terms

**Figure 3.3**   Partial sums of the Fourier series of the square wave

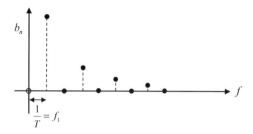

**Figure 3.4**  The coefficients $b_n$ of the Fourier series of the square wave

Note the behaviour of the partial sums near points of discontinuity displaying what is known as the 'overshoot' or *Gibbs' phenomenon*, which will be discussed shortly. The above Fourier series can be represented in the frequency domain, where it appears as a *line* spectrum as shown in Figure 3.4.

We now note some aspects of Fourier series.

### *Convergence of the Fourier Series*

We have assumed that a periodic function may be represented by a Fourier series. Now we state (without proof) the conditions (known as the Dirichlet conditions, see Oppenheim *et al.* (1997) for more details) under which a Fourier series representation is possible. The *sufficient conditions* are follows. If a bounded periodic function with period $T_P$ is piecewise continuous (with a finite number of maxima, minima and discontinuities) in the interval $-T_P/2 < t \leq T_P/2$ and has a left and right hand derivative at each point $t_0$ in that interval, then its Fourier series converges. The sum is $x(t_0)$ if $x$ is continuous at $t_0$. If $x$ is not continuous at $t_0$, then the sum is the average of the left and right hand limits of $x$ at $t_0$. In the square wave example above, at $t = 0$ the Fourier series converges to

$$\frac{1}{2}\left[\lim_{t \to 0+} x(t) + \lim_{t \to 0-} x(t)\right] = \frac{1}{2}[1 - 1] = 0$$

### *Gibbs' Phenomenon*[M3.1]

When a function is approximated by a partial sum of a Fourier series, there will be a significant error in the vicinity of a discontinuity, no matter how many terms are used for the partial sum. This is known as Gibbs' phenomenon.

Consider the square wave in the previous example. As illustrated in Figure 3.5, near the discontinuity the continuous terms of the series struggle to simulate a sudden jump. As the number of terms in the partial sum is increased, the 'ripples' are squashed towards the point of the discontinuity, but the overshoot does not reduce to zero. In this example it turns out that a lower bound on the overshoot is about 9% of the height of the discontinuity (Oppenheim *et al.*, 1997).

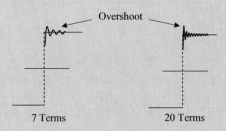

**Figure 3.5** Illustrations of Gibbs' phenomenon

## Differentiation and Integration of Fourier Series

If $x(t)$ satisfies the Dirichlet conditions, then its Fourier series may be integrated term by term. Integration 'smoothes' jumps and so results in a series whose convergence is enhanced. Satisfaction of the Dirichlet conditions by $x(t)$ does not justify differentiation term by term. But, if periodic $x(t)$ is continuous and its derivative, $\dot{x}(t)$, satisfies the Dirichlet conditions, then the Fourier series of $\dot{x}(t)$ may be obtained from the Fourier series of $x(t)$ by differentiating term by term. Note, however, that these are general guidelines only. Each situation should be considered carefully. For example, the integral of a periodic function for which $a_0 \neq 0$ (mean value of the signal is not zero) is no longer periodic.

## Even and Odd Functions

A function $x(t)$ is even if $x(t) = x(-t)$, as shown for example in Figure 3.6.
A function $x(t)$ is odd if $x(t) = -x(-t)$, as shown for example in Figure 3.7.
Any function $x(t)$ may be expressed as the sum of even and odd functions, i.e.

$$x(t) = \frac{1}{2}[x(t) + x(-t)] + \frac{1}{2}[x(t) - x(-t)] = x_e(t) + x_o(t) \tag{3.18}$$

If $x(t)$ and $y(t)$ are two functions, then the following four properties hold:

1. If $x(t)$ is odd and $y(t)$ is odd, then $x(t) \cdot y(t)$ is even.
2. If $x(t)$ is odd and $y(t)$ is even, then $x(t) \cdot y(t)$ is odd.
3. If $x(t)$ is even and $y(t)$ is odd, then $x(t) \cdot y(t)$ is odd.
4. If $x(t)$ is even and $y(t)$ is even, then $x(t) \cdot y(t)$ is even.

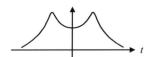

**Figure 3.6** An example of an even function

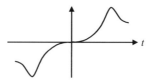

**Figure 3.7**   An example of an odd function

Also:

1. If $x(t)$ is odd, then $\int_{-a}^{a} x(t)dt = 0$.
2. If $x(t)$ is even, then $\int_{-a}^{a} x(t)dt = 2 \int_{0}^{a} x(t)dt$.

### Fourier Series of Odd and Even Functions

If $x(t)$ is an odd periodic function with period $T_P$, then

$$x(t) = \sum_{n=1}^{\infty} \left[ b_n \sin\left(\frac{2\pi n t}{T_P}\right) \right]$$

It is a series of odd terms only with a zero mean value, i.e. $a_n = 0$, $n = 0, 1, 2, \ldots$. If $x(t)$ is an even periodic function with period $T_P$, then

$$x(t) = \frac{a_0}{2} + \sum_{n=1}^{\infty} \left[ a_n \cos\left(\frac{2\pi n t}{T_P}\right) \right]$$

It is now a series of even terms only, i.e. $b_n = 0$, $n = 1, 2, \ldots$.

   We now have a short 'mathematical aside' to introduce the delta function, which turns out to be very convenient in signal analysis.

## 3.2  THE DELTA FUNCTION

The Dirac delta function is denoted by $\delta(t)$, and is sometimes called the unit impulse function. Mathematically, it is defined by

$$\delta(t) = 0 \quad \text{for } t \neq 0, \quad \text{and} \quad \int_{-\infty}^{\infty} \delta(t)dt = 1 \tag{3.19}$$

This is not an ordinary function, but is classified as a 'generalized' function. We may consider this function as a very narrow and tall spike at $t = 0$ as illustrated in Figure 3.8. Then, Figure 3.8 can be expressed by Equation (3.20), where the integration of the function is $\int_{-\infty}^{\infty} \delta_\varepsilon(t)dt = 1$. This is a unit *impulse*:

$$\delta_\varepsilon(t) = \frac{1}{\varepsilon} \quad \text{for } -\frac{\varepsilon}{2} < t < \frac{\varepsilon}{2}$$

$$= 0 \quad \text{otherwise} \tag{3.20}$$

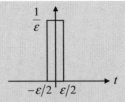

**Figure 3.8**  Model of a unit impulse

Now, if we visualize the spike as infinitesimally narrow, i.e. $\delta(t) = \lim_{\varepsilon \to 0} \delta_\varepsilon(t)$, then we may represent it on a graph as shown in Figure 3.9, i.e. an arrow whose height indicates the magnitude of the impulse.

**Figure 3.9**  Graphical representation of the delta function

An alternative interpretation of the unit impulse is to express it in terms of the unit step function $u(t)$ defined as

$$\begin{aligned} u(t) &= 1 \quad \text{for } t > 0 \\ &= 0 \quad \text{for } t < 0 \end{aligned} \tag{3.21}$$

Since Equation (3.20) can be obtained by using two unit step functions appropriately, i.e. $\delta_\varepsilon(t) = (1/\varepsilon)\,[u\,(t + \varepsilon/2) - u\,(t - \varepsilon/2)]$, the Dirac delta function and the unit step function have the following relationship, which is the Dirac delta function as the derivative of the unit step function:

$$\delta(t) = \lim_{\varepsilon \to 0} \delta_\varepsilon(t) = \frac{d}{dt} u(t) \tag{3.22}$$

Note that the concept of Equation (3.22) makes it possible to deal with differentiating functions that contain discontinuities.

### Properties of the Delta Function

A shifted delta function: if a delta function is located at $t = a$, then it can be written as $\delta(t - a)$. Some useful properties of the delta function are:

1. $\delta(t) = \delta(-t)$, i.e. a delta function is an even function.
2. *Sifting property:* if $x(t)$ is an 'ordinary' function, then the integral of the product of the ordinary function and a shifted delta function is

$$\int_{-\infty}^{\infty} x(t)\delta(t - a)dt = x(a) \tag{3.23}$$

i.e. the delta function 'sifts out' the value of the ordinary function at $t = a$.

The result (3.23) is justified in Figure 3.10 which shows the product of an ordinary function and a shifted $\delta_\varepsilon(t)$, i.e.

$$I_\varepsilon = \int_{-\infty}^{\infty} x(t)\delta_\varepsilon(t-a)dt \qquad (3.24)$$

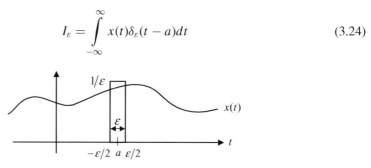

**Figure 3.10**   Graphical illustration of the sifting property

We evaluate this integral and then let $\varepsilon \to 0$ to obtain Equation (3.23), i.e.

$$I_\varepsilon = \frac{1}{\varepsilon} \int_{a-\varepsilon/2}^{a+\varepsilon/2} x(t)dt \qquad (3.25)$$

This is the average height of $x(t)$ within the range $a - \varepsilon/2 < t < a + \varepsilon/2$ and we write this as $x(a + \theta\varepsilon/2)$, $|\theta| < 1$. So $I_\varepsilon = x(a + \theta\varepsilon/2)$, from which $\lim_{\varepsilon \to 0} I_\varepsilon = x(a)$. This justifies Equation (3.23).

3. $\quad \displaystyle\int_{-\infty}^{\infty} e^{\pm j2\pi at} dt = \delta(a), \quad \text{or} \quad \int_{-\infty}^{\infty} e^{\pm jat} dt = 2\pi\,\delta(a) \qquad (3.26)$

The justification of this property is given below, where the delta function is described in terms of the limiting form of a tall narrow 'sinc function' as shown in Figure 3.11:

$$\int_{-\infty}^{\infty} e^{\pm j2\pi at} dt = \lim_{M \to \infty} \int_{-M}^{M} (\cos 2\pi at \pm j \sin 2\pi at)dt = \lim_{M \to \infty} \int_{-M}^{M} (\cos 2\pi at)dt$$

$$= \lim_{M \to \infty} 2 \left. \frac{\sin 2\pi at}{2\pi a} \right|_0^M = \lim_{M \to \infty} 2M \frac{\sin 2\pi aM}{2\pi aM} = \delta(a) \qquad (3.27)$$

Note that it can be verified that the integral of the function in Figure 3.11 is unity (see Appendix A).

4. $\delta(at) = \dfrac{1}{|a|}\delta(t)$, where $a$ is an arbitrary constant $\qquad (3.28)$

5. $\quad \displaystyle\int_{-\infty}^{\infty} f(t)\delta^{(n)}(t-a)dt = (-1)^n f^{(n)}(a)$, where $(n)$ denotes the $n$th derivative $\qquad (3.29)$

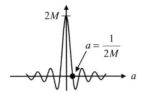

**Figure 3.11**  Representation of the delta function using a sinc function

## 3.3 FOURIER SERIES AND THE DELTA FUNCTION

As already noted, we can differentiate discontinuous functions if we introduce delta functions. Let us consider the Fourier series of derivatives of discontinuous periodic functions. Consider an example of a discontinuous function $x(t)$ as shown in Figure 3.12, whose Fourier series is given as

$$x(t) = \frac{1}{2} + \frac{1}{\pi} \sum_{n=1}^{\infty} \frac{1}{n} \sin\left(\frac{2\pi nt}{T}\right)$$

(Note that this is an odd function offset with a d.c. component.)

Differentiating the function in the figure gives

$$-\frac{1}{T} + \sum_{n=-\infty}^{\infty} \delta(t - nT)$$

and differentiating the Fourier series term by term gives

$$\frac{2}{T} \sum_{n=1}^{\infty} \cos\left(\frac{2\pi nt}{T}\right)$$

Equating these gives

$$\sum_{n=-\infty}^{\infty} \delta(t - nT) = \frac{1}{T} + \frac{2}{T} \sum_{n=1}^{\infty} \cos\left(\frac{2\pi nt}{T}\right) \qquad (3.30)$$

This is a periodic 'train' of impulses, and it has a Fourier series representation whose coefficients are constant $(2/T)$ for all frequencies except for the d.c. component. The periodic train of impulses (usually written $\delta_T(t)$ or $i(t)$) is drawn as in Figure 3.13, and will be used in sampling theory later.

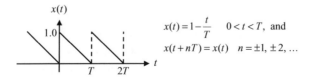

$$x(t) = 1 - \frac{t}{T} \qquad 0 < t < T, \text{ and}$$

$$x(t + nT) = x(t) \qquad n = \pm1, \pm2, \dots$$

**Figure 3.12**  An example of a discontinuous periodic function

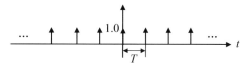

**Figure 3.13**   The periodic train of impulses

So far we have used sines and cosines explicitly in the Fourier series. These functions can be combined using $e^{\pm j\theta} = \cos\theta \pm j\sin\theta$ and so sines and cosines can be replaced by complex exponentials. This is the basis of the complex form of the Fourier series.

## 3.4 THE COMPLEX FORM OF THE FOURIER SERIES

It is often convenient to express a Fourier series in terms of $e^{\pm j\omega_1 nt}$ ($\omega_1 = 2\pi/T_p$). Note that

$$\cos\theta = \frac{1}{2}\left(e^{j\theta} + e^{-j\theta}\right) \quad \text{and} \quad \sin\theta = \frac{1}{2j}\left(e^{j\theta} - e^{-j\theta}\right)$$

so the Fourier series defined in Equation (3.3)

$$x(t) = \frac{a_0}{2} + \sum_{n=1}^{\infty}\left[a_n\cos\left(\frac{2\pi nt}{T_P}\right) + b_n\sin\left(\frac{2\pi nt}{T_P}\right)\right]$$

becomes

$$x(t) = \frac{a_0}{2} + \sum_{n=1}^{\infty}\left[\frac{a_n}{2}\left(e^{j2\pi nt/T_P} + e^{-j2\pi nt/T_P}\right) + \frac{b_n}{2j}\left(e^{j2\pi nt/T_P} - e^{-j2\pi nt/T_P}\right)\right]$$

$$= \frac{a_0}{2} + \sum_{n=1}^{\infty}\frac{1}{2}\left(a_n + \frac{b_n}{j}\right)e^{j2\pi nt/T_P} + \sum_{n=1}^{\infty}\frac{1}{2}\left(a_n - \frac{b_n}{j}\right)e^{-j2\pi nt/T_P}$$

$$= \frac{a_0}{2} + \sum_{n=1}^{\infty}\frac{a_n - jb_n}{2}e^{j2\pi nt/T_P} + \sum_{n=1}^{\infty}\frac{a_n + jb_n}{2}e^{-j2\pi nt/T_P} \tag{3.31}$$

Let $c_0 = a_0/2$, $c_n = (a_n - jb_n)/2$, so $c_n^* = (a_n + jb_n)/2$, i.e.

$$x(t) = c_0 + \sum_{n=1}^{\infty}c_n e^{j2\pi nt/T_P} + \sum_{n=1}^{\infty}c_n^* e^{-j2\pi nt/T_P} \tag{3.32}$$

Substituting for $a_n$ and $b_n$ from Equation (3.4) gives

$$c_0 = \frac{1}{T_P}\int_0^{T_P}x(t)dt, \quad c_n = \frac{1}{T_P}\int_0^{T_P}x(t)e^{-j2\pi nt/T_P}dt, \quad c_n^* = \frac{1}{T_P}\int_0^{T_P}x(t)e^{j2\pi nt/T_P}dt = c_{-n}$$

$$\tag{3.33}$$

Note that the *negative frequency* terms $(c_{-n})$ are introduced, so that

$$\sum_{n=1}^{\infty} c_n^* e^{-j2\pi nt/T_P} = \sum_{n=1}^{\infty} c_{-n} e^{-j2\pi nt/T_P}$$

in Equation (3.32). Thus, we obtain the very important results:

$$x(t) = \sum_{n=-\infty}^{\infty} c_n e^{j2\pi nt/T_P} \tag{3.34}$$

$$c_n = \frac{1}{T_P} \int_0^{T_P} x(t) e^{-j2\pi nt/T_P} dt \tag{3.35}$$

Note that the 'basic elements' are now *complex* exponentials and the Fourier coefficients $c_n$ are also complex, representing both *amplitude* and *phase* information. Note also that the notion of 'negative frequencies', i.e. $f_n = n/T_P$, $n = 0, \pm 1, \pm 2, \ldots$, has been introduced by the *algebraic manipulation* in Equation (3.33).

Referring to Equation (3.12)

$$x(t) = \frac{a_0}{2} + \sum_{n=1}^{\infty} M_n \cos(2\pi nf_1 t + \phi_n)$$

the relationship between the coefficients is given by

$$\left. \begin{aligned} c_n &= \frac{a_n - jb_n}{2} \quad \text{so } |c_n| = \frac{1}{2}\sqrt{a_n^2 + b_n^2} = \frac{M_n}{2} \\ \text{and} \quad \arg c_n &= \tan^{-1}\left(-\frac{b_n}{a_n}\right) = \phi_n \end{aligned} \right\} \quad \text{for } n \neq 0 \tag{3.36}$$

All previous discussions on Fourier series still hold except that now manipulations are considerably easier using the complex form. We note a generalization of the concept of orthogonality of functions for the complex case. *Complex-valued* functions $u_n(t)$ are orthogonal if

$$\int_0^{T_P} u_n(t) u_m^*(t) dt = 0 \quad \text{for } m \neq n \tag{3.37}$$

This is easily verified by using $u_n(t) = e^{j2\pi nt/T_P}$. Also, when $n = m$, the integral is $T_P$.

## 3.5 SPECTRA

We now introduce the notion of the spectrum of a process. We shall refer to the complex representation of Equation (3.34) using positive and negative frequencies and also Equation (3.12) using only positive frequencies. Referring to Equation (3.34) first, a plot of the magnitude $|c_n|$ versus frequency $f$ (or $\omega$) is called the amplitude spectrum of the periodic function $x(t)$. A plot of the phase angle $\arg c_n$ versus frequency is called the phase spectrum. These are *not* continuous curves but take values only at discrete values for $f = n/T_P$, $n = 0$, $\pm 1$, $\pm 2$, .... We can draw these spectra as in Figures 3.14 and 3.15 respectively.

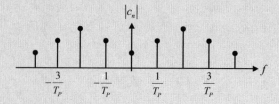

**Figure 3.14**  Amplitude spectrum of a Fourier series (a *line spectrum* and an *even function*)

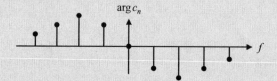

**Figure 3.15**  Phase spectrum of a Fourier series (a *line spectrum* and an *odd function*)

If we did not want to include negative frequencies, we could plot $M_n$, $\phi_n$ (Equation (3.12)) versus $n$ above (note that $M_n = 2|c_n|$ for $n \neq 0$).

As an example, consider a periodic function that can be depicted as in Figure 3.16. A calculation will give the coefficients as

$$c_n = \frac{(Ad/T)\sin(n\pi d/T)}{(n\pi d/T)}, \arg c_n = 0$$

Since this function is even the phase spectrum is zero for all frequency components. If, for example, $T = 1/4$, $d = 1/20$, then the amplitude spectrum is as given in Figure 3.17.

If the function is shifted to the right by $d/2$, then the function is depicted as in Figure 3.18. Then, $|c_n|$ is *unchanged* but the phase components are changed, so that $\arg c_n = -n\pi(d/T)$ rads.

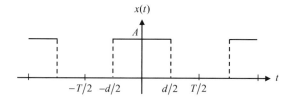

**Figure 3.16**  A periodic function of a rectangular pulse

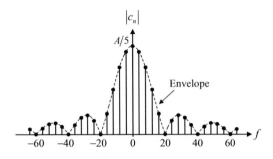

**Figure 3.17**   Amplitude spectrum of the function given in Figure 3.16

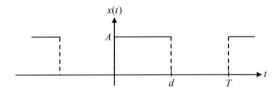

**Figure 3.18**   A periodic function of a rectangular pulse shifted to the right by $d/2$

---

***Parseval's Theorem – The Power Spectrum***

Suppose $x(t)$ is interpreted as a voltage. Then the instantaneous power dissipated across a 1 ohm resistor is $x^2(t)$, and the average power dissipated across the resistor is

$$\frac{1}{T_P} \int_0^{T_P} x^2(t)dt$$

Now, using Equation (3.34),

$$x(t) = \sum_{n=-\infty}^{\infty} c_n e^{j2\pi nt/T_P}$$

the voltage squared is

$$x^2(t) = x(t) \cdot x^*(t) = \sum_{n=-\infty}^{\infty} c_n e^{j2\pi nt/T_P} \cdot \sum_{m=-\infty}^{\infty} c_m^* e^{-j2\pi mt/T_P}$$

Thus, the average power can be written as

$$\frac{1}{T_P} \int_0^{T_P} x^2(t)dt = \frac{1}{T_P} \sum_{n=-\infty}^{\infty} \sum_{m=-\infty}^{\infty} c_n c_m^* \int_0^{T_P} e^{j2\pi(n-m)t/T_P} dt \qquad (3.38)$$

By the property of orthogonality, this is reduced to the following form known as (one form of) *Parseval's theorem*:

$$\frac{1}{T_P}\int_0^{T_P} x^2(t)dt = \sum_{n=-\infty}^{\infty} |c_n|^2 \tag{3.39}$$

This has a 'physical' interpretation. It indicates that the average power of the signal $x(t)$ may be regarded as the sum of power associated with individual frequency components. The power spectrum is $|c_n|^2$ and may be drawn (typically) as in Figure 3.19. It is a decomposition of the power of the process over frequency. Note that the power spectrum is *real* valued and *even* (and there is *no* phase information).

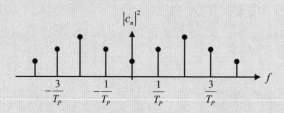

**Figure 3.19**  An example of a power spectrum (Compare with Figure 3.14)

If we wished to restrict ourselves to positive frequencies only, we could fold the left hand portion over to double the values at frequencies $f = n/T_P, n = 1, 2, \ldots$. The name 'periodogram' is sometimes given to this power decomposition.

## 3.6 SOME COMPUTATIONAL CONSIDERATIONS[M3.2]

When calculating the Fourier coefficients of a periodic signal which we *measure*, it is important to be able to identify the period of the signal. For example, if $x(t)$ has the form shown in Figure 3.20, the period is $T_P$.

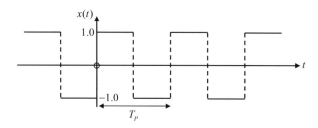

**Figure 3.20**  A periodic signal with a period $T_P$

If we *acquire* (record or measure) exactly one period, we can calculate Fourier coefficients correctly (subject to 'computational errors') from the formula given by Equation (3.35), i.e.

$$c_n = \frac{1}{T_P} \int_0^{T_P} x(t)e^{-j2\pi nt/T_P} dt$$

If we acquire $rT_P$ seconds of data and use the formula

$$c_n = \frac{1}{rT_P} \int_0^{rT_P} x(t)e^{-j2\pi nt/T_P} dt \tag{3.40}$$

then if $r$ is an integer we can obtain the same Fourier coefficients spaced at frequencies $1/T_P$ along the frequency axis.

However, if $r$ is not an integer and we use the formula

$$c_n = \frac{1}{rT_P} \int_0^{rT_P} x(t)e^{-j\frac{2\pi n}{rT_P}t} dt \tag{3.41}$$

then the Fourier coefficients need careful consideration. For example, if we use the period of $1.5T_P$ (note that $r$ is no longer an integer), then we are *assuming* that the signal is as shown in Figure 3.21 (compare this with the true signal in Figure 3.20).

Clearly, the Fourier coefficients change (we see immediately that there is a *non-zero mean value*) and frequencies are present at every $1/1.5T_P$ Hz.

In practice, if the period is not known, then it is necessary to capture a large number of periods so that 'end effects' are small. This point should be noted since computational methods of obtaining Fourier coefficients often restrict the data set of $N$ points where $N = 2^M$, i.e. a power of two ($M$ is an integer). This means we may analyse a non-integer number of periods. These features are now demonstrated for a square wave.

For the square wave shown in Figure 3.20 (a period of $T_P$), the theoretical Fourier coefficient $c_n$ has magnitude

$$
\left.
\begin{aligned}
|c_n| &= \frac{2}{n\pi} && \text{for } n = \text{odd} \\
&= 0 && \text{for } n = 0, \text{ even}
\end{aligned}
\right\} \tag{3.42}
$$

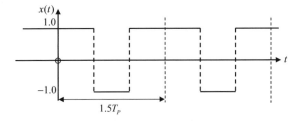

**Figure 3.21** A periodic signal with a period of $1.5T_P$

If $T_P = 1$ second, frequency components $|c_n|$ occur at every $1/T_P = 1$ Hz. Figure 3.22(a) shows $|c_n|$ plotted (for $n \geq 0$ only) up to 10 Hz for an exact single period using Equation (3.35) (or using Equation (3.41) with $r = 1$). Figure 3.22(b) shows $x(t)$ for *two* periods using Equation (3.41), where $r = 2$. By comparing these two figures, it can be seen that the Fourier coefficients are exactly the same except that Figure 3.22(b) has more 'zeros' due to the fact that we calculate the coefficients at every $1/2T_P = 0.5$ Hz.

Figures 3.22(c) and (d) show changes in the amplitude spectra when a non-integer number of periods is taken (i.e. non-integer $r$) and Equation (3.41) is used. In Figure 3.22(c), 1.5 periods are used. Note that it appears that there are no frequency components at 1, 3, 5, ... Hz as indicated in Figures 3.22(a) and (b). For Figure 3.22(d), 3.5 periods are taken. The increased 'density' of the line frequencies shows maxima near the true values (also refer to Figure 3.22(e) where 10.5 periods are taken). Note that in Figures 3.22(c)–(e) the amplitudes have decreased. This follows since there is an increased density of frequency components in the decomposition. Recall Parseval's identity (theorem)

$$\frac{1}{T_P} \int_0^{T_P} x^2(t)dt = \sum_{n=-\infty}^{\infty} |c_n|^2$$

and note the $1/T_P$ on the left hand side. For the square wave considered, the average power is always unity when an integer number of periods is included. When a non-integer number of periods is included the increased density of frequency components means that the amplitudes change.

### Some Comments on the Computation of Fourier Coefficients

This is an introductory comment on the computation of Fourier coefficients, which will be expanded later in Chapter 6. We address the problem of performing the following integral using digital techniques:

$$c_k = \frac{1}{T_P} \int_0^{T_P} x(t)e^{-j2\pi kt/T_P} dt \tag{3.43}$$

Consider an arbitrary signal measured for $T_P$ seconds as shown in Figure 3.23. Suppose the signal is sliced as shown in the figure, and the values of $x(t)$ at $N$ discrete points $x(n\Delta)$, $n = 0$, 1, 2, ..., $N-1$, are known, each point separated by a time interval $\Delta$, say. Then a logical and simple approximation to the integral of Equation (3.43) is

$$c_k \approx \frac{1}{N\Delta} \sum_{n=0}^{N-1} x(n\Delta)e^{-j\frac{2\pi kn\Delta}{N\Delta}} \cdot \Delta \tag{3.44}$$

or

$$c_k \approx \frac{1}{N} \sum_{n=0}^{N-1} x(n\Delta)e^{-j\frac{2\pi kn}{N}} = \frac{X_k}{N} \text{ (say)} \tag{3.45}$$

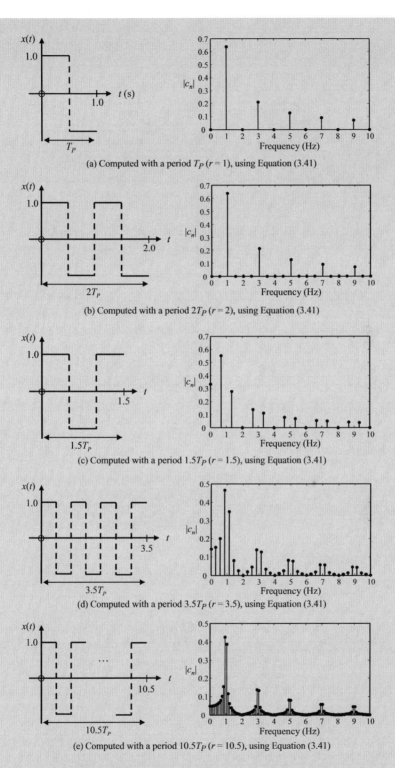

(a) Computed with a period $T_P$ ($r = 1$), using Equation (3.41)

(b) Computed with a period $2T_P$ ($r = 2$), using Equation (3.41)

(c) Computed with a period $1.5T_P$ ($r = 1.5$), using Equation (3.41)

(d) Computed with a period $3.5T_P$ ($r = 3.5$), using Equation (3.41)

(e) Computed with a period $10.5T_P$ ($r = 10.5$), using Equation (3.41)

**Figure 3.22**    Fourier coefficients of a square wave

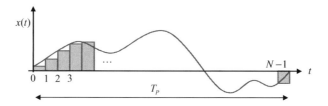

**Figure 3.23**  An arbitrary signal measured for $T_P$ seconds

The expression $X_k$ on the right hand side of Equation (3.45) is called a discrete Fourier transform (DFT) and we note some important points that will be further discussed later:

1. It is a *finite* summation.
2. Do *not* assume $X_k/N \equiv c_k$. In fact $X_k$ turns out to be periodic (proved below), i.e. $X_k = X_{k+rN}$, where $r$ is an integer, though it would seem reasonable to expect that, if $\Delta$ is 'sufficiently small', then $c_k \approx X_k/N$ for at least some 'useful' range of values for $k$.
   *Proof of the periodic nature of the DFT*: From

$$X_k = \sum_{n=0}^{N-1} x_n e^{-j\frac{2\pi}{N}nk}$$

   (note the change in notation $x_n \equiv x(n\Delta)$), substitute $k$ by $k + rN$ ($r$ is an integer). Then the equation becomes

$$X_{k+rN} = \sum_{n=0}^{N-1} x_n e^{-j\frac{2\pi}{N}n(k+rN)} = \sum_{n=0}^{N-1} x_n e^{-j\frac{2\pi}{N}nk} \underbrace{e^{-j2\pi nr}}_{=1.0}$$

   and thus $X_{k+rN} = X_k$.
3. The DFT relationship

$$X_k = \sum_{n=0}^{N-1} x_n e^{-j\frac{2\pi}{N}nk} \tag{3.46}$$

has the inverse relationship (IDFT)

$$x_n = \frac{1}{N} \sum_{k=0}^{N-1} X_k e^{j\frac{2\pi}{N}nk} \tag{3.47}$$

So, although $X_k$ may not provide enough information to allow the continuous time series $x(t)$ to be obtained, it is important to realize that it does permit the discrete values of the series $x_n$ to be regained exactly.

*Proof of the inverse DFT (IDFT) relationship*: Starting from

$$X_k = \sum_{n=0}^{N-1} x_n e^{-j\frac{2\pi}{N}nk}$$

multiply both sides by $e^{j(2\pi/N)sk}$ and sum over $k$ ($s$ is an integer, $0 \le s \le N - 1$). This yields

$$\sum_{k=0}^{N-1} X_k e^{j\frac{2\pi}{N}sk} = \sum_{k=0}^{N-1}\sum_{n=0}^{N-1} x_n e^{-j\frac{2\pi}{N}nk} e^{j\frac{2\pi}{N}sk} = \sum_{k=0}^{N-1}\sum_{n=0}^{N-1} x_n e^{j\frac{2\pi}{N}(s-n)k} = \sum_{n=0}^{N-1} x_n \sum_{k=0}^{N-1} e^{j\frac{2\pi}{N}(s-n)k}$$

Consider the second summation. Let $s - n = m$ (integer); then we get $\sum_{k=0}^{N-1} e^{j\frac{2\pi}{N}mk}$.
(a) If $m = 0$, this is $N$.
(b) If $m \neq 0$, this is a 'geometric series' with common ratio $e^{j\frac{2\pi}{N}m}$ and the sum is

$$S_{N-1} = \frac{1 - \left(e^{j\frac{2\pi}{N}m}\right)^N}{1 - e^{j\frac{2\pi}{N}m}} = 0$$

Thus

$$\sum_{k=0}^{N-1} X_k e^{j\frac{2\pi}{N}sk} = Nx_s, \quad \text{and so} \quad x_s = \frac{1}{N}\sum_{k=0}^{N-1} X_k e^{j\frac{2\pi}{N}sk}$$

or more usually

$$x_n = \frac{1}{N}\sum_{k=0}^{N-1} X_k e^{j\frac{2\pi}{N}nk}$$

4. Some authors have defined the DFT in related but different ways, e.g.

$$X_k = \frac{1}{N}\sum_{n=0}^{N-1} x_n e^{-j\frac{2\pi}{N}nk} \tag{3.48}$$

Clearly such differences are ones of scale only. We shall use Equations (3.46) and (3.47) since these are widely adopted as 'standard' in signal processing.

5. $N$ is of course arbitrary above, but is often chosen to be a power of two ($N = 2^M$, $M$ an integer) owing to the advent of efficient Fourier transform algorithms called fast Fourier transforms (FFTs).

6. We have introduced the DFT as an approximation for calculating Fourier coefficients. However, we shall see that a formal body of theory has been constructed for 'discrete-time' systems in which the properties are exact and must be considered to stand on their own. Analogies with continuous-time theory are not always useful and in some cases are confusing.

## 3.7 BRIEF SUMMARY

1. A periodic signal of period $T_P$ may be expressed (Equations (3.34) and (3.35)) by

$$x(t) = \sum_{n=-\infty}^{\infty} c_n e^{j2\pi nt/T_P} \quad \text{with} \quad c_n = \frac{1}{T_P} \int_0^{T_P} x(t) e^{-j2\pi nt/T_P} dt$$

2. The plots of $|c_n|$ versus frequency and arg $c_n$ versus frequency are amplitude and phase (*line*) spectra of the Fourier decomposition.
3. The average power of a periodic signal is described by Equation (3.39), i.e.

$$\frac{1}{T_P} \int_0^{T_P} x^2(t) dt = \sum_{n=-\infty}^{\infty} |c_n|^2 \quad \text{Parseval's theorem (identity)}$$

A plot of $|c_n|^2$ versus frequency is called a power spectrum (or a periodogram).
4. The DFT and IDFT relationships for discrete data are defined by Equations (3.46) and (3.47),

$$X_k = \sum_{n=0}^{N-1} x_n e^{-j\frac{2\pi}{N}nk} \quad \text{and} \quad x_n = \frac{1}{N} \sum_{k=0}^{N-1} X_k e^{j\frac{2\pi}{N}nk}$$

The Fourier coefficients $c_k$ are approximated by $c_k \approx X_k/N$ if an integer number of periods is taken and only for a *restricted range* of $k$.

We now include some MATLAB examples illustrating the material covered.

## 3.8 MATLAB EXAMPLES

*Example 3.1:* **Illustration of the convergence of the Fourier series and Gibbs' phenomenon** (see Section 3.1)

Consider Equation (3.17),

$$x(t) = \frac{4}{\pi} \left[ \sin \omega_1 t + \frac{1}{3} \sin 3\omega_1 t + \frac{1}{5} \sin 5\omega_1 t + \cdots \right]$$

In this MATLAB example, we compare the results of 3, 7 and 20 partial sums in Equation (3.17).

| Line | MATLAB code | Comments |
|------|-------------|----------|
| 1<br>2 | clear all<br>t=[0:0.001:1]; | Define the time variable (vector) t from 0 to 1 second with a step size of 0.001. |
| 3 | x=[]; x_tmp=zeros(size(t)); | Define an empty matrix x, and define the vector x_tmp having the same size as the vector t. All the elements of x_tmp are zeros. |
| 4 | for n=1:2:39 | Start a 'for' loop where n are 1, 3, 5,..., 39 (n = 39 implies the 20 partial sums). |
| 5 | x_tmp=x_tmp+4/pi*(1/n*sin(2*pi*n*t)); | MATLAB expression of Equation (3.17), and the result of each partial sum is stored in the vector x_tmp. |
| 6 | x=[x; x_tmp]; | Each row of matrix x has a corresponding partial sum of Equation (3.17). For example, the second row of x corresponds to the sum of two terms (i.e. n=3). |
| 7 | end | End of the 'for' loop. |
| 8<br>9 | plot(t,x(3,:),t,x(7,:),t,x(20,:))<br>xlabel('\itt\rm (seconds)');<br>ylabel('\itx\rm(\itt\rm)') | Plot the results of only 3, 7 and 20 partial sums against the time variable. |
| 10 | grid on | |

## Results

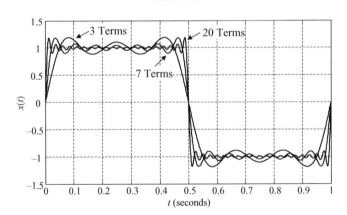

**Comments:** The square wave is better represented as the number of terms is increased, but its errors remain near the discontinuities.

*Example 3.2:* **Fourier coefficients of a square wave** (Figure 3.20, i.e. $T_P = 1$ second and the amplitude is '1') This is examined for various values of $r$ in Equation (3.41) below (see Section 3.6):

$$c_n = \frac{1}{rT_P} \int_0^{rT_P} x(t)e^{-j\frac{2\pi n}{rT_P}t}dt$$

Case 1: $r$ is an integer number. We chose $r = 3$ for this example; however, readers may choose any arbitrary positive integer number. The Fourier coefficients are calculated up to 10 Hz.

| Line | MATLAB code | Comments |
|------|-------------|----------|
| 1<br>2 | clear all<br>r=3; cn=[]; | Define a parameter r for the number of periods, and the empty matrix cn for the Fourier coefficients. |
| 3<br>4 | for n=1:10*r<br>    temp1=0; temp2=0; | Define a 'for' loop for the Fourier coefficients up to 10 Hz, and set temporary variables. |
| 5<br>6<br>7<br>8 | for k = 1:2:2*r<br>    tmp_odd = exp(-i*(k/r)*n*pi);<br>    temp1=temp1+tmp_odd;<br>end | This nested 'for' loop calculates the integral in Equation (3.41) for the intervals of $x(t) = 1$ in Figure 3.20, and stores the result in the variable temp1. |
| 9<br>10<br>11<br>12 | for k = 2:2:2*r-1<br>    tmp_even = -exp(-i*(k/r)*n*pi);<br>    temp2=temp2+tmp_even;<br>end | Another nested 'for' loop, which calculates the integral in Equation (3.41) for the intervals of $x(t) = -1$ in Figure 3.20, and stores the result in the variable temp2. |
| 13 | temp = -1/2 + temp1 + temp2<br>-1/2*exp(-i*2*n*pi); | This completes the calculation of the integral in Equation (3.41). |
| 14 | cn = [cn; i*temp/(pi*n)]; | 'i*temp/(pi*n)' is the final calculation of Equation (3.41) for each value of n. As a result, cn is a '30 × 1' vector, and each row of the vector cn contains the complex-valued Fourier coefficients. |
| 15 | end | End of the 'for' loop. |
| 16 | stem([0:1/r:n/r],[0; abs(cn)], 'o', 'filled') | Plot the result using the 'stem' command. [0:1/r:n/r] defines the frequencies (horizontal axis) from 0 Hz to 10 Hz at every 1/3 Hz.<br>[0; abs(cn)] is the modulus of the Fourier coefficient at each frequency. Note that the value of zero is added for 0 Hz.<br>The result is the amplitude spectrum. |
| 17<br>18 | xlabel('Frequency (Hz)')<br>ylabel('Modulus (\mid\itc_n\rm\mid)') | Insert labels for each axis.<br>'\mid' is for '|', '\it' is for italic font, 'c_n' is for $c_n$, and '\rm' is for normal font. |

**Results**

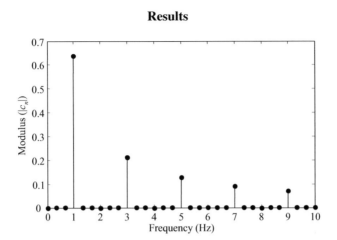

**Comments:** Compare this graph with Figures 3.22(a) and (b) and with the next case.

Case 2: $r$ is not an integer number. We chose $r = 7.5$ for this example; however, readers may choose any arbitrary positive integer number $+ 0.5$. The Fourier coefficients are calculated up to 10 Hz.

| Line | MATLAB code | Comments |
|---|---|---|
| 1<br>2 | clear all<br>r=7.5; r2=ceil(r); cn=[]; | 'ceil' command rounds the element to the nearest integer towards infinity, so in this case, r2 has a value of 8. |
| 3<br>4 | for n=1:10*r<br>  temp1=0; temp2=0; | Same as previous case. |
| 5<br>6<br>7<br>8 | for k = 1:2:2*r2-3<br>  tmp_odd = exp(-i*(k/r)*n*pi);<br>  temp1=temp1+tmp_odd;<br>end | Except for 'k= 1:2:2*r2-3', it is the same script as in the previous case, i.e. it calculates the integral in Equation (3.41) for the intervals of $x(t) = 1$. |
| 9<br>10<br>11<br>12 | for k = 2:2:2*r2-1<br>  tmp_even=-exp(-i*(k/r)*n*pi);<br>  temp2=temp2+tmp_even;<br>end | Same script as in previous case, except for 'k=2:2:2*r2-1'. It is for the intervals of $x(t) = -1$. |
| 13 | temp=-1/2 + temp1 + temp2<br>+1/2*exp(-i*(2*r/r)*n*pi); | This completes the calculation of the integral in Equation (3.41). |
| 14 | cn = [cn; i*temp/(pi*n)]; | Same as in previous case, but now cn is a '75 × 1' vector. |
| 15 | end | End of the 'for' loop. |

| 16 | stem([0:1/r:n/r],[0.5/r; abs(cn)], 'o', 'filled') | Frequencies are from 0 to 10 Hz at every 1/7.5 Hz. 0.5/r is added for 0 Hz (note the non-zero mean value). |
| 17 | xlabel('Frequency (Hz)') | Same as in previous case. |
| 18 | ylabel('Modulus (\mid\itc_n\rm\mid)') | |

## Results

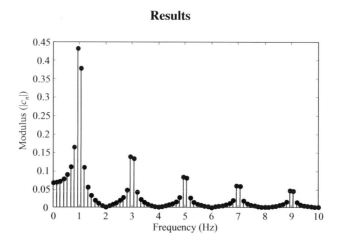

**Comments:** Compare this graph with Figures 3.22(c)–(e) and with the previous case.

# 4

# Fourier Integrals (Fourier Transform) and Continuous-Time Linear Systems

## Introduction

This chapter introduces the central concept for signal representation, the Fourier integral. All classes of signals may be accommodated from this as a starting point – periodic, almost periodic, transient and random – though each relies on a rather different perspective and interpretation.

In addition, the concept of convolution is introduced which allows us to describe linear filtering and interpret the effect of data truncation (windowing). We begin with a derivation of the Fourier integral.

## 4.1 THE FOURIER INTEGRAL

We shall extend Fourier analysis to non-periodic phenomena. The basic change in the representation is that the discrete summation of the Fourier series becomes a continuous summation, i.e. an integral form. To demonstrate this change, we begin with the Fourier series representation of a periodic signal as in Equation (4.1), where the interval of integration is defined from $-T_P/2$ to $T_P/2$ for convenience:

$$x(t) = \sum_{n=-\infty}^{\infty} c_n e^{j2\pi nt/T_P} \quad \text{where} \quad c_n = \frac{1}{T_P} \int_{-T_P/2}^{T_P/2} x(t) e^{-j2\pi nt/T_P} dt \qquad (4.1)$$

As an example, visualize a periodic signal $x(t)$ as having the form below, in Equation (4.2) and Figure 4.1:

$$
\begin{aligned}
x(t) &= 0 & -T_P/2 < t < -1 \\
&= 1 & -1 < t < 1 \quad (T_P/2 > 1) \\
&= 0 & 1 < t < T_P/2
\end{aligned}
\qquad (4.2)
$$

*Fundamentals of Signal Processing for Sound and Vibration Engineers*
K. Shin and J. K. Hammond.    © 2008 John Wiley & Sons, Ltd

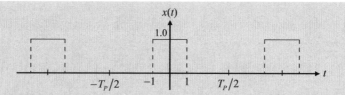

**Figure 4.1**   An example of a periodic signal with a period $T_P$

Now let $T_P$ become large. As this happens we are left with a single 'pulse' near $t = 0$ and the others get further and further away. We examine what happens to the Fourier representation under these conditions.

The fundamental frequency $f_1 = 1/T_P$ becomes smaller and smaller and all other frequencies ($nf_1 = f_n$, say), being multiples of the fundamental frequency, are *more densely packed* on the frequency axis. Their separation is $1/T_P = \Delta f$ (say). So, as $T_P \rightarrow \infty$, $\Delta f \rightarrow 0$, i.e. the integral form of $c_n$ in Equation (4.1) becomes

$$c_n = \frac{1}{T_P} \int_{-T_P/2}^{T_P/2} x(t)e^{-j2\pi nt/T_P}dt \quad \rightarrow \quad c_n = \lim_{\substack{T_P \rightarrow \infty \\ (\Delta f \rightarrow 0)}} \Delta f \int_{-T_P/2}^{T_P/2} x(t)e^{-j2\pi f_n t}dt \quad (4.3)$$

If the integral is finite the Fourier coefficients $c_n \rightarrow 0$ as $\Delta f \rightarrow 0$ (i.e. the more frequency components there are, the smaller are their amplitudes). To avoid this rather unhelpful result, it is more desirable to form the ratio $c_n/\Delta f$, and so it can be rewritten as

$$\lim_{\Delta f \rightarrow 0} \left( \frac{c_n}{\Delta f} \right) = \lim_{T_P \rightarrow \infty} \int_{-T_P/2}^{T_P/2} x(t)e^{-j2\pi f_n t}dt \quad (4.4)$$

Assuming that the limits exist, we write this as

$$X(f_n) = \int_{-\infty}^{\infty} x(t)e^{-j2\pi f_n t}dt \quad (4.5)$$

Since $\Delta f \rightarrow 0$, the frequencies $f_n$ in the above representation become a continuum, so we write $f$ instead of $f_n$. From Equation (4.4), $X(f_n)$ which is now expressed as $X(f)$ is an amplitude divided by bandwidth or *amplitude density* which is called the Fourier integral or the Fourier transform of $x(t)$, written as

$$X(f) = \int_{-\infty}^{\infty} x(t)e^{-j2\pi ft}dt \quad (4.6)$$

Now consider the corresponding change in the representation of $x(t)$ as the sum of sines and cosines, i.e. $x(t) = \sum_{n=-\infty}^{\infty} c_n e^{j2\pi nt/T_P}$. Using the above results,

$$\lim_{\Delta f \rightarrow 0} \left( \frac{c_n}{\Delta f} \right) = X(f_n)$$

$x(t)$ can be rewritten as

$$x(t) = \lim_{\Delta f \to 0} \sum_{n=-\infty}^{\infty} X(f_n)\Delta f \cdot e^{j2\pi nt/T_P} \tag{4.7}$$

which can be represented in a continuous form as

$$x(t) = \int_{-\infty}^{\infty} X(f)e^{j2\pi ft}df \tag{4.8}$$

Equations (4.6) and (4.8) are called the Fourier integral pair.

## Comments on the Fourier Integral

1. Interpretation and appearance of the Fourier transform: $X(f)$ is a (*complex*) amplitude density. From the representation $x(t) = \int_{-\infty}^{\infty} X(f)e^{j2\pi ft}df$, we see that $|X(f)|\,df$ represents the contribution in magnitude (to $x(t)$) of the frequency components in a narrow band near frequency $f$. Since $X(f)$ is complex, we may write

$$X(f) = X_{Re}(f) + jX_{Im}(f) = |X(f)|\,e^{j\phi(f)} \tag{4.9}$$

where $|X(f)|$ is the magnitude (or amplitude) spectrum and $\phi(f)$ is the phase spectrum. When $x(t)$ is in volts, $|X(f)|$ is in volts/Hz.

If $x(t)$ is real valued, $X_{Re}(f)$ is an *even* function and $X_{Im}(f)$ is an *odd* function, and also $|X(f)|$ is an *even* function while $\phi(f)$ is an *odd* function. A typical display of $X(f)$ may look like that shown in Figure 4.2. An alternative way of displaying $X(f)$ is to use the 'polar (or Nyquist) diagram' as shown in Figure 4.3, where the positive frequency $(+f)$ is drawn clockwise and the negative frequency $(-f)$ is drawn anti-clockwise. Note the relationship between the 'magnitude/phase' pair with the 'real/imaginary' pair in these figures.

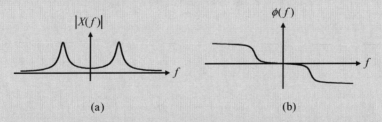

(a)  (b)

**Figure 4.2** Typical display of $X(f)$: (a) magnitude spectrum, (b) phase spectrum

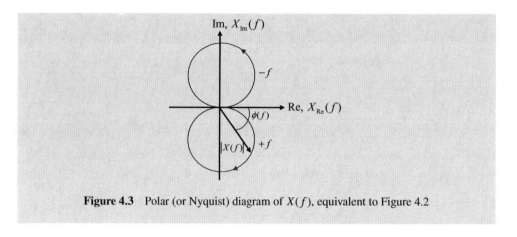

**Figure 4.3**   Polar (or Nyquist) diagram of $X(f)$, equivalent to Figure 4.2

2. We have chosen to derive Equations (4.6) and (4.8) using $f$(Hz). Often $\omega$ is used and alternatives to the above equations are

$$
x(t) = \frac{1}{2\pi} \int_{-\infty}^{\infty} X(\omega)e^{j\omega t}\,d\omega \quad \text{and} \quad X(\omega) = \int_{-\infty}^{\infty} x(t)e^{-j\omega t}\,dt \tag{4.10}
$$

$$
x(t) = \int_{-\infty}^{\infty} X(\omega)e^{j\omega t}\,d\omega \quad \text{and} \quad X(\omega) = \frac{1}{2\pi} \int_{-\infty}^{\infty} x(t)e^{-j\omega t}\,dt \tag{4.11}
$$

$$
x(t) = \frac{1}{\sqrt{2\pi}} \int_{-\infty}^{\infty} X(\omega)e^{j\omega t}\,d\omega \quad \text{and} \quad X(\omega) = \frac{1}{\sqrt{2\pi}} \int_{-\infty}^{\infty} x(t)e^{-j\omega t}\,dt \tag{4.12}
$$

So, the definition used must be noted carefully. Equations (4.10) are a common alternative which we shall use when necessary, and Equations (4.11) and (4.12) are not used in this book.

3. The inversion of the Fourier pair is often accomplished using the delta function. In order to be able to do this we need to use the properties of the delta function. Recall Equation (3.26), i.e.

$$
\int_{-\infty}^{\infty} e^{\pm j2\pi a t}\,dt = \delta(a), \quad \text{or} \quad \int_{-\infty}^{\infty} e^{\pm j a t}\,dt = 2\pi\delta(a)
$$

We now demonstrate the inversion. We start with $x(t) = \int_{-\infty}^{\infty} X(f)e^{j2\pi ft}\,df$, multiply both sides by $e^{-j2\pi gt}$ and integrate with respect to $t$. Then, we obtain

$$
\int_{-\infty}^{\infty} x(t)e^{-j2\pi gt}\,dt = \int_{-\infty}^{\infty}\int_{-\infty}^{\infty} X(f)e^{j2\pi(f-g)t}\,df\,dt = \int_{-\infty}^{\infty} X(f) \int_{-\infty}^{\infty} e^{j2\pi(f-g)t}\,dt\,df
$$

$$
\tag{4.13}
$$

Using the property of the delta function (Equation (3.26)), the inner integral term on the right hand side of Equation (4.13) can be written as $\int_{-\infty}^{\infty} e^{j2\pi(f-g)t} dt = \delta(f - g)$, and so the right hand side of Equation (4.13) becomes

$$\int_{-\infty}^{\infty} X(f)\delta(f - g)df = X(g) \tag{4.14}$$

Hence, equating this with the left hand side of Equation (4.13) gives $X(g) = \int_{-\infty}^{\infty} x(t)e^{-j2\pi gt} dt$, which proves the inverse.

Similarly, $x(t)$ can be obtained via *inversion* of $X(f)$ using the delta function. That is, we start with $X(f) = \int_{-\infty}^{\infty} x(t)e^{-j2\pi ft} dt$, multiply both sides by $e^{j2\pi ft_1}$ and integrate with respect to $f$:

$$\int_{-\infty}^{\infty} X(f)e^{j2\pi ft_1}df = \int_{-\infty}^{\infty}\int_{-\infty}^{\infty} x(t)e^{-j2\pi ft}e^{j2\pi ft_1} dt df = \int_{-\infty}^{\infty} x(t) \int_{-\infty}^{\infty} e^{j2\pi f(t_1-t)}df dt$$

$$= \int_{-\infty}^{\infty} x(t)\delta(t_1 - t)dt = x(t_1) \tag{4.15}$$

4. The sufficient conditions for the existence of a Fourier integral are usually given as

$$\int_{-\infty}^{\infty} |x(t)|dt < \infty \tag{4.16}$$

but we shall transform functions failing to satisfy this condition using delta functions (see example (d) in Section 4.3).

## 4.2 ENERGY SPECTRA

Using an electrical analogy, if $x(t)$ is the voltage across a unit resistor then the *total energy* dissipated in the resistor is $\int_{-\infty}^{\infty} x^2(t)dt$. This may be decomposed into a frequency distribution from the relationship given in Equation (4.17), which is a form of Parseval's theorem:

$$\int_{-\infty}^{\infty} x^2(t)dt = \int_{-\infty}^{\infty} |X(f)|^2 df \tag{4.17}$$

This can be proved using the delta function, as given below:

$$\int_{-\infty}^{\infty} x^2(t)dt = \int_{-\infty}^{\infty} x(t)x^*(t)dt = \int_{-\infty}^{\infty}\int_{-\infty}^{\infty}\int_{-\infty}^{\infty} X(f_1)e^{j2\pi f_1 t}X^*(f_2)e^{-j2\pi f_2 t} dt df_1 df_2$$

$$= \int_{-\infty}^{\infty}\int_{-\infty}^{\infty} X(f_1)X^*(f_2)\delta(f_1 - f_2)df_1 df_2 = \int_{-\infty}^{\infty} |X(f_1)|^2 df_1 \tag{4.18}$$

Note that we are using *energy* here, whereas for *Fourier series* we talked of *power* (power spectra). The quantity $|X(f)|^2$ is an *energy spectral density* (energy per unit bandwidth) since it must be multiplied by a bandwidth $df$ to give energy. It is a measure of the decomposition of the energy of the process over frequency.

## 4.3 SOME EXAMPLES OF FOURIER TRANSFORMS

Some examples are given below, which help to understand the properties of the Fourier transform:

(a) The Fourier transform of the Dirac delta function $\delta(t)$ is

$$F\{\delta(t)\} = \int_{-\infty}^{\infty} \delta(t)e^{-j2\pi ft}\,dt = e^{-j2\pi f\cdot 0} = 1 \qquad (4.19)$$

where $F\{\}$ denotes the Fourier transform (shown in Figure 4.4). Note that the sifting property of the delta function is used (see Equation (3.23)).

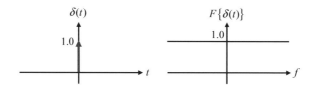

**Figure 4.4**   Dirac delta function and its Fourier transform

(b) For an exponentially decaying symmetric function

$$x(t) = e^{-\lambda|t|}, \quad \lambda > 0 \qquad (4.20)$$

$$X(f) = \int_{-\infty}^{\infty} x(t)e^{-j2\pi ft}\,dt = \int_{-\infty}^{\infty} e^{-\lambda|t|}e^{-j2\pi ft}\,dt$$

$$= \int_{-\infty}^{0} e^{\lambda t}e^{-j2\pi ft}\,dt + \int_{0}^{\infty} e^{-\lambda t}e^{-j2\pi ft}\,dt = \frac{2\lambda}{\lambda^2 + 4\pi^2 f^2} \qquad (4.21)$$

The time history and the transform are shown in Figure 4.5.

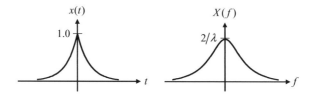

**Figure 4.5**   Time domain and frequency domain graphs of example (b)

Observations:

(i) The time history is symmetric with respect to $t = 0$ (i.e. an even function), so the transform is entirely real (i.e. a cosine transform) and the phase is zero, i.e. a so-called zero-phase signal.

(ii) The parameter $\lambda$ controls the shape of the signal and its transform. In the frequency domain the transform falls to $1/2$ of its value at $f = 0$ at a frequency $f = \lambda/2\pi$. So if $\lambda$ is large then $x(t)$ is narrow in the time domain, but wide in the frequency domain and vice versa. This is an example of the so-called inverse spreading property of the Fourier transform, i.e. the wider in one domain, then the narrower in the other.

(c) For an exponentially decaying function

$$x(t) = e^{-\alpha t} \quad t \geq 0, \quad \alpha > 0$$
$$= 0 \quad\quad t < 0 \tag{4.22}$$

$$X(f) = \int_{-\infty}^{\infty} x(t)e^{-j2\pi ft}\,dt = \int_{0}^{\infty} e^{-\alpha t}e^{-j2\pi ft}\,dt = \int_{0}^{\infty} e^{-(\alpha+j2\pi f)t}\,dt$$

$$= \frac{1}{\alpha + j2\pi f} = |X(f)|\,e^{j\phi(f)} \tag{4.23}$$

where

$$|X(f)| = \frac{1}{\sqrt{\alpha^2 + 4\pi^2 f^2}} \quad \text{and} \quad \phi(f) = \tan^{-1}\left(-\frac{2\pi f}{\alpha}\right)$$

The time and frequency domains are shown in Figure 4.6.

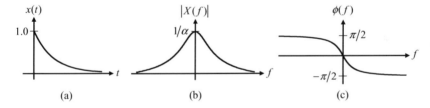

**Figure 4.6**    Time domain and frequency domain graphs of example (c): (a) time domain, (b) magnitude spectrum, (c) phase spectrum

(d) For a sine function

$$x(t) = A\sin(2\pi pt) \tag{4.24}$$

$$X(f) = \int_{-\infty}^{\infty} x(t)e^{-j2\pi ft}\,dt = \int_{-\infty}^{\infty} A\sin 2\pi pt \cdot e^{-j2\pi ft}\,dt = \int_{-\infty}^{\infty} \frac{A}{2j}\left(e^{j2\pi pt} - e^{-j2\pi pt}\right)e^{-j2\pi ft}\,dt$$

$$= \frac{A}{2j}\int_{-\infty}^{\infty} e^{-j2\pi(f-p)t} - e^{-j2\pi(f+p)t}\,dt = \frac{A}{2j}[\delta(f-p) - \delta(f+p)] \tag{4.25}$$

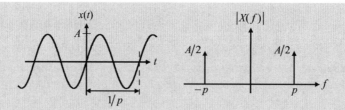

**Figure 4.7**  Time domain and frequency domain graphs of example (d)

In this example, $X(f)$ is non-zero at $f = p$ and $f = -p$ only, and is imaginary valued (the sine function is an odd function). The phase components are arg $X(p) = -\pi/2$ and arg $X(-p) = \pi/2$. This shows that a distinct frequency component results in spikes in the amplitude density.

(e) For a rectangular pulse

$$x(t) = a \quad |t| < b$$
$$= 0 \quad |t| > b \tag{4.26}$$

$$X(f) = \int_{-\infty}^{\infty} x(t)e^{-j2\pi ft}dt = \int_{-b}^{b} ae^{-j2\pi ft}dt$$

$$= \frac{2ab\,\sin(2\pi fb)}{2\pi fb} \tag{4.27}$$

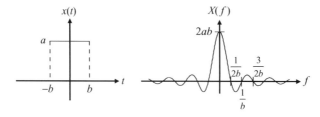

**Figure 4.8**  Time domain and frequency domain graphs of example (e)

The expression for $X(f)$ has been written so as to highlight the term $\sin(2\pi fb)/2\pi fb$, i.e. $\sin(x)/x$, which is the so-called sinc function which is unity at $x = 0$, and thereafter is an amplitude-modulated oscillation, where the modulation is $1/x$. The width (in time) of $x(t)$ is $2b$ and the distance to the first zero crossing in the frequency domain is $1/2b$ (as shown in Figure 4.8). This once again demonstrates the inverse spreading property.

For the case $a = 1$, then as $b \to \infty$, $X(f)$ is more and more concentrated around $f = 0$ and becomes taller and taller. In fact, $\lim_{b \to \infty} 2b\,\sin(2\pi fb)/2\pi fb$ is another way of expressing the delta function $\delta(f)$, as we have seen in Chapter 3 (see Equation (3.27)).

(f) For a damped symmetrically oscillating function

$$x(t) = e^{-a|t|}\cos 2\pi f_0 t, \quad a > 0 \tag{4.28}$$

$$X(f) = \int_{-\infty}^{\infty} x(t)e^{-j2\pi ft}\,dt = \int_{-\infty}^{\infty} e^{-a|t|}\cos 2\pi f_0 t\, e^{-j2\pi ft}\,dt = \int_{-\infty}^{\infty} e^{-a|t|}\frac{1}{2}\left(e^{j2\pi f_0 t} + e^{-j2\pi f_0 t}\right)e^{-j2\pi ft}\,dt$$

$$= \frac{1}{2}\int_{-\infty}^{\infty} e^{-a|t|}e^{-j2\pi(f-f_0)t}\,dt + \frac{1}{2}\int_{-\infty}^{\infty} e^{-a|t|}e^{-j2\pi(f+f_0)t}\,dt$$

$$= \frac{a}{a^2 + [2\pi(f-f_0)]^2} + \frac{a}{a^2 + [2\pi(f+f_0)]^2} \tag{4.29}$$

The time and frequency domains are shown in Figure 4.9.

**Figure 4.9** Time domain and frequency domain graphs of example (f)

(g) For a damped oscillating function

$$x(t) = e^{-at}\sin 2\pi f_0 t, \quad t \geq 0 \text{ and } a > 0 \tag{4.30}$$

$$X(f) = \int_{-\infty}^{\infty} x(t)e^{-j2\pi ft}\,dt = \int_{0}^{\infty} e^{-at}\sin 2\pi f_0 t\, e^{-j2\pi ft}\,dt = \int_{0}^{\infty} e^{-at}\frac{1}{2j}\left(e^{j2\pi f_0 t} - e^{-j2\pi f_0 t}\right)e^{-j2\pi ft}\,dt$$

$$= \frac{1}{2j}\int_{0}^{\infty} e^{-[a+j2\pi(f-f_0)]t}\,dt - \frac{1}{2j}\int_{0}^{\infty} e^{-[a+j2\pi(f+f_0)]t}\,dt = \frac{2\pi f_0}{(2\pi f_0)^2 + (a + j2\pi f)^2} \tag{4.31}$$

The time and frequency domains are shown in Figure 4.10.

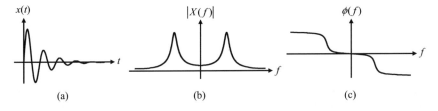

(a)          (b)          (c)

**Figure 4.10** Time domain and frequency domain graphs of example (g): (a) time domain, (b) magnitude spectrum, (c) phase spectrum

(h) For the Gaussian pulse

$$x(t) = e^{-at^2}, \quad a > 0 \tag{4.32}$$

$$X(\omega) = \sqrt{\pi/a} \cdot e^{-\omega^2/4a} \tag{4.33}$$

i.e. $X(\omega)$ is also a Gaussian pulse. Proof of Equation (4.33) is given below. We shall use $X(\omega)$ instead of $X(f)$ for convenience.

We start from

$$X(\omega) = \int_{-\infty}^{\infty} e^{-at^2} e^{-j\omega t} dt = \int_{-\infty}^{\infty} e^{-a(t^2 + j\omega t/a)} dt$$

and multiply by $e^{-\omega^2/4a} \cdot e^{\omega^2/4a}$ to complete the square, i.e. so that

$$X(\omega) = e^{-\omega^2/4a} \cdot \int_{-\infty}^{\infty} e^{-a(t^2 + j\omega t/a - \omega^2/4a^2)} dt = e^{-\omega^2/4a} \cdot \int_{-\infty}^{\infty} e^{-a[t + j(\omega/2a)]^2} dt$$

Now, let $y = [t + j(\omega/2a)]$; then finally we have

$$X(\omega) = e^{-\omega^2/4a} \cdot \int_{-\infty}^{\infty} e^{-ay^2} dy = \sqrt{\pi/a} \cdot e^{-\omega^2/4a}$$

The time and frequency domains are shown in Figure 4.11.

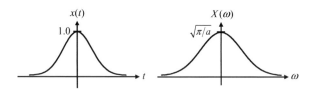

**Figure 4.11**   Time domain and frequency domain graphs of example (h)

(i) For a unit step function

$$u(t) = 1 \quad t > 0$$
$$= 0 \quad t < 0 \tag{4.34}$$

The unit step function is not defined at $t = 0$, i.e. it has a discontinuity at this point. The Fourier transform of $u(t)$ is given by

$$F\{u(t)\} = \frac{1}{2}\delta(f) + \frac{1}{j2\pi f} \tag{4.35}$$

where $F\{\}$ denotes the Fourier transform (shown in Figure 4.12). The derivation of this result requires the use of the delta function and some properties of the Fourier transform. Details of the derivation can be found in Hsu (1970), if required. Also, note the presence of the delta function at $f = 0$, which indicates a d.c. component.

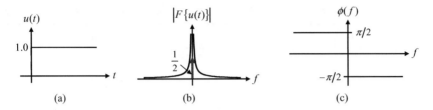

**Figure 4.12** Time domain and frequency domain graphs of example (i): (a) time domain, (b) magnitude spectrum, (c) phase spectrum

(j) For the Fourier transform of a *periodic* function

If $x(t)$ is periodic with a period $T_P$, then

$$x(t) = \sum_{n=-\infty}^{\infty} c_n e^{j2\pi nt/T_P}$$

(i.e. Equation (3.34)). The Fourier transform of this equation gives

$$X(f) = \int_{-\infty}^{\infty} \sum_{n=-\infty}^{\infty} c_n e^{j2\pi nt/T_P} e^{-j2\pi ft} dt = \sum_{n=-\infty}^{\infty} c_n \int_{-\infty}^{\infty} e^{-j2\pi(f-n/T_P)t} dt$$

$$= \sum_{n=-\infty}^{\infty} c_n \delta(f - n/T_P) \tag{4.36}$$

This shows that the Fourier transform of a periodic function is a series of delta functions scaled by $c_n$, and located at multiples of the fundamental frequency, $1/T_P$.

Note that, in examples (a), (b), (e), (f) and (h), arg $X(f) = 0$ owing to the evenness of $x(t)$. Some useful Fourier transform pairs are given in Table 4.1.

## 4.4 PROPERTIES OF FOURIER TRANSFORMS

We now list some important properties of Fourier transforms. Here $F\{x(t)\}$ denotes $X(f)$.

(a) Time scaling:

$$F\{x(at)\} = \frac{1}{|a|} X(f/a) \tag{4.37a}$$

or

$$F\{x(at)\} = \frac{1}{|a|} X(\omega/a) \tag{4.37b}$$

where $a$ is a real constant. The proof is given below.

**Table 4.1**  Some Fourier transform (integral) pairs

| No. | Time function $x(t)$ | Fourier transform $X(f)$ | Fourier transform $X(\omega)$ |
|---|---|---|---|
| 1 | $\delta(t)$ | $1$ | $1$ |
| 2 | $1$ | $\delta(f)$ | $2\pi\,\delta(\omega)$ |
| 3 | $A$ | $A\delta(f)$ | $2\pi A\delta(\omega)$ |
| 4 | $u(t)$ | $\dfrac{1}{2}\delta(f)+\dfrac{1}{j2\pi f}$ | $\pi\,\delta(\omega)+\dfrac{1}{j\omega}$ |
| 5 | $\delta(t-t_0)$ | $e^{-j2\pi f t_0}$ | $e^{-j\omega t_0}$ |
| 6 | $e^{j2\pi f_0 t}$ or $e^{j\omega_0 t}$ | $\delta(f-f_0)$ | $2\pi\,\delta(\omega-\omega_0)$ |
| 7 | $\cos(2\pi f_0 t)$ or $\cos(\omega_0 t)$ | $\dfrac{1}{2}[\delta(f-f_0)+\delta(f+f_0)]$ | $\pi[\delta(\omega-\omega_0)+\delta(\omega+\omega_0)]$ |
| 8 | $\sin(2\pi f_0 t)$ or $\sin(\omega_0 t)$ | $\dfrac{1}{2j}[\delta(f-f_0)-\delta(f+f_0)]$ | $\dfrac{\pi}{j}[\delta(\omega-\omega_0)-\delta(\omega+\omega_0)]$ |
| 9 | $e^{-\alpha|t|}$ | $\dfrac{2\alpha}{\alpha^2+4\pi^2 f^2}$ | $\dfrac{2\alpha}{\alpha^2+\omega^2}$ |
| 10 | $\dfrac{1}{\alpha^2+t^2}$ | $\dfrac{\pi}{\alpha}e^{-\alpha2\pi|f|}$ | $\dfrac{\pi}{\alpha}e^{-\alpha|\omega|}$ |
| 11 | $x(t)=e^{-\alpha t}u(t)$ | $\dfrac{1}{\alpha+j2\pi f}$ | $\dfrac{1}{\alpha+j\omega}$ |
| 12 | $\begin{aligned}x(t)&=A\quad |t|<T\\&=0\quad |t|>T\end{aligned}$ | $2AT\dfrac{\sin(2\pi f T)}{2\pi f T}$ | $2AT\dfrac{\sin(\omega T)}{\omega T}$ |
| 13 | $2Af_0\dfrac{\sin(2\pi f_0 t)}{2\pi f_0 t}$ or $A\dfrac{\sin(\omega_0 t)}{\pi t}$ | $\begin{aligned}X(f)&=A\quad |f|<f_0\\&=0\quad |f|>f_0\end{aligned}$ | $\begin{aligned}X(\omega)&=A\quad |\omega|<\omega_0\\&=0\quad |\omega|>\omega_0\end{aligned}$ |
| 14 | $\displaystyle\sum_{n=-\infty}^{\infty}c_n e^{j2\pi n f_0 t}$ or $\displaystyle\sum_{n=-\infty}^{\infty}c_n e^{jn\omega_0 t}$ | $\displaystyle\sum_{n=-\infty}^{\infty}c_n\delta(f-nf_0)$ | $2\pi\displaystyle\sum_{n=-\infty}^{\infty}c_n\delta(\omega-n\omega_0)$ |
| 15 | $\mathrm{sgn}(t)$ | $\dfrac{1}{j\pi f}$ | $\dfrac{2}{j\omega}$ |
| 16 | $\dfrac{1}{t}$ | $-j\pi\,\mathrm{sgn}(f)$ | $-j\pi\,\mathrm{sgn}(\omega)$ |

For $a>0$, the Fourier transform is $F\{x(at)\}=\int_{-\infty}^{\infty}x(at)e^{-j2\pi f t}dt$. Let $at=\tau$; Then

$$F\{x(at)\}=\frac{1}{a}\int_{-\infty}^{\infty}x(\tau)e^{-j2\pi(f/a)\tau}d\tau=\frac{1}{a}X(f/a)$$

Similarly for $a<0$,

$$F\{x(at)\}=\frac{1}{a}\int_{\infty}^{-\infty}x(\tau)e^{-j2\pi(f/a)\tau}d\tau$$

thus

$$F\{x(at)\} = -\frac{1}{a} \int_{-\infty}^{\infty} x(\tau)e^{-j2\pi(f/a)\tau} d\tau = \frac{1}{|a|}X(f/a)$$

That is, time scaling results in frequency scaling, again demonstrating the inverse spreading relationship.

(b) Time reversal:

$$\boxed{F\{x(-t)\} = X(-f) \quad (= X^*(f), \quad \text{for } x(t) \text{ real})} \tag{4.38a}$$

or

$$F\{x(-t)\} = X(-\omega) \tag{4.38b}$$

*Proof:* We start from $F\{x(-t)\} = \int_{-\infty}^{\infty} x(-t)e^{-j2\pi ft} dt$, let $-t = \tau$, then obtain

$$F\{x(-t)\} = -\int_{\infty}^{-\infty} x(\tau)e^{j2\pi f\tau} d\tau = \int_{-\infty}^{\infty} x(\tau)e^{-j2\pi(-f)\tau} d\tau = X(-f)$$

Note that if $x(t)$ is real, then $x^*(t) = x(t)$. In this case,

$$X(-f) = \int_{-\infty}^{\infty} x(t)e^{-j2\pi(-f)t} dt = \int_{-\infty}^{\infty} x^*(t)e^{j2\pi ft} dt = X^*(f)$$

This is called the conjugate symmetry property.

It is interesting to note that the Fourier transform of $X(-\omega)$ is $x(t)$, i.e. $F\{X(-\omega)\} = x(t)$, and similarly $F\{X(\omega)\} = x(-t)$.

(c) Time shifting:

$$\boxed{F\{x(t - t_0)\} = e^{-j2\pi ft_0} X(f)} \tag{4.39a}$$

or

$$F\{x(t - t_0)\} = e^{-j\omega t_0} X(\omega) \tag{4.39b}$$

*Proof:* We start from $F\{x(t - t_0)\} = \int_{-\infty}^{\infty} x(t - t_0)e^{-j2\pi ft} dt$, let $t - t_0 = \tau$, then obtain

$$F\{x(t - t_0)\} = \int_{-\infty}^{\infty} x(\tau)e^{-j2\pi f(t_0+\tau)} d\tau = e^{-j2\pi ft_0} \int_{-\infty}^{\infty} x(\tau)e^{-j2\pi f\tau} d\tau = e^{-j2\pi ft_0} X(f)$$

This important property is expanded upon in Section 4.5.

(d) Modulation (or multiplication) property:

(i)

$$\boxed{F\{x(t)e^{j2\pi f_0t}\} = X(f - f_0)} \tag{4.40a}$$

or

$$F\{x(t)e^{j\omega_0 t}\} = X(\omega - \omega_0) \tag{4.40b}$$

This property is usually known as the 'frequency shifting' property.

*Proof*:

$$F\left\{x(t)e^{j2\pi f_0 t}\right\} = \int_{-\infty}^{\infty} x(t)e^{j2\pi f_0 t}e^{-j2\pi f t}\,dt = \int_{-\infty}^{\infty} x(t)e^{-j2\pi(f-f_0)t}\,dt = X(f - f_0)$$

(ii)

$$\boxed{F\{x(t)\cos(2\pi f_0 t)\} = \frac{1}{2}[X(f - f_0) + X(f + f_0)]}$$ (4.41a)

or

$$F\{x(t)\cos(\omega_0 t)\} = \frac{1}{2}[X(\omega - \omega_0) + X(\omega + \omega_0)]$$ (4.41b)

This characterizes 'amplitude modulation'. For communication systems, usually $x(t)$ is a low-frequency signal, and $\cos(2\pi f_0 t)$ is a high-frequency carrier signal.

*Proof*:

$$F\{x(t)\cos(2\pi f_0 t)\} = F\left\{\frac{1}{2}x(t)e^{j2\pi f_0 t} + \frac{1}{2}x(t)e^{-j2\pi f_0 t}\right\}$$

$$= \frac{1}{2}F\left\{x(t)e^{j2\pi f_0 t}\right\} + \frac{1}{2}F\left\{x(t)e^{-j2\pi f_0 t}\right\}$$

$$= \frac{1}{2}[X(f - f_0) + X(f + f_0)]$$

(e) Differentiation:

$$\boxed{F\{\dot{x}(t)\} = j2\pi f X(f)}\text{ (if } x(t) \to 0 \text{ as } t \to \pm\infty)$$ (4.42a)

or

$$F\{\dot{x}(t)\} = j\omega X(\omega)$$ (4.42b)

*Proof*:

$$F\{\dot{x}(t)\} = \int_{-\infty}^{\infty} \dot{x}(t)e^{-j2\pi f t}\,dt = x(t)e^{-j2\pi f t}\Big|_{-\infty}^{\infty} + j2\pi f \int_{-\infty}^{\infty} x(t)e^{-j2\pi f t}\,dt$$

Since $x(t) \to 0$ as $t \to \pm\infty$, the first part of the right hand side diminishes. Thus

$$F\{\dot{x}(t)\} = j2\pi f \int_{-\infty}^{\infty} x(t)e^{-j2\pi f t}\,dt = j2\pi f X(f)$$

(f) The Fourier transform of the 'convolution' of two functions:

$$\boxed{F\{h(t) * x(t)\} = H(f)X(f)}$$ (4.43)

where the convolution of the two functions $h(t)$ and $x(t)$ is defined as

$$h(t) * x(t) = \int_{-\infty}^{\infty} h(\tau)x(t - \tau)\,d\tau$$ (4.44)

The property of Equation (4.43) is very important in linear system theory and is explained fully in Section 4.7.

*Proof:* Let $t - \tau = v$. Then

$$F\{h(t) * x(t)\} = \int\limits_{-\infty}^{\infty} \int\limits_{-\infty}^{\infty} h(\tau)x(t-\tau)e^{-j2\pi ft}d\tau dt$$

$$= \int\limits_{-\infty}^{\infty} \int\limits_{-\infty}^{\infty} h(\tau)x(v)e^{-j2\pi f(\tau+v)}d\tau dv$$

$$= \int\limits_{-\infty}^{\infty} h(\tau)e^{-j2\pi f\tau}d\tau \int\limits_{-\infty}^{\infty} x(v)e^{-j2\pi fv}dv = H(f)X(f)$$

(g) The Fourier transform of the 'product' of two functions:

$$F\{x(t)w(t)\} = \int\limits_{-\infty}^{\infty} X(g)W(f-g)dg = X(f) * W(f) \qquad (4.45)$$

This is also a very important property, and will be examined in detail in Section 4.11.

*Proof:* We start from $F\{x(t)w(t)\} = \int_{-\infty}^{\infty} x(t)w(t)e^{-j2\pi ft}dt$. If $x(t)$ and $w(t)$ both have Fourier representations, then the right hand side is

$$\int\limits_{-\infty}^{\infty} x(t)w(t)e^{-j2\pi ft}dt = \int\limits_{-\infty}^{\infty} \int\limits_{-\infty}^{\infty} \int\limits_{-\infty}^{\infty} X(f_1)e^{j2\pi f_1 t}W(f_2)e^{j2\pi f_2 t} \cdot e^{-j2\pi ft}df_1 df_2 dt$$

$$= \int\limits_{-\infty}^{\infty} X(f_1) \int\limits_{-\infty}^{\infty} W(f_2) \int\limits_{-\infty}^{\infty} e^{-j2\pi(f-f_1-f_2)t}dt df_2 df_1$$

$$= \int\limits_{-\infty}^{\infty} X(f_1) \int\limits_{-\infty}^{\infty} W(f_2)\delta(f-f_1-f_2)df_2 df_1$$

$$= \int\limits_{-\infty}^{\infty} X(f_1)W(f-f_1)df_1 = X(f) * W(f)$$

## 4.5  THE IMPORTANCE OF PHASE

In many cases, we sometimes only draw the magnitude spectral density, $|X(f)|$, and not the phase spectral density, $\arg X(f) = \phi(f)$. However, in order to reconstruct a signal we need both. An infinite number of different-looking signals may have the same magnitude spectra – it is their phase structure that differs. We now make a few general comments:

1. A symmetrical signal has a real-valued transform, i.e. its phase is zero. We saw this property in examples given in Section 4.3.
2. A *pure delay* imposed on a signal results in a *linear phase change* to the transform (see property (c) in Section 4.4). An example of this is illustrated in Figure 4.13.

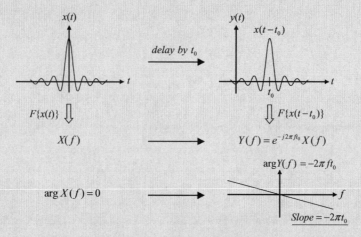

**Figure 4.13**   The effect of a pure delay on a zero-phase signal

   The *slope of the phase curve* gives the *delay*, i.e. $d\phi/df = -2\pi t_0$, or $d\phi/d\omega = -t_0$. Specifically, the quantity $-d\phi/d\omega = t_0$ is known as the *group delay* of the signal. In the above case, the delay is the same for all frequencies due to the pure delay (i.e. there is *no dispersion*). The reason for the term group delay is given in Section 4.8.
3. If the phase curve is nonlinear, i.e. $-d\phi/d\omega$ is a nonlinear function of $\omega$, then the signal shape is altered.

## 4.6 ECHOES[M4.1]

If a signal $y(t)$ contains a pure echo (a scaled replica of the main signal), it may be modelled as

$$y(t) = x(t) + ax(t - t_0) \qquad (4.46)$$

where $x(t)$ is the main signal and $ax(t - t_0)$ is the echo, $a$ is the amplitude of the echo, and $t_0$ is called the 'epoch' of the echo (i.e. the time delay of the echo relative to the main signal). A typical example may be illustrated as shown in Figure 4.14, and the Fourier transform of $y(t)$ is

$$Y(f) = (1 + ae^{-j2\pi f t_0})X(f) \qquad (4.47)$$

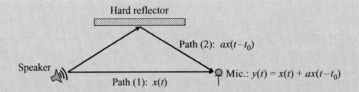

**Figure 4.14** Example of a signal containing a pure echo

The term $(1 + ae^{-j2\pi f t_0})$ is a function of frequency and has an oscillatory form in both magnitude and phase. This describes the effect of the echo on the main signal, and may be illustrated as shown in Figure 4.15. The magnitude of $Y(f)$ is $\sqrt{(1 + a^2 + 2a\cos 2\pi f t_0)}\,|X(f)|$ where an oscillatory form is imposed on $|X(f)|$ due to the echo. Thus, such a 'rippling' appearance in energy (or power) spectra may indicate the existence of an echo. However, additional echoes and dispersion result in more complicated features. The autocorrelation function can also be used to detect the time delays of echoes in a signal (the correlation function will be discussed in Part II of this book), but are usually limited to wideband signals (e.g. a pulse-like signal). Another approach to analysing such signals is 'cepstral analysis' (Bogert *et al.*, 1963) later generalized as homomorphic deconvolution (Oppenheim and Schafer, 1975).

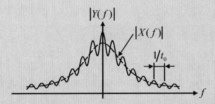

**Figure 4.15** Effect of a pure echo

## 4.7 CONTINUOUS-TIME LINEAR TIME-INVARIANT SYSTEMS AND CONVOLUTION

Consider the input–output relationship for a linear time-invariant (LTI) system as shown in Figure 4.16.

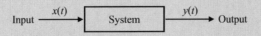

**Figure 4.16** A continuous LTI system

We now define the terms 'linear' and 'time-invariant'.

## *Linearity*

Let $y_1(t)$ and $y_2(t)$ be the responses of the system to inputs $x_1(t)$ and $x_2(t)$, respectively. If the system is linear it satisfies the properties in Figure 4.17, where $a$ is an arbitrary constant.

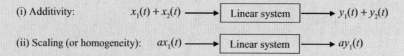

(i) Additivity: $\quad\quad\quad x_1(t) + x_2(t) \longrightarrow$ Linear system $\longrightarrow y_1(t) + y_2(t)$

(ii) Scaling (or homogeneity): $\quad a x_1(t) \longrightarrow$ Linear system $\longrightarrow a y_1(t)$

**Figure 4.17**   Properties of a linear system

Or the two properties can be combined to give a more general expression that is known as the 'superposition property' (Figure 4.18), where $a_1$ and $a_2$ are arbitrary constants.

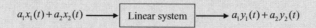

$a_1 x_1(t) + a_2 x_2(t) \longrightarrow$ Linear system $\longrightarrow a_1 y_1(t) + a_2 y_2(t)$

**Figure 4.18**   Superposition property of a linear system

## *Time Invariance*

A time-invariant system may be illustrated as in Figure 4.19, such that if the input is shifted by $t_0$, then the response will also be shifted by the same amount of time.

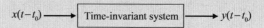

$x(t - t_0) \longrightarrow$ Time-invariant system $\longrightarrow y(t - t_0)$

**Figure 4.19**   Property of a time-invariant system

## Mathematical Characterization of an LTI System

Very commonly LTI systems are described in differential equation form. The forced vibration of a single-degree-of-freedom system is a typical example, which may be expressed as

$$m\ddot{y}(t) + c\dot{y}(t) + ky(t) = x(t) \tag{4.48}$$

where $x(t)$ is the input and $y(t)$ is the output of the system.

Relating $y(t)$ to $x(t)$ in the time domain then requires the solution of the differential equation. Transformation (Laplace and Fourier) techniques allow a 'systems approach' with the input/response relationships described by transfer functions or frequency response functions.

We shall use a general approach to linear system characterization that does not require a differential equation format. We could characterize a system in terms of its response to specific inputs, e.g. a step input or a harmonic input, but we shall find that the response to an ideal impulse (the Dirac delta function) turns out to be very helpful - even though such an input is a mathematical idealization.

We define the response of a linear system to a unit impulse at $t = 0$ (i.e. $\delta(t)$) to be $h(t)$. See Figure 4.20. In the figure, it can be seen that the system only responds after the impulse, i.e. we assume that the system is *causal*, in other words $h(t) = 0$ for $t < 0$. For a causal system, the output $y(t)$ at the present time, say $t = t_1$, is dependent upon only the past and present values of the input $x(t)$, i.e. $x(t)$ for $t \leq t_1$, and does not depend on the future values of $x(t)$.

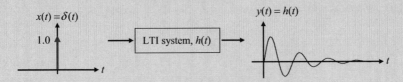

**Figure 4.20** Impulse response of a system

We shall now show how the concept of the ideal impulse response function $h(t)$ can be used to describe the system response to any input. We start by noting that for a time-invariant system, the response to a delayed impulse $\delta(t - t_1)$ is a delayed impulse response $h(t - t_1)$.

Consider an arbitrary input signal $x(t)$ split up into elemental impulses as given in Figure 4.21. The impulse at time $t_1$ is $x(t_1)\Delta t_1$. Because the system is linear, the response to this impulse at time $t$ is $h(t - t_1)x(t_1)\Delta t_1$. Now, adding all the responses to such impulses, the total response of $y(t)$ at time $t$ (the present) becomes

$$y(t) \approx \sum h(t - t_1)x(t_1)\Delta t_1 \tag{4.49}$$

and by letting $\Delta t_1 \to 0$ this results in

$$y(t) = \int_{-\infty}^{t} h(t - t_1)x(t_1)dt_1 \tag{4.50}$$

Note that the upper limit is $t$ because we assume that the system is causal. Using the substitution $t - t_1 = \tau$ $(-dt_1 = d\tau)$, the expression can be written in an alternative

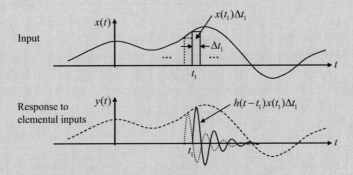

**Figure 4.21** The response of a system to elemental inputs

form as given by Equation (4.51a), i.e. the convolution integral has the commutative property

$$y(t) = \int_{-\infty}^{t} h(t - t_1)x(t_1)dt_1 = -\int_{\infty}^{0} h(\tau)x(t - \tau)d\tau = \int_{0}^{\infty} h(\tau)x(t - \tau)d\tau \qquad (4.51a)$$

or simply

$$y(t) = x(t) * h(t) = h(t) * x(t) \qquad (4.51b)$$

As depicted in Figure 4.22, we see $h(\tau)$ in its role as a 'memory' or weighting function.

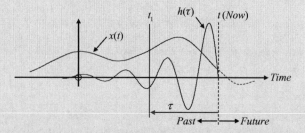

**Figure 4.22** The impulse response function as a 'memory'

If the input $x(t)$ is zero for $t < 0$, the response of a causal system is

$$y(t) = \int_{0}^{t} h(\tau)x(t - \tau)d\tau = \int_{0}^{t} h(t - \tau)x(\tau)d\tau \qquad (4.52)$$

And, if the system is *non-causal*, i.e. the system also responds to future inputs, the convolution integrals are

$$y(t) = \int_{-\infty}^{\infty} h(\tau)x(t - \tau)d\tau = \int_{-\infty}^{\infty} h(t - \tau)x(\tau)d\tau \qquad (4.53)$$

An example of convolution operation of a causal input and a causal LTI system is illustrated in Figure 4.23.

We note that, obviously,

$$h(t) = h(t) * \delta(t) = \int_{-\infty}^{\infty} h(\tau)\delta(t - \tau)d\tau \qquad (4.54)$$

The convolution integral also satisfies 'associative' and 'distributive' properties, i.e.

Associative: $[x(t) * h_1(t)] * h_2(t) = x(t) * [h_1(t) * h_2(t)]$      (4.55)

Distributive: $x(t) * [h_1(t) + h_2(t)] = x(t) * h_1(t) + x(t) * h_2(t)$      (4.56)

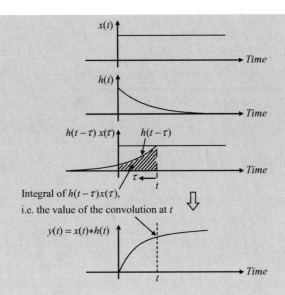

**Figure 4.23**　Illustrations of a convolution operation

### The Frequency Response Function

Consider the steady state response of a system to a harmonic excitation, i.e. let $x(t) = e^{j2\pi ft}$. Then the convolution integral becomes

$$y(t) = \int_0^\infty h(\tau)x(t-\tau)d\tau = \int_0^\infty h(\tau)e^{j2\pi f(t-\tau)}d\tau = e^{j2\pi ft}\underbrace{\int_0^\infty h(\tau)e^{-j2\pi f\tau}d\tau}_{H(f)} \qquad (4.57)$$

The system response to frequency $f$ is embodied in $H(f) = \int_0^\infty h(\tau)e^{-j2\pi f\tau}d\tau$, which is the system 'frequency response function (FRF)'.

The expression of the convolution operation in the time domain is very much simplified when the integral transform (Laplace or Fourier transform) is taken. If the response is $y(t) = \int_0^\infty h(\tau)x(t-\tau)d\tau$, then taking the Fourier transform gives

$$Y(f) = \int_{-\infty}^\infty \int_0^\infty h(\tau)x(t-\tau)e^{-j2\pi ft}d\tau dt$$

Let $t - \tau = u$; then

$$Y(f) = \int_0^\infty h(\tau)e^{-j2\pi f\tau}d\tau \int_{-\infty}^\infty x(u)e^{-j2\pi fu}du$$

Thus,

$$Y(f) = H(f)X(f) \qquad (4.58)$$

The convolution operation becomes a 'product' (see property (f) in Section 4.4). $H(f)$ is the Fourier transform of the impulse response function and is the frequency response function of the system. Sometimes, Equation (4.58) is used to 'identify' a system if the input and response are all available, i.e. $H(f) = Y(f)/X(f)$. Following on from this the relationship between the input and output *energy spectra* is

$$|Y(f)|^2 = |H(f)|^2 |X(f)|^2 \qquad (4.59)$$

If the Laplace transform is taken (the Laplace transform will be discussed further in Section 5.1), then by a similar argument as for the Fourier transform, it becomes

$$Y(s) = H(s)X(s) \qquad (4.60)$$

where $s = \sigma + j\omega$ is complex. The ratio $Y(s)/X(s) = H(s)$ is called the transfer function of the system. The relationships between the impulse response function, the frequency response function and the transfer function are depicted in Figure 4.24. Note that $H(\omega)$ can be obtained by $H(s)$ on the imaginary axis in the $s$-plane, i.e. the Fourier transform can be considered as the Laplace transform taking the values on the imaginary axis only (see Section 5.1).

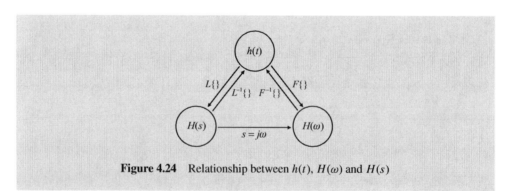

**Figure 4.24**   Relationship between $h(t)$, $H(\omega)$ and $H(s)$

## Examples of Systems

### *Example 1*

Reconsider the simple acoustic problem in Figure 4.25, with input $x(t)$ and response $y(t)$.
   The relationship between $x(t)$ and $y(t)$ may be modelled as

$$y(t) = ax(t - \Delta_1) + bx(t - \Delta_2) \qquad (4.61)$$

The impulse response function relating $x(t)$ to $y(t)$ is

$$h(t) = a\delta(t - \Delta_1) + b\delta(t - \Delta_2) \qquad (4.62)$$

and is illustrated in Figure 4.26.

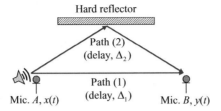

**Figure 4.25**　A simple acoustic example

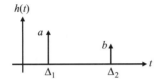

**Figure 4.26**　Impulse response function for Example 1

The frequency response function is

$$H(\omega) = \int_{-\infty}^{\infty} h(t)e^{-j\omega t}\,dt = ae^{-j\omega\Delta_1} + be^{-j\omega\Delta_2} = ae^{-j\omega\Delta_1}\left(1 + \frac{b}{a}e^{-j\omega(\Delta_2-\Delta_1)}\right) \qquad (4.63)$$

If we let $\Delta = \Delta_2 - \Delta_1$, then the modulus of $H(\omega)$ is

$$|H(\omega)| = a\sqrt{\left(1 + \frac{b^2}{a^2} + \frac{2b}{a}\cos\omega\Delta\right)} \qquad (4.64)$$

This has an oscillatory form in frequency (compare this with the case depicted in Figure 4.15). The phase component arg $H(\omega)$ also has an oscillatory behaviour as expected from Equation (4.63). These characteristics of the frequency response function are illustrated in Figure 4.27, where $H(\omega)$ is represented as a vector on a polar diagram.

Next, applying the Laplace transform to $h(t)$, the transfer function is

$$H(s) = \int_{-\infty}^{\infty} h(t)e^{-st}\,dt = ae^{-s\Delta_1} + be^{-s\Delta_2} \qquad (4.65)$$

Now we shall examine the *poles* and *zeros* in the $s$-plane. From Equation (4.65), it can be seen that there are no poles. Zeros are found, such that $H(s) = 0$ when $ae^{-s\Delta_1} = -be^{-s\Delta_2}$, i.e. at

$$e^{s\Delta} = -\frac{b}{a} \qquad (4.66)$$

where $\Delta = \Delta_2 - \Delta_1$. Let $s = \sigma + j\omega$ so that Equation (4.66) can be written as

$$e^{\sigma\Delta}e^{j\omega\Delta} = \frac{b}{a}e^{\pm j(\pi + 2k\pi)} \qquad (4.67)$$

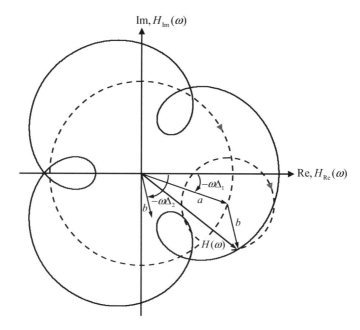

**Figure 4.27**    Polar diagram of $H(\omega)$ for Example 1 (where $\Delta_1 = 1$, $\Delta_2 = 4$ and $a/b = 2$)

where $k$ is an integer. Since $e^{\sigma \Delta} = b/a$ and $\omega \Delta = \pm j\pi(2k + 1)$, zeros are located at

$$\sigma = \frac{1}{\Delta} \ln \left( \frac{b}{a} \right), \quad \omega = \pm j\frac{\pi}{\Delta}(2k + 1) \tag{4.68}$$

and are depicted in Figure 4.28.

In the figure, the corresponding oscillatory nature of the modulus of the frequency response function is seen, as it is in the phase. However, the phase has a superimposed linear

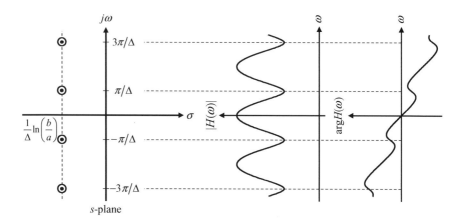

**Figure 4.28**    Representation in the $s$-plane and its corresponding $H(\omega)$ for Example 1

component due to the delay $\Delta_1$ of the first 'spike' in the impulse response function (see Figure 4.26).

## *Example 2*

Consider the single-degree-of-freedom mechanical system as given in Equation (4.48), which can be rewritten in the following form:

$$\ddot{y}(t) + 2\zeta\omega_n\dot{y}(t) + \omega_n^2 y(t) = \frac{1}{m}x(t) \tag{4.69}$$

where $\omega_n = \sqrt{k/m}$ and $\zeta = c/2\sqrt{km}$. The impulse response function can be obtained from $\ddot{h}(t) + 2\zeta\omega_n\dot{h}(t) + \omega_n^2 h(t) = (1/m)\delta(t)$, and assuming that the system is underdamped (i.e. $0 < \zeta < 1$), the impulse response function is

$$h(t) = \frac{1}{m\omega_d}e^{-\zeta\omega_n t}\sin\omega_d t \tag{4.70}$$

where $\omega_d = \omega_n\sqrt{1 - \zeta^2}$, and is illustrated in Figure 4.29.

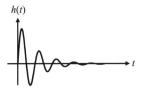

*h(t)*

**Figure 4.29** Impulse response function for Example 2

The corresponding frequency response function and transfer function are

$$H(\omega) = \frac{1/m}{\omega_n^2 - \omega^2 + j2\zeta\omega_n\omega} \tag{4.71}$$

$$H(s) = \frac{1/m}{s^2 + 2\zeta\omega_n s + \omega_n^2} \tag{4.72}$$

Note that there are only poles in the $s$-plane for this case as shown in Figure 4.30.

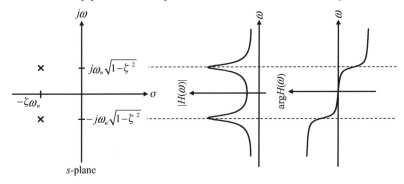

**Figure 4.30** Representation in the $s$-plane and its corresponding $H(\omega)$ for Example 2

## 4.8 GROUP DELAY[1] (DISPERSION)[M4.2]

We have seen that a pure delay results in a linear phase component. We now interpret nonlinear phase characteristics. Suppose we have a system $H(\omega) = A(\omega)e^{j\phi(\omega)}$, where $A(\omega)(= |H(\omega)|)$ is amplitude and $\phi(\omega)$ is phase. Consider a *group of frequencies* near $\omega_k$ in the range from $\omega_k - B$ to $\omega_k + B$ ($B \ll \omega_k$), i.e. a narrow-band element of $H(\omega)$, approximated by $H_k(\omega)$, as shown in Figure 4.31, i.e. $H(\omega) \approx \sum_k H_k(\omega)$, and $H_k(\omega) = |H_k(\omega)|\, e^{j\,\arg H_k(\omega)} = A(\omega_k)e^{j\phi(\omega)}$. The phase $\phi(\omega)$ may be linearly approximated over the narrow frequency interval (by applying the Taylor expansion) such that $\phi(\omega) \approx \phi(\omega_k) + (\omega - \omega_k)\phi'(\omega_k)$ as shown in Figure 4.32. Then, $H_k(\omega)$ has the form of an ideal band-pass filter with a linear phase characteristic.

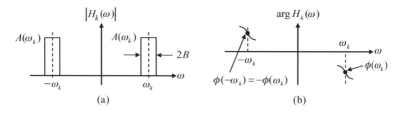

**Figure 4.31**  Narrow-band frequency components of $H_k(\omega)$: (a) magnitude, (b) phase

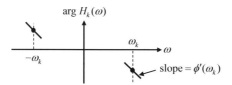

**Figure 4.32**  Linear approximation of arg $H_k(\omega)$

Now, based on the representation $H(\omega) \approx \sum_k A(\omega_k)e^{j[\phi(\omega_k)+(\omega-\omega_k)\phi'(\omega_k)]}$ we shall inverse transform this to obtain a corresponding expression for the impulse response function. We start by noting that the 'equivalent' low-pass filter can be described as in Figure 4.33(a) whose corresponding time signal is $2A(\omega_k)B \sin[B(t + \phi'(\omega_k))]/[\pi B(t + \phi'(\omega_k))]$ (see Equation (4.39b) and No. 13 of Table 4.1). Now, consider the Fourier transform of a cosine function with a phase, i.e. $F\{\cos(\omega_k t + \phi(\omega_k))\} = \pi[e^{j\phi(\omega_k)}\delta(\omega - \omega_k) + e^{-j\phi(\omega_k)}\delta(\omega + \omega_k)]$ as shown in Figure 4.33(b). In fact, $H_k(\omega)$ can be obtained by taking the convolution of Figures 4.33(a) and (b) in the frequency domain. This may be justified by noting that the frequency domain convolution described in Equation (4.45) can be rewritten as

$$X(f) * W(f) = \int_{-\infty}^{\infty} X(g)W(f-g)dg = \int_{-\infty}^{\infty} |X(g)|\, e^{j\phi_X(g)}\, |W(f-g)|\, e^{j\phi_W(f-g)}dg$$

$$= \int_{-\infty}^{\infty} |X(g)| \cdot |W(f-g)|\, e^{j[\phi_X(g)+\phi_W(f-g)]}dg \tag{4.73}$$

---

[1] See Zadeh and Desoer (1963); Papoulis (1977).

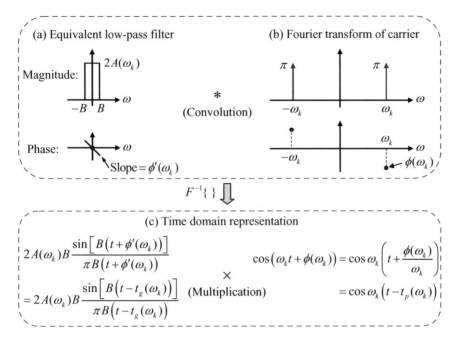

**Figure 4.33** Frequency and time domain representation of $H_k(\omega)$

Thus, the frequency domain convolution (Equation (4.73)) may be interpreted in the form that the resultant magnitude is the running sum of the multiplication of two magnitude functions while the resultant phase is the running sum of the addition of two phase functions.

Since the convolution in the frequency domain results in the multiplication in the time domain (see Equation (4.45)) as depicted in Figure 4.33(c), the inverse Fourier transform of $H_k(\omega)$ becomes

$$F^{-1}\{H_k(\omega)\} \approx 2A(\omega_k)B\frac{\sin\left[B\left(t + \phi'(\omega_k)\right)\right]}{\pi B\left(t + \phi'(\omega_k)\right)}\cos\left(\omega_k t + \phi(\omega_k)\right) \tag{4.74}$$

and finally, the inverse Fourier transform of $H(\omega)$ is

$$h(t) = F^{-1}\{H(\omega)\} \approx \sum_k \underbrace{2A(\omega_k)B\frac{\sin\left[B\left(t - t_g(\omega_k)\right)\right]}{\pi B\left(t - t_g(\omega_k)\right)}}_{\text{envelope}}\underbrace{\cos\omega_k(t - t_p(\omega_k))}_{\text{carrier}} \tag{4.75}$$

where $t_g$ and $t_p$ are the 'group delay' and 'phase delay' respectively, and are defined by Equations (4.76) and (4.77). The relationship between these two properties is illustrated in Figure 4.34.

$$t_g(\omega) = -\frac{d\phi(\omega)}{d\omega} \tag{4.76}$$

$$t_p(\omega) = -\frac{\phi(\omega)}{\omega} \tag{4.77}$$

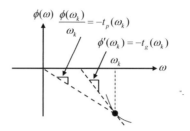

**Figure 4.34**  Illustrations of group delay and phase delay in the frequency domain

Note that each signal component given in Equation (4.75) is an amplitude modulation signal where the 'envelope' is delayed by $t_g$, while the 'carrier' is delayed by $t_p$. This is illustrated in Figure 4.35. As shown in the figure, the *phase delay* gives the time delay of each sinusoidal component while the *group delay* can be interpreted as the time delay of the amplitude envelope (or the group of sinusoidal components within a small frequency band centred at $\omega_k$). The delays are a continuous function of $\omega$, i.e. they may have different values at different frequencies. This deviation of the group delay away from a *constant* indicates the *degree of nonlinearity* of the phase. If a system has a non-constant group delay, each frequency component in the input is delayed differently, so the shape of output signal will be different from the input. This phenomenon is called the *dispersion*. In our simple acoustic models (e.g. Figure 4.25), a single path is non-dispersive, but the inclusion of an echo results in a nonlinear phase characteristic. Most structural systems exhibit dispersive characteristics.

In the case of a pure delay, the group delay and the phase delay are the same as shown in Figure 4.36 (compare the carrier signal with that in Figure 4.35 where the group delay and the phase delay are different).

Directly allied concepts in sound and vibration are the group velocity and the phase velocity of a wave, which are defined by

$$\text{Group velocity of a wave: } v_g = \frac{d\omega}{dk} \tag{4.78}$$

$$\text{Phase velocity of a wave: } v_p = \frac{\omega}{k} \tag{4.79}$$

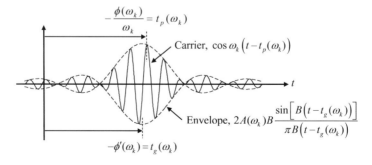

**Figure 4.35**  Illustrations of group delay and phase delay in the time domain

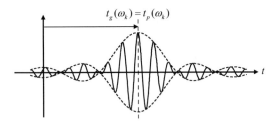

**Figure 4.36** The case of pure delay (group delay and phase delay are the same)

where $\omega$ is the wave's angular frequency, and $k = 2\pi/\lambda$ is the angular wave number ($\lambda$ is the wavelength in the medium). The group velocity and the phase velocity are the same for a non-dispersive wave. Since velocity is distance divided by time taken, the group delay is related to the group velocity of a wave and the phase delay to the phase velocity.

## 4.9 MINIMUM AND NON-MINIMUM PHASE SYSTEMS

### All-pass Filter

We shall now consider the phase characteristics of a special filter (system). Suppose we have a filter with transfer function

$$H(s) = \frac{s-a}{s+a} \tag{4.80}$$

The pole–zero map on the $s$-plane is shown in Figure 4.37.

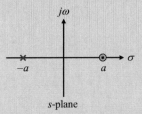

**Figure 4.37** The pole–zero map of Equation (4.80)

Equation (4.80) may be rewritten as

$$H(s) = 1 - \frac{2a}{s+a} \tag{4.81}$$

Then, taking the inverse Laplace transform gives the impulse response function

$$h(t) = \delta(t) - 2ae^{-at} \tag{4.82}$$

which is depicted in Figure 4.38.

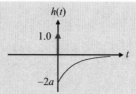

**Figure 4.38**   Impulse response function of the all-pass filter

The corresponding frequency response function is

$$H(\omega) = \frac{j\omega - a}{j\omega + a} \tag{4.83}$$

Thus, the modulus of $H(\omega)$ is

$$|H(\omega)| = \frac{\sqrt{\omega^2 + a^2}}{\sqrt{\omega^2 + a^2}} = 1 \tag{4.84}$$

This implies that there is no amplitude distortion through this filter. So, it is called the 'all-pass filter'. But note that the phase of the filter is *nonlinear* as given in Equation (4.85) and Figure 4.39. So, the all-pass filter distorts the shape of the input signal.

$$\arg H(\omega) = \arg(j\omega - a) - \arg(j\omega + a) = \pi - 2\tan^{-1}\left(\frac{\omega}{a}\right) \quad (\omega \geq 0) \tag{4.85}$$

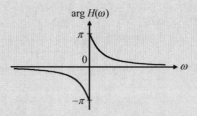

**Figure 4.39**   Phase characteristic of the all-pass filter

From Equation (4.85), the group delay of the all-pass system is

$$-\frac{d}{d\omega}\left(\arg H(\omega)\right) = \frac{2}{a\left(1 + \omega^2/a^2\right)} \tag{4.86}$$

Note that the group delay is always positive as shown in Figure 4.40.

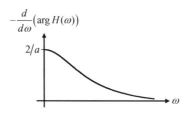

**Figure 4.40**   Group delay of the all-pass filter (shown for $\omega \geq 0$)

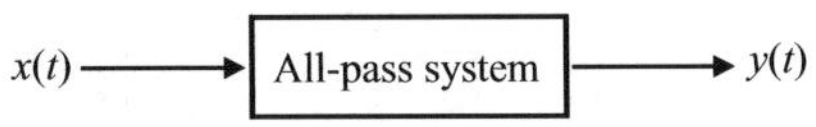

**Figure 4.41** Input–output relationship of the all-pass system

Now, suppose that the response of an all-pass system to an input $x(t)$ is $y(t)$ as in Figure 4.41.

Then, the following properties are obtained:

(i) $$\int_{-\infty}^{\infty} |x(t)|^2 \, dt = \int_{-\infty}^{\infty} |y(t)|^2 \, dt \tag{4.87}$$

(ii) $$\int_{-\infty}^{t_0} |x(t)|^2 \, dt \geq \int_{-\infty}^{t_0} |y(t)|^2 \, dt \tag{4.88}$$

The first Equation (4.87) follows directly from Parseval's theorem. The second Equation (4.88) implies that the energy 'build-up' in the input is more rapid than in the output, and the proof is as follows. Let $y_1(t)$ be the output of the system to the input

$$x_1(t) = x(t), \quad t \leq t_0$$
$$= 0 \qquad t > t_0$$

Then for $t \leq t_0$,

$$y_1(t) = \int_{-\infty}^{t} h(t - \tau)x_1(\tau)d\tau = \int_{-\infty}^{t} h(t - \tau)x(\tau)d\tau = y(t) \tag{4.89}$$

Applying Equation (4.87) to input $x_1(t)$ and output $y_1(t)$, then

$$\int_{-\infty}^{t_0} |x_1(t)|^2 \, dt = \int_{-\infty}^{\infty} |y_1(t)|^2 \, dt = \int_{-\infty}^{t_0} |y_1(t)|^2 \, dt + \int_{t_0}^{\infty} |y_1(t)|^2 \, dt \tag{4.90}$$

Thus, Equation (4.88) follows because $x(t) = x_1(t)$ and $y(t) = y_1(t)$ for $t \leq t_0$.

---

### Minimum and Non-minimum Phase Systems

A stable causal system has all its poles in the left half of the $s$-plane. This is referred to as BIBO (Bounded Input/Bounded Output) stable, i.e. the output will be bounded for every bounded input to the system. For the time domain condition for BIBO stability, the necessary and sufficient condition is $\int_{-\infty}^{\infty} |h(t)| \, dt < \infty$. We now assume that the system is causal and satisfies the BIBO stability criterion. Then, systems may be classified by the structure of the poles and zeros as follows: a system with all its poles and zeros in the left half of the $s$-plane is a minimum phase system; a system with all its zeros in the right half of the $s$-plane is a maximum phase system; a system with some zeros in the left and some in the right half plane is a mixed phase (or non-minimum phase) system. The meaning of 'minimum phase' will be explained shortly.

Consider the following (stable) maximum phase system which has poles and a zero as shown in Figure 4.42:

$$H(s) = \frac{s - a}{s^2 + 2\zeta\omega_n s + \omega_n^2} \tag{4.91}$$

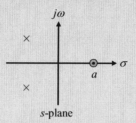

Figure 4.42 The pole–zero map of Equation (4.91)

This may be expressed as

$$H(s) = \left(\frac{s + a}{s^2 + 2\zeta\omega_n s + \omega_n^2}\right)\left(\frac{s - a}{s + a}\right) = H_{\text{min}}(s)H_{\text{ap}}(s) \tag{4.92}$$

where $H_{\text{min}}(s)$ is the minimum phase system with $|H_{\text{min}}(\omega)| = |H(\omega)|$, and $H_{\text{ap}}(s)$ is the all-pass system with $|H_{\text{ap}}(\omega)| = 1$. This decomposition is very useful when dealing with 'inverse' problems (Oppenheim *et al.*, 1999). Note that the direct inversion of the system, $H^{-1}(s)$, has a pole in the right half of the $s$-plane, so the system is unstable. On the other hand, the inverse of a minimum phase system, $H_{\text{min}}^{-1}(s)$, is always stable.

The term 'minimum phase' may be explained by comparing two systems, $H_1(s) = (s + a)/D(s)$ and $H_2(s) = (s - a)/D(s)$. Both systems have the same pole structure but the zeros are at $-a$ and $a$ respectively, so the phase of the system is

$$\arg H_1(\omega) = \tan^{-1}\left(\frac{\omega}{a}\right) - \arg D(\omega) \tag{4.93}$$

$$\arg H_2(\omega) = \pi - \tan^{-1}\left(\frac{\omega}{a}\right) - \arg D(\omega) \tag{4.94}$$

Comparing $\tan^{-1}(\omega/a)$ and $\pi - \tan^{-1}(\omega/a)$, it can be easily seen that $\arg H_1(\omega) < \arg H_2(\omega)$ as shown in Figure 4.43.

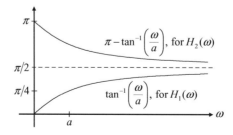

Figure 4.43 Phase characteristics of $H_1(\omega)$ and $H_2(\omega)$

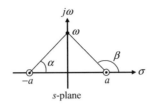

**Figure 4.44** Phase characteristics of $H_1(s)$ and $H_2(s)$

Or, the angles in the $s$-plane show that $\alpha < \beta$ as shown in Figure 4.44. This implies that $H_1(s)$ is minimum phase, since 'phase of $H_1(s) <$ phase of $H_2(s)$'.

It follows that, if $H(s)$ is a stable transfer function with zeros anywhere and $H_{min}(s)$ is a minimum phase system with $|H(\omega)| = |H_{min}(\omega)|$, then the group delay of $H(s)$, $-d \arg H(\omega)/d\omega$, is larger than $-d \arg H_{min}(\omega)/d\omega$. Also, if input $x(t)$ is applied to arbitrary system $H(s)$ giving response $y(t)$ and to $H_{min}(s)$ giving response $y_{min}(t)$, then for any $t_0$ the following energy relationship is given:

$$\int_{-\infty}^{t_0} |y(t)|^2 \, dt \geq \int_{-\infty}^{t_0} |y_{min}(t)|^2 \, dt \qquad (4.95)$$

As a practical example, consider the cantilever beam excited by a shaker as shown in Figure 4.45. Let the signal from the force transducer be the input $x(t)$, and the signals from the accelerometers be the outputs $y_1(t)$ and $y_2(t)$ for positions 1 and 2 respectively. Also, let $H_1(\omega)$ and $H_2(\omega)$ be the frequency response functions between $x(t)$ and $y_1(t)$, and between $x(t)$ and $y_2(t)$ respectively.

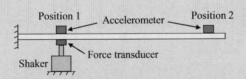

**Figure 4.45** Cantilever beam excited by a shaker

If the input and the output are *collocated* (i.e. measured at the same point) the frequency response function $H_1(\omega)$ is minimum phase, and if they are *non-collocated* the frequency response function $H_2(\omega)$ is non-minimum phase (Lee, 2000). Typical characteristics of the accelerance frequency response functions $H_1(\omega)$ and $H_2(\omega)$ are shown in Figure 4.46. Note that the minimum phase system $H_1(\omega)$ shows distinct anti-resonances with a phase response over $0 \leq \arg H_1(\omega) \leq \pi$.

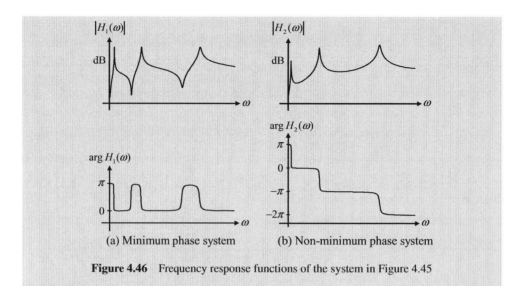

**Figure 4.46**   Frequency response functions of the system in Figure 4.45

## 4.10 THE HILBERT TRANSFORM[M4.3-4.5]

Consider the input–output relationship as described in Figure 4.47.

$$x(t) \longrightarrow \boxed{h(t) = \frac{1}{\pi t}} \longrightarrow y(t) = \hat{x}(t)$$

**Figure 4.47**   Input–output relationship of the 90° phase shifter

The output of the system is the convolution of $x(t)$ with $1/\pi t$:

$$\hat{x}(t) = h(t) * x(t) = \frac{1}{\pi t} * x(t) \qquad (4.96)$$

This operation is called the Hilbert transform. Note that $h(t)$ is a non-causal filter with a singularity at $t = 0$. The Fourier transform of the above convolution operation can be written as

$$\hat{X}(\omega) = H(\omega)X(\omega) \qquad (4.97)$$

where $H(\omega)$ is the Fourier transform of $1/\pi t$, which is given by (see No. 16 of Table 4.1)

$$H(\omega) = -j\,\text{sgn}(\omega) = \begin{cases} -j & \text{for } \omega > 0 \\ j & \text{for } \omega < 0 \\ 0 & \text{for } \omega = 0 \end{cases} \qquad (4.98a)$$

or

$$H(\omega) = \begin{cases} e^{-j(\pi/2)} & \text{for } \omega > 0 \\ e^{j(\pi/2)} & \text{for } \omega < 0 \\ 0 & \text{for } \omega = 0 \end{cases} \qquad (4.98b)$$

From Equation (4.98), it can be seen that

$$|H(\omega)| = 1 \text{ for all } \omega, \text{ except } \omega = 0 \tag{4.99}$$

$$\arg H(\omega) = \begin{cases} -\pi/2 & \text{for } \omega > 0 \\ \pi/2 & \text{for } \omega < 0 \end{cases} \tag{4.100}$$

Thus, the Hilbert transform is often referred to as a 90° phase shifter. For example, the Hilbert transform of $\cos \omega_0 t$ is $\sin \omega_0 t$, and that of $\sin \omega_0 t$ is $-\cos \omega_0 t$.

The significance of the Hilbert transform is that it is used to form the so called 'analytic signal' or 'pre-envelope signal'. An analytic signal is a *complex* time signal whose real part is the original signal $x(t)$ and where imaginary part is the Hilbert transform of $x(t)$, i.e. $\hat{x}(t)$. Thus, the analytic signal $a_x(t)$ is defined as

$$a_x(t) = x(t) + j\hat{x}(t) \tag{4.101}$$

The Fourier transform of analytic signal $F\{a_x(t)\}$ is zero for $\omega < 0$, and is $2X(\omega)$ for $\omega > 0$ and $X(\omega)$ for $\omega = 0$. Since the analytic signal is complex, it can be expressed as

$$a_x(t) = A_x(t)e^{j\phi_x(t)} \tag{4.102}$$

where $A_x(t) = \sqrt{x^2(t) + \hat{x}^2(t)}$ is the *instantaneous amplitude*, and $\phi_x(t) = \tan^{-1}(\hat{x}(t)/x(t))$ is the *instantaneous phase*. The time derivative of the unwrapped instantaneous phase $\omega_x(t) = \dot{\phi}_x(t) = d\phi_x(t)/dt$ is called the *instantaneous frequency*. For a trivial case $x(t) = \cos \omega_0 t$, the analytic signal is $a_x(t) = e^{j\omega_0 t}$ where $A_x(t) = 1$ and $\omega_x(t) = \omega_0$, i.e. both are constants as expected. These concepts of instantaneous amplitude, phase and frequency are particularly useful for amplitude-modulated and frequency-modulated signals.

To visualize these concepts, consider the following amplitude-modulated signal[M4.3]

$$x(t) = m(t) \cos \omega_c t = (A_c + A_m \sin \omega_m t) \cos \omega_c t \tag{4.103}$$

where $\omega_c > \omega_m$. We note that if $m(t)$ is *band-limited* and has a maximum frequency less than $\omega_c$, the Hilbert transform of $x(t) = m(t) \cos \omega_c t$ is $\hat{x}(t) = m(t) \sin \omega_c t$. Then, using the relationship between Equations (4.101) and (4.102), the analytic signal can be written as

$$a_x(t) = A_x(t)e^{j\phi_x(t)} = (A_c + A_m \sin \omega_m t) e^{j\omega_c t} \tag{4.104}$$

and the corresponding $A_x(t)$, $\phi_x(t)$ and $\omega_x(t)$ are as shown in Figure 4.48.

In sound and vibration engineering, a practical application of the Hilbert transform related to amplitude modulation/demodulation is 'envelope analysis' (Randall, 1987), where the demodulation refers to a technique that extracts the modulating components, e.g. extracting $A_m \sin \omega_m t$ from Equation (4.103). Envelope analysis is used for the early detection of a machine fault. For example, a fault in an outer race of a rolling bearing may generate a series of

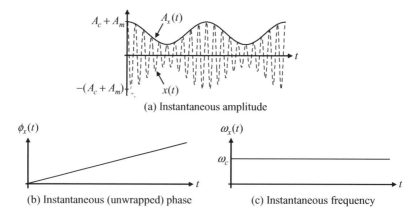

(a) Instantaneous amplitude

(b) Instantaneous (unwrapped) phase          (c) Instantaneous frequency

**Figure 4.48**    Analytic signal associated with the amplitude-modulated signal

burst signals at a regular interval. Such burst signals decay very quickly and contain relatively small energies, thus the usual Fourier analysis may not reveal the repetition frequency of the bursts. However, it may be possible to detect this frequency component by forming the analytic signal and then applying Fourier analysis to the envelope $A_x(t)$.

## Examples

### *Example 1:* **Estimation of damping from time domain records of an oscillator** [M4.4]

Suppose we have a free response of a damped single-degree-of-freedom system as below:

$$x(t) = Ae^{-\zeta\omega_n t}\sin(\omega_d t + \phi) \quad t \geq 0 \tag{4.105}$$

where $\omega_d = \omega_n\sqrt{1 - \zeta^2}$. The analytic signal for this may be approximated as

$$a_x(t) = A_x(t)e^{j\phi_x(t)} \approx \left(Ae^{-\zeta\omega_n t}\right)e^{j(\omega_d t + \phi - \pi/2)} \quad t \geq 0 \tag{4.106}$$

Since $\ln A_x(t) \approx \ln A - \zeta\omega_n t$, the damping ratio $\zeta$ can be estimated from the plot of $\ln A_x(t)$ versus time, provided that the natural frequency $\omega_n$ is known. This is demonstrated in MATLAB Example 4.4. However, as shown in MATLAB Example 4.4, it must be noted that $A_x(t)$ and $\phi_x(t)$ are usually distorted, especially at the beginning and the last parts of the signal. This undesirable phenomenon occurs from the following: (i) the modulating component $Ae^{-\zeta\omega_n t}$ is not band-limited, (ii) the non-causal nature of the filter $(h(t) = 1/\pi t)$, and (iii) practical windowing effects (truncation in the frequency domain). Thus, the part of the signal near $t = 0$ must be avoided in the estimation of the damping characteristic. The windowing effect is discussed in the next section.

## *Example 2:* Frequency modulation[M4.5]

Now, to demonstrate another feature of the analytic signal, we consider the frequency modulated signal as given below:

$$x(t) = A_c \cos(\omega_c t + A_m \sin \omega_m t) \tag{4.107}$$

This can be written as $x(t) = A_c [\cos \omega_c t \cos (A_m \sin \omega_m t) - \sin \omega_c t \sin (A_m \sin \omega_m t)]$ which consists of two amplitude-modulated signals, i.e. $x(t) = m_1(t) \cos \omega_c t - m_2(t) \sin \omega_c t$, where $m_1(t)$ and $m_2(t)$ may be approximated as band-limited (Oppenheim *et al.*, 1999). So, for $A_m \omega_m \ll \omega_c$, the analytic signal associated with Equation (4.107) may be approximated as

$$a_x(t) = A_x(t) e^{j \phi_x(t)} \approx A_c e^{j(\omega_c t + A_m \sin \omega_m t)} \tag{4.108}$$

and the corresponding $A_x(t)$, $\phi_x(t)$ and $\omega_x(t)$ are as shown in Figure 4.49. Note that the instantaneous frequency is $\omega_x(t) = d\phi_x(t)/dt = \omega_c + \omega_m A_m \cos \omega_m t$, as can be seen in Figure 4.49(c).

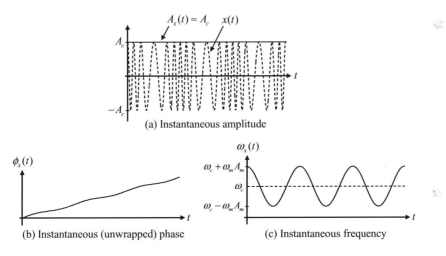

(a) Instantaneous amplitude

(b) Instantaneous (unwrapped) phase

(c) Instantaneous frequency

**Figure 4.49**   Analytic signal associated with the frequency-modulated signal

From this example, we have seen that it may be possible to examine how the frequency contents of a signal vary with time by forming an analytic signal. We have seen two methods of relating the temporal and frequency structure of a signal. First, based on the Fourier transform we saw how group delay relates how groups of frequencies are delayed (shifted) in time, i.e. the group delays are time dependent. Second, we have seen a 'non-Fourier' type of representation of a signal as $A(t) \cos \phi(t)$ (based on the analytic signal derived using the Hilbert transform). This uses the concepts of amplitude modulation and instantaneous phase and frequency.

These two approaches are different and only under certain conditions do they give similar results (for signals with large bandwidth–time product – see the uncertainty principle in the next section). These considerations are fundamental to many of the time–frequency analyses of signals. Readers may find useful information on time–frequency methods in two review papers (Cohen, 1989; Hammond and White, 1996).

## 4.11  THE EFFECT OF DATA TRUNCATION (WINDOWING)[M4.6–4.9]

Suppose $x(t)$ is a deterministic signal but is known only for $-T/2 \leq t \leq T/2$, as shown in Figure 4.50.

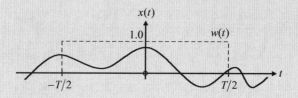

**Figure 4.50**   Truncated data with a rectangular window $w(t)$

In effect, we are observing the data through a *window* $w(t)$ where

$$w(t) = 1 \quad |t| < T/2$$
$$= 0 \quad |t| > T/2 \tag{4.109}$$

so that we see the truncated data $x_T(t) = x(t)w(t)$.

If we Fourier transform $x_T(t)$ (in an effort to get $X(f)$) we obtain the Fourier transform of the product of two signals $x(t)$ and $w(t)$ as (see Equation (4.45))

$$X_T(f) = F\{x(t)w(t)\} = \int_{-\infty}^{\infty} X(g)W(f - g)dg = X(f) * W(f) \tag{4.110}$$

i.e. the Fourier transform of the product of two time signals is the convolution of their Fourier transforms. $W(f)$ is called the *spectral window*, and is $W(f) = T\sin(\pi f T)/\pi f T$ for the rectangular window. Owing to this convolution operation in the frequency domain, the window (which need not be restricted to the rectangular data window) results in *bias* or *truncation error*. Recall the shape of $W(f)$ for the rectangular window as in Figure 4.51.

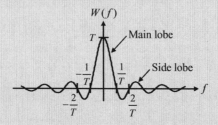

**Figure 4.51**   Fourier transform of the rectangular window $w(t)$

The convolution integral indicates that the shape of $X(g)$ is distorted, such that it broadens the true Fourier transform. The distortion due to the main lobe is sometimes called *smearing*, and the distortion caused by the side lobes is called *leakage* since the frequency components of $X(g)$ at values other than $g = f$ 'leak' through the side lobes to contribute to the value of $X_T(f)$ at $f$. For example, consider a sinusoidal signal $x(t) = \cos(2\pi pt)$ whose Fourier transform is $X(f) = \frac{1}{2}[\delta(f + p) + \delta(f - p)]$. Then the Fourier transform of the truncated signal $x_T(t)$ is

$$X_T(f) = \int_{-\infty}^{\infty} X(g)W(f - g)dg = \frac{1}{2}\int_{-\infty}^{\infty} [\delta(g + p) + \delta(g - p)] W(f - g)dg$$

$$= \frac{1}{2}[W(f + p) + W(f - p)] \qquad (4.111)$$

This shows that the delta functions (in the frequency domain) are replaced by the shape of the spectral window. The 'theoretical' and 'achieved (windowed)' spectra are illustrated in Figure 4.52 (compare $X(f)$ and $X_T(f)$ for both shape and magnitude).

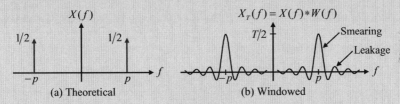

**Figure 4.52**   Fourier transform of a cosine wave

If two or more closely spaced sinusoidal components are present in a signal, then they may not easily be resolved in the frequency domain because of the distortion (especially due to the main lobe). A rough guide as to the effect of this rectangular window is obtained from Figure 4.53 (shown for $f > 0$ only).

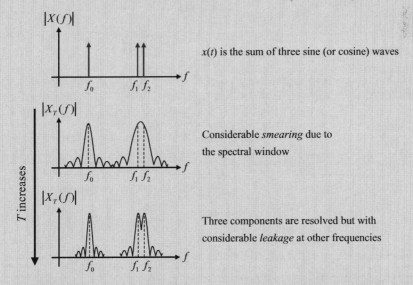

**Figure 4.53**   Effects of windowing on the modulus of the Fourier transform

In fact, in order to get two separate peaks of frequencies $f_1$, $f_2$ given in this example it is necessary to use a data length $T$ of order $T \geq 2/(f_2 - f_1)$ (i.e. $f_2 - f_1 \geq 2/T$) for the rectangular window. Note that the rectangular window is considered a 'poor' window with respect to the side lobes, i.e. the side lobes are large and decay slowly. The highest side lobe is 13 dB below the peak of the main lobe, and the asymptotic roll-off is 6 dB/octave. This results from the sharp corners of the rectangular window. However, the main lobe of the rectangular window is narrower than any other windows.

MATLAB examples are given at the end of the chapter. Since we are using sinusoidal signals in MATLAB Examples 4.6 and 4.7, it is interesting to compare this windowing effect with the computational considerations for a periodic signal given in Section 3.6 (and with MATLAB Example 3.2).

A wide variety of windows are available, each with its own frequency characteristics. For example, by tapering the windows to zero, the side lobes can be reduced but the main lobe is wider than that of the rectangular window, i.e. increased smearing. To see this effect, consider the following two window functions:

1.  A 20 % cosine tapered window (at each side, 10 % of the data record is tapered):

$$w_C(t) = 1 \qquad\qquad |t| < 4T/10$$

$$= \cos^2 \frac{5\pi t}{T} \qquad -T/2 \leq t \leq -4T/10, \quad 4T/10 \leq t \leq T/2$$

$$= 0 \qquad\qquad |t| > T/2 \qquad\qquad\qquad (4.112)$$

2.  A Hann (Hanning) window (full cosine tapered window):

$$w_H(t) = \cos^2 \frac{\pi t}{T} \quad |t| < T/2$$

$$= 0 \qquad\quad |t| > T/2 \qquad\qquad\qquad (4.113)$$

These window functions are sometimes called the Tukey window, and are shown in Figure 4.54. Note that the cosine tapered window has a narrower bandwidth and so better frequency resolution whilst the Hann window has smaller side lobes and sharper roll-off, giving improved leakage suppression.

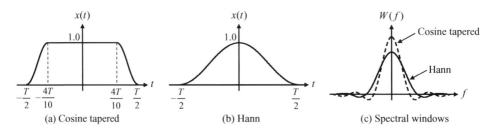

**Figure 4.54**  Effect of tapering window

Window 'carpentry' is used to design windows to reduce leakage at the expense of main lobe width in Fourier transform calculations, i.e. to obtain windows with small side lobes. One 'trades' the side lobe reduction for 'bandwidth', i.e. by tapering the window smoothly to zero, the side lobes are greatly reduced, but the price paid is a much wider main lobe. The frequency characteristic of a window is often presented in dB normalized to unity gain (0 dB) at zero frequency, e.g. as shown in Figure 4.55 for the rectangular window (in general, $A = 1$).

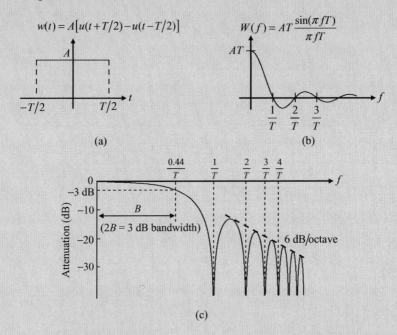

$$w(t) = A\left[u(t + T/2) - u(t - T/2)\right]$$

$$W(f) = AT \frac{\sin(\pi f T)}{\pi f T}$$

(a)                                                                                (b)

(c)

**Figure 4.55**   Rectangular window and its frequency characteristic

The rectangular window may be good for separating closely spaced sinusoidal components, but the leakage is the price to pay. Some other commonly used windows and their spectral properties (for $f \geq 0$ only) are shown in Figure 4.56. The Hann window is a good general purpose window, and has a moderate frequency resolution and a good side lobe roll-off characteristic. Through MATLAB Examples 4.6–4.9, the frequency characteristics of the rectangular window and the Hann window are compared. Another widely used window is the Hamming window (a Hann window sitting on a small rectangular base). It has a low level of the first few side lobes, and is used for speech signal processing. The frequency characteristics of these window functions are compared in Figure 4.57.

We now note a few general comments on windows:

1. The ability to pick out peaks (resolvability) depends on the data widow *width* as well as the *shape*.
2. The windows in Figure 4.56 (and others except the rectangular window) are not generally applicable to transient waveforms where a significant portion of the information is lost by

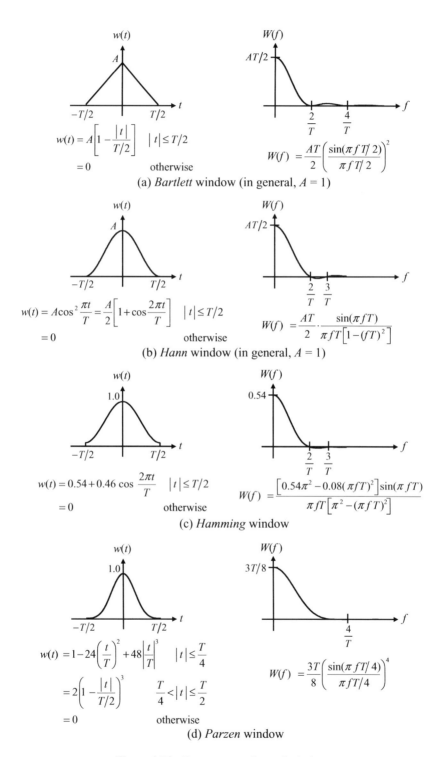

$$w(t) = A\left[1 - \frac{|t|}{T/2}\right] \quad |t| \le T/2$$

$$= 0 \qquad \text{otherwise}$$

$$W(f) = \frac{AT}{2}\left(\frac{\sin(\pi f T/2)}{\pi f T/2}\right)^2$$

(a) *Bartlett* window (in general, $A = 1$)

$$w(t) = A\cos^2\frac{\pi t}{T} = \frac{A}{2}\left[1 + \cos\frac{2\pi t}{T}\right] \quad |t| \le T/2$$

$$= 0 \qquad \text{otherwise}$$

$$W(f) = \frac{AT}{2} \cdot \frac{\sin(\pi f T)}{\pi f T\left[1 - (fT)^2\right]}$$

(b) *Hann* window (in general, $A = 1$)

$$w(t) = 0.54 + 0.46 \cos\frac{2\pi t}{T} \quad |t| \le T/2$$

$$= 0 \qquad \text{otherwise}$$

$$W(f) = \frac{\left[0.54\pi^2 - 0.08(\pi f T)^2\right]\sin(\pi f T)}{\pi f T\left[\pi^2 - (\pi f T)^2\right]}$$

(c) *Hamming* window

$$w(t) = 1 - 24\left(\frac{t}{T}\right)^2 + 48\left|\frac{t}{T}\right|^3 \quad |t| \le \frac{T}{4}$$

$$= 2\left(1 - \frac{|t|}{T/2}\right)^3 \quad \frac{T}{4} < |t| \le \frac{T}{2}$$

$$= 0 \qquad \text{otherwise}$$

$$W(f) = \frac{3T}{8}\left(\frac{\sin(\pi f T/4)}{\pi f T/4}\right)^4$$

(d) *Parzen* window

**Figure 4.56**   Some commonly used windows

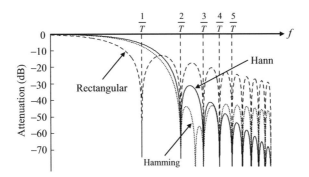

**Figure 4.57** Frequency characteristics of some windows

windowing.[M4.9] (The exponential window is sometimes used for exponentially decaying signals such as responses to impact hammer tests.)

3. A correction factor (scaling factor) should be applied to the window functions to account for the loss of 'energy' relative to a rectangular window as follows:

$$\text{Scaling factor} = \sqrt{\frac{\int_{-T/2}^{T/2} w_{rect}^2(t)dt}{\int_{-T/2}^{T/2} w^2(t)dt}} \qquad (4.114)$$

where $w_{rect}(t)$ is the rectangular window, and $w(t)$ is the window function applied on the signal. For example, the scaling factor for the Hann window is $\sqrt{8/3}$. This correction factor is used in MATLAB Examples 4.7–4.9. This correction is more readily interpreted in relation to stationary random signals and will be commented upon again in that context with a more general formula for the estimation of the power spectral density.

4. For the data windows, we define two 'bandwidths' of the windows, namely (a) *3 dB bandwidth*; (b) *noise bandwidth*. The 3 dB bandwidth is the width of the power transmission characteristic at the 3 dB points, i.e. where there are 3 dB points below peak amplification, as shown in Figure 4.58.

 The (equivalent) noise bandwidth is the width of an ideal filter with the same peak power gain that accumulates the same power from a white noise source, as shown in Figure 4.59 (Harris, 1978).

5. The properties of some commonly used windows are summarised in Table 4.2. More comprehensive discussions on window functions can be found in Harris (1978).

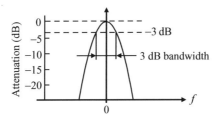

**Figure 4.58** The 3 dB bandwidth

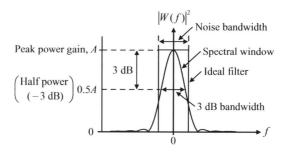

**Figure 4.59**   Noise bandwidth

**Table 4.2**   Properties of some window functions

| Window (length $T$) | Highest side lobe (dB) | Asymptotic roll-off (dB/octave) | 3 dB bandwidth | Noise bandwidth | First zero crossing (freq.) |
|---|---|---|---|---|---|
| Rectangular | −13.3 | 6 | $0.89\dfrac{1}{T}$ | $1.00\dfrac{1}{T}$ | $\dfrac{1}{T}$ |
| Bartlett (triangle) | −26.5 | 12 | $1.28\dfrac{1}{T}$ | $1.33\dfrac{1}{T}$ | $\dfrac{2}{T}$ |
| Hann(ing) (Tukey or cosine squared) | −31.5 | 18 | $1.44\dfrac{1}{T}$ | $1.50\dfrac{1}{T}$ | $\dfrac{2}{T}$ |
| Hamming | −43 | 6 | $1.30\dfrac{1}{T}$ | $1.36\dfrac{1}{T}$ | $\dfrac{2}{T}$ |
| Parzen | −53 | 24 | $1.82\dfrac{1}{T}$ | $1.92\dfrac{1}{T}$ | $\dfrac{4}{T}$ |

### The Uncertainty Principle (Bandwidth–Time Product)

As can be seen from the Fourier transform of a rectangular pulse (see Figure 4.8), i.e. Equation (4.27), $X(f) = 2ab \sin(2\pi f b)/2\pi fb$, a property of the Fourier transform of a signal is that the narrower the signal description in one domain, the wider its description in the other. An extreme example is a delta function $\delta(t)$ whose Fourier transform is a constant. Another example is a sinusoidal function $\cos(2\pi f_0 t)$ whose Fourier transform is $\frac{1}{2}[\delta(f - f_0) + \delta(f + f_0)]$. This fundamental property of signals is generalized by the so-called *uncertainty principle*.

Similar to Heisenberg's uncertainty principle in quantum mechanics, the uncertainty principle in Fourier analysis is that the product of the spectral bandwidth and the time duration of a signal must be greater than a certain value. Consider a signal $x(t)$ with finite energy, such that $\|x\|^2 = \int_{-\infty}^{\infty} x^2(t)dt < \infty$, and its Fourier transform $X(\omega)$. We define the following:

$$\bar{t} = \frac{1}{\|x\|^2} \int_{-\infty}^{\infty} t x^2(t)dt \tag{4.115a}$$

$$(\Delta t)^2 = \frac{1}{\|x\|^2} \int_{-\infty}^{\infty} (t - \bar{t})^2 x^2(t)dt \tag{4.115b}$$

where $\bar{t}$ is the centre of gravity of the area defined by $x^2(t)$, i.e. the measure of location, and the *time dispersion* $\Delta t$ is the measure of the spread of $x(t)$. Similarly, on the frequency scale, $\|X\|^2 = \int_{-\infty}^{\infty} |X(\omega)|^2 d\omega$, and we define

$$\bar{\omega} = \frac{1}{\|X\|^2} \int_{-\infty}^{\infty} \omega |X(\omega)|^2 d\omega \tag{4.116a}$$

$$(\Delta\omega)^2 = \frac{1}{\|X\|^2} \int_{-\infty}^{\infty} (\omega - \bar{\omega})^2 |X(\omega)|^2 d\omega \tag{4.116b}$$

where $\bar{\omega}$ is the measure of location on the frequency scale, and $\Delta\omega$ is called the *spectral bandwidth*, which is the measure of spread of $X(\omega)$. Note that for a real signal $x(t)$, $\bar{\omega}$ is equal to zero since $|X(\omega)|^2$ is even. Using *Schwartz's inequality*

$$\int_{-\infty}^{\infty} |f(t)|^2 dt \cdot \int_{-\infty}^{\infty} |g(t)|^2 dt \geq \left| \int_{-\infty}^{\infty} f(t)g(t)dt \right|^2 \tag{4.117}$$

and Parseval's theorem, it can be shown that (Hsu, 1970)

$$\Delta\omega \cdot \Delta t \geq \frac{1}{2} \tag{4.118}$$

or, if the spectral bandwidth is defined in hertz,

$$\Delta f \cdot \Delta t \geq \frac{1}{4\pi} \tag{4.119}$$

Thus, the bandwidth–time (*BT*) product of a signal has a lower bound of $1/2$ . For example, the *BT* product of the rectangular window is $\Delta\omega \cdot \Delta t = 2\pi$ (or $\Delta f \cdot \Delta t = 1$), and the Gaussian pulse $e^{-at^2}$ has the 'minimum *BT* product' of $\Delta\omega \cdot \Delta t = 1/2$ (recall that the Fourier transform of a Gaussian pulse is another Gaussian pulse, see Equation (4.33)). For the proof of these results, see Hsu (1970).

The inequality above points out a difficulty (or a limitation) in the *Fourier-based* time–frequency analysis methods. That is, if we want to obtain a 'local' Fourier transform then increasing the 'localization' in the time domain results in poorer resolution in the frequency domain, and vice versa. In other words, we cannot achieve arbitrarily fine 'resolution' in both the time and frequency domains at the same time.

Sometimes, the concept of the above inverse spreading property can be very useful to understand principles of noise control. For example, when the impact between two solid bodies produces a significant noise, the most immediate remedy may be to increase the impact duration by adding some resilient material. This increase of time results in narrower frequency bandwidth, i.e. removes the high-frequency noise, and reduces the total noise level. This is illustrated in Figure 4.60 assuming that the force is a half-sine pulse. Note that the impulse

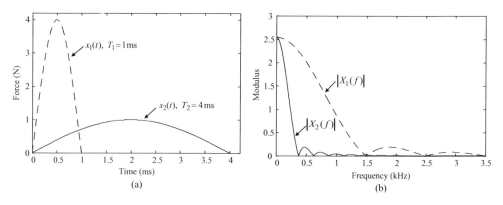

**Figure 4.60**   Interpretation of impact noise

(the area under the force curve, $x_i(t)$) is the same for both cases, i.e.

$$\int_0^{T_1} x_1(t)dt = \int_0^{T_2} x_2(t)dt$$

However, the total energy of the second impulse is much smaller, i.e.

$$\int_{-\infty}^{\infty} |X_1(f)|^2 df \gg \int_{-\infty}^{\infty} |X_2(f)|^2 df$$

as shown in Figure 4.60(b). Also note that, for each case, Parseval's theorem is satisfied, i.e.

$$\int_0^{T_i} x_i^2(t)dt = \int_{-\infty}^{\infty} |X_i(f)|^2 df$$

## 4.12 BRIEF SUMMARY

1. A deterministic aperiodic signal may be expressed by

$$x(t) = \int_{-\infty}^{\infty} X(f)e^{j2\pi ft}df \text{ and } X(f) = \int_{-\infty}^{\infty} x(t)e^{-j2\pi ft}dt : \text{Fourier integral pair}$$

2. Then, the *energy* spectral density of $x(t)$ is $|X(f)|^2$ and satisfies

$$\int_{-\infty}^{\infty} x^2(t)dt = \int_{-\infty}^{\infty} |X(f)|^2 df : \text{ Parseval's theorem}$$

3. The input–output relationship for an LTI system is expressed by the convolution integral,

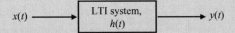

i.e. $y(t) = h(t) * x(t) = \int_{-\infty}^{\infty} h(\tau)x(t - \tau)d\tau$, and in the frequency domain $Y(f) = H(f)X(f)$.

4. A pure delay preserves the shape of the original shape, and gives a constant value of group delay $-d\phi/d\omega = t_0$. A non-constant group delay indicates the degree of nonlinearity of the phase.

5. A minimum phase system has all its poles and zeros in the left half of the $s$-plane, and is especially useful for inverse problems.

6. The analytic signal $a_x(t) = A_x(t)e^{j\phi_x(t)}$ provides the concepts of instantaneous amplitude, instantaneous phase and instantaneous frequency.

7. If a signal is truncated such that $x_T(t) = x(t)w(t)$, then $X_T(f) = \int_{-\infty}^{\infty} X(g)W(f - g)dg$.

8. Data windows $w(t)$ introduce 'leakage' and distort the Fourier transform. Both the *width* and *shape* of the window dictate the resolvability of closely spaced frequency components. A 'scale factor' should be employed when a window is used.

9. The uncertainty principle states that the product of the spectral bandwidth and the time extent of a signal is $\Delta\omega \cdot \Delta t \geq 1/2$. This indicates the fundamental limitations of the Fourier-based analyses.

## 4.13 MATLAB EXAMPLES

### Example 4.1: The effect of an echo

Consider a signal with a pure echo, $y(t) = x(t) + ax(t - t_0)$ as given in Equation (4.46), where the main signal is $x(t) = e^{-\lambda|t|}$ (see Equation (4.20) and Figure 4.5). For this example, the parameters $a = 0.2$, $\lambda = 300$ and $t_0 = 0.15$ are chosen. Readers may change these values to examine the effects for various cases.

| Line | MATLAB code | Comments |
|---|---|---|
| 1 | clear all | Define time variable from $-5$ to 5 seconds |
| 2 | fs=500; t=-5:1/fs:5; | with sampling rate fs $= 500$. |
| 3 | lambda=300; t0=0.15; a=0.2; | Assign values for the parameters of the signal. |
| 4 | x=exp(-lambda*abs(t)); | Expression of the main signal, $x(t)$. This is for the comparison with $y(t)$. |

| 5 | y=x+a*exp(-lambda*abs(t-t0)); | Expression of the signal, $y(t)$. |
|---|---|---|
| 6 | X=fft(x); Y=fft(y); | Fourier transforms of signals $x(t)$ and $y(t)$. In fact, this is the discrete Fourier transform (DFT) which will be discussed in Chapter 6. |
| 7 | N=length(x); | Define the frequency variables for both |
| 8 | fp=0:fs/N:fs/2; % for the positive frequency | positive and negative frequencies. (The |
| 9 | fn=-fs/N:-fs/N:-fs/2; | frequency spacing of the DFT will also be |
| | % for the negative frequency | discussed in Chapter 6.) The command |
| 10 | f=[fliplr(fn) fp]; | 'fliplr' flips the vector (or matrix) in the left/right direction. |
| 11 | plot(f,fftshift(abs(X)/fs), 'r:') | Plot the magnitude of $X(f)$, i.e. $|X(f)|$ |
| 12 | xlabel('Frequency (Hz)'); ylabel('Modulus') | (dashed line)[2], and hold the graph. The |
| 13 | hold on | command 'fftshift' shifts the zero frequency component to the middle of the spectrum. Note that the magnitude is scaled by '1/fs', and the reason for doing this will also be found in Chapter 6. |
| 14 | plot(f,fftshift(abs(Y)/fs)) | Plot the magnitude $|Y(f)|$ on the same |
| 15 | hold off | graph, and release the graph. Compare this with $|X(f)|$. |

## Results

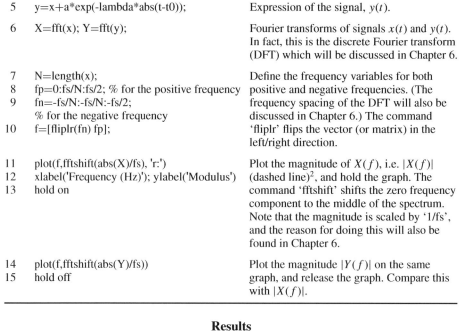

---

*Example 4.2:* **Appearances of envelope and carrier signals**

This is examined for the cases of $t_p = t_g$, $t_p < t_g$ and $t_p > t_g$ in Equation (4.75), i.e.

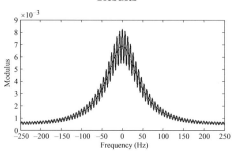

$$x(t) = 2AB\underbrace{\frac{\sin\left(B(t-t_g)\right)}{\pi B(t-t_g)}}_{\text{envelope}} \underbrace{\cos\omega_k(t-t_p)}_{\text{carrier}}$$

---

[2] It is dotted line in the MATLAB code. However, dashed lines are used for generating figures. So, the dashed line in the comments denotes the 'dotted line' in the corresponding MATLAB code. This applies to all MATLAB examples in this book.

| Line | MATLAB code | Comments |
|---|---|---|
| 1<br>2 | clear all<br>B=1; | Define the frequency band in rad/s. |
| 3<br>4 | A=3;<br>wk=6; | Select the amplitude A arbitrary, and define the carrier frequency, wk such that wk $\gg$ B. |
| 5<br>6 | tg=5;<br>tp=5; % tp=4.7 (for tp < tg),<br>      % tp=5.3 (for tp > tg) | Define the group delay tg, and the phase delay tp.<br>In this example, we use tp=5 for tp = tg, tp=4.7 for tp < tg, and tp=5.3 for tp > tg. Try with different values. |
| 7 | t=0:0.03:10; | Define the time variable. |
| 8 | x=2*A*B*sin(B*(t-tg))./(pi*B*<br>(t-tg)).*cos(wk*(t-tp)); | Expression of the above equation. This is the actual time signal. |
| 9 | xe=2*A*B*sin(B*(t-tg))./(pi*B*(t-tg)); | Expression of the 'envelope' signal. |
| 10<br><br>11 | plot(t,x); xlabel('Time (s)');<br>ylabel('\itx\rm(\itt\rm)')<br>hold on | Plot the actual amplitude-modulated signal, and hold the graph. |
| 12<br>13<br>14 | plot(t, xe, 'g:', t, -xe, 'g:')<br>hold off<br>grid on | Plot the envelope signal with the dashed line, and release the graph. |

**Results**

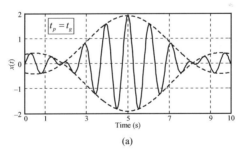

(a)

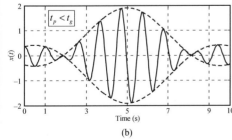

(b)

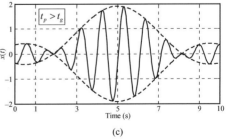

(c)

*Example 4.3:* **Hilbert transform: amplitude-modulated signal** (see Equation (4.103))

$$x(t) = (A_c + A_m \sin \omega_m t) \cos \omega_c t = (A_c + A_m \sin 2\pi f_m t) \cos 2\pi f_c t$$

For this example, the parameters $A_c = 1$, $A_m = 0.5$, $f_m = 1$ and $f_c = 10$ are chosen.

| Line | MATLAB code | Comments |
|---|---|---|
| 1 | clear all | Define parameters and the time variable. |
| 2 | Ac=1; Am=0.5; fm=1; fc=10; | |
| 3 | t=0:0.001:3; | |
| 4 | x=(Ac+Am*cos(2*pi*fm*t)).*cos(2*pi*fc*t); | Expression of the amplitude-modulated signal, $x(t)$. |
| 5 | a=hilbert(x); | Create the analytic signal. Note that, in MATLAB, the function 'hilbert' creates the analytic signal, not $\hat{x}(t)$. |
| 6 | fx=diff(unwrap(angle(a)))./diff(t)/(2*pi); | This is an approximate derivative, which computes the instantaneous frequency in Hz. |
| 7 | figure(1) | Plot the instantaneous amplitude |
| 8 | plot(t, abs(a), t, x, 'g:') | $A_x(t)$. |
| 9 | axis([0 3 -2 2]) | Note that $A_x(t)$ estimates well the |
| 10 | xlabel('Time (s)'); ylabel('\itA_x\rm(\itt\rm)') | envelope of the signal, $A_c + A_m \sin 2\pi f_m t = 1 + 0.5 \sin 2\pi \cdot 1 \cdot t$. |
| 11 | figure(2) | Plot the instantaneous (unwrapped) |
| 12 | plot(t, unwrap(angle(a))) | phase $\phi_x(t)$, which increases linearly |
| 13 | axis([0 3 0 200]) | with time. |
| 14 | xlabel('Time (s)'); ylabel('\it\phi_x\rm(\itt\rm)') | |
| 15 | figure(3) | Plot the instantaneous frequency, |
| 16 | plot(t(2:end),fx) | where $f_x(t) = \omega_x(t)/2\pi$. |
| 17 | axis([0 3 8 12]) | Note that $f_x(t)$ estimates $f_c = 10$ |
| 18 | xlabel('Time (s)'); ylabel('\itf_x\rm(\itt\rm)') | reasonably well, except small regions at the beginning and end. |

**Results**

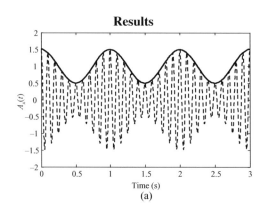

Time (s)

(a)

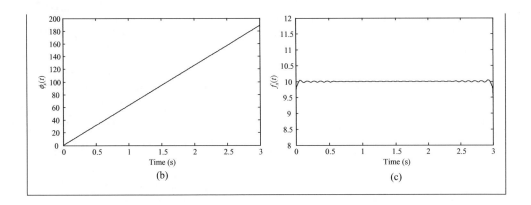

(b)                                                                       (c)

---

*Example 4.4:* **Hilbert transform: estimation of damping coefficient** (see Equation (4.106))

Suppose we have a signal represented as Equation (4.105), i.e.

$$x(t) = Ae^{-\zeta \omega_n t} \sin(\omega_d t + \phi) = Ae^{-\zeta 2\pi f_n t} \sin(\omega_d t + \phi)$$

and, for this example, the parameters $A = 1$, $\zeta = 0.01$, $f_n = 10$ and $\phi = 0$ are chosen.

| Line | MATLAB code | Comments |
|---|---|---|
| 1 | clear all | Define parameters and the time variable. |
| 2 | A=1; zeta=0.01; fn=10; wn=2*pi*fn; | |
| 3 | wd=wn*sqrt(1-zeta^2); phi=0; t=0:0.001:6; | |
| 4 | x=A*exp(-zeta*wn*t).*sin(wd*t+phi); | Expression of the signal (Equation (4.105)). |
| 5 | a=hilbert(x); | Create the analytic signal. |
| 6 | ax=log(abs(a)); | Compute ln $A_x(t)$. Note that 'log' in MATLAB denotes the natural logarithm. |
| 7 | figure(1) | Plot the instantaneous amplitude $A_x(t)$. |
| 8 | plot(t, abs(a), t, x, 'g:'); axis([0 6 -1.5 1.5]) | Note that, in this figure (Figure (a) below), |
| 9 | xlabel('Time (s)'); | the *windowing effect* (truncation in the |
|  | ylabel('\itA_x\rm(\itt\rm)') | frequency domain – MATLAB uses the FFT-based algorithm, see MATLAB help window for details) and the *non-causal* component are clearly visible. |
| 10 | figure(2) | Plot ln $A_x(t)$ versus time. The figure shows |
| 11 | plot(t, ax); axis([0 6 -6 1]) | a linearly decaying characteristic over the |
| 12 | xlabel('Time (s)'); | range where the windowing effects are not |
|  | ylabel('ln\itA_x\rm(\itt\rm)') | significant. |

| 13 | p=polyfit(t(1000:4000), ax(1000:4000), 1); | 'polyfit' finds the coefficients of a polynomial that fits the data in the least squares sense. In this example, we use a polynomial of degree 1 (i.e. linear regression). Also, we use the data set in the well-defined region only (i.e. 1 to 4 seconds). |
| 14 | format long | 'format long' displays the number with 15 digits. |
| 15 | zeta_est=-p(1)/wn | The first element of the vector p represents the slope of the graph in Figure (b) below. Thus, the $\zeta$ can be estimated by dividing $-p(1)$ by the natural frequency $\omega_n$. |

## Results

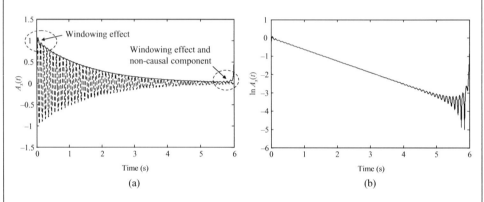

(a)                                                   (b)

The variable 'zeta_est' returns the value '0.00999984523039' which is very close to the true value $\zeta = 0.01$.

---

***Example 4.5:* Hilbert transform: frequency-modulated signal** (see Equation (4.107))

$$x(t) = A_c \cos(\omega_c t + A_m \sin \omega_m t) = A_c \cos(2\pi f_c t + A_m \sin 2\pi f_m t)$$

For this example, the parameters $A_c = 1$, $A_m = 4$, $f_m = 1$ and $f_c = 8$ are chosen.

| Line | MATLAB code | Comments |
|------|-------------|----------|
| 1 | clear all | Note that we define a much finer time variable for a better approximation of the derivative (see Line 6 of the MATLAB code). |
| 2 | Ac=1; Am=4; fm=1; fc=8; | |
| 3 | t=0:0.0001:4; | |
| 4 | x=Ac*cos(2*pi*fc*t + Am*sin(2*pi*fm*t)); | Expression of the frequency-modulated signal, $x(t)$. |

| | | |
|---|---|---|
| 5 | a=hilbert(x); | Create the analytic signal. |
| 6 | fx=diff(unwrap(angle(a)))./diff(t)/(2*pi); | Compute the instantaneous frequency in Hz. |
| 7<br>8<br>9 | figure(1)<br>plot(t, abs(a), t, x, 'g:'); axis([0 4 -1.5 1.5])<br>xlabel('Time (s)'); ylabel('\itA_x\rm(\itt\rm)') | Plot the instantaneous amplitude $A_x(t)$.<br>Note that the envelope is $A_x(t) \approx A_c = 1$. |
| 10<br>11<br>12 | figure(2)<br>plot(t, unwrap(angle(a))); axis([0 4 0 220])<br>xlabel('Time (s)');<br>ylabel('\it\phi_x\rm(\itt\rm)') | Plot the instantaneous (unwrapped) phase $\phi_x(t)$. |
| 13<br>14<br>15 | figure(3)<br>plot(t(2:end),fx); axis([0 4 0 13])<br>xlabel('Time (s)'); ylabel('\itf_x\rm(\itt\rm)') | Plot the instantaneous frequency, where $f_x(t) = \omega_x(t)/2\pi$.<br>Note that $f_x(t) = f_c + f_m A_m$<br>$\cos 2\pi f_m t = 8 + 4\cos 2\pi \cdot 1 \cdot t$. |

## Results

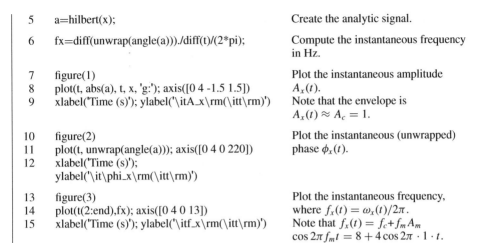

(a)

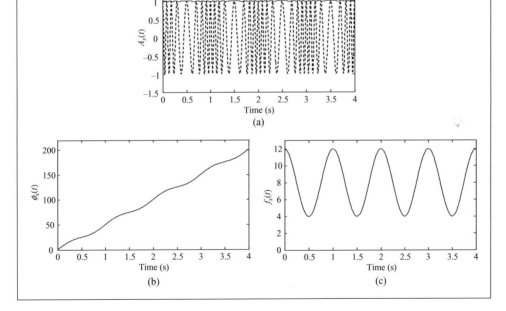

(b)                                                                (c)

---

*Example 4.6:* **Effects of windowing on the modulus of the Fourier transform**

Case 1: Rectangular window (data truncation)
Consider the following signal with three sinusoidal components:

$$x(t) = A_1 \sin 2\pi f_1 t + A_2 \sin 2\pi f_2 t + A_3 \sin 2\pi f_3 t$$

Amplitudes are $A_1 = A_2 = A_3 = 2$, which gives the magnitude '1' for each sinusoidal component in the frequency domain. The frequencies are chosen as $f_1 = 10$, $f_2 = 20$ and $f_3 = 21$.

| Line | MATLAB code | Comments |
|---|---|---|
| 1<br>2 | clear all<br>f1=10; f2=20; f3=21; fs=60; | Define frequencies. The sampling rate is chosen as 60 Hz. |
| 3 | T=0.6; % try different values: 0.6, 0.8, 1.0,<br>1.5, 2, 2.5, 3, 4 | Define the window length 0.6 s. In this example, we use various lengths to demonstrate the effect of windowing. |
| 4 | t=0:1/fs:T-1/fs; | Define time variable from 0 to T-1/fs seconds. The subtraction by 1/fs is introduced in order to make 'exact' periods of the sinusoids (see Chapter 6 for more details of DFT properties). |
| 5 | x=2*sin(2*pi*f1*t) +<br>2*sin(2*pi*f2*t)+2*sin(2*pi*f3*t); | Description of the above equation. |
| 6<br>7<br>8 | N=length(x);<br>X=fft(x);<br>f=fs*(0:N-1)/N; | Perform DFT using the 'fft' function of MATLAB. Calculate the frequency variable (see Chapter 6). |
| 9<br>10<br>11 | Xz=fft([x zeros(1,2000-N)]); %zero padding<br>Nz=length(Xz);<br>fz=fs*(0:Nz-1)/Nz; | Perform '2000-point' DFT by adding zeros at the end of the time sequence 'x'. This procedure is called the 'zero padding' (see the comments below). Calculate new frequency variable accordingly. |
| 12<br>13<br>14<br>15<br>16<br>17<br>18 | figure(1)<br>stem(f(1:N/2+1), abs(X(1:N/2+1)/fs/T), 'r:')<br>axis([0 30 0 1.2])<br>xlabel('Frequency (Hz)'); ylabel('Modulus')<br>hold on<br>plot(fz(1:Nz/2+1), abs(Xz(1:Nz/2+1)/fs/T))<br>hold off; grid on | Plot the modulus of the DFT (from 0 to fs/2 Hz). Note that the DFT coefficients are divided by the sampling rate fs in order to make its amplitude the same as the Fourier integral (see Chapter 6). Also note that, since the time signal is *periodic*, it is further divided by 'T' in order to compensate for its amplitude, and to make it same as the Fourier series coefficients (see Chapter 6 and Chapter 3, Equation (3.45)).<br>The DFT without zero padding is drawn as the dashed stem lines with circles, and the DFT with zero padding is drawn as a solid line. Two graphs are drawn in the same figure. |

**Comments:**

1. Windowing with the rectangular window is just the truncation of the signal (i.e. from 0 to T seconds). The results are shown next together with MATLAB Example 4.7.
2. **Zero padding**: Padding 'zeros' at the end of the time sequence improves the appearance in the frequency domain since the spacing between frequencies is reduced. In

other words, zero padding in the time domain results in interpolation in the frequency domain (Smith, 2003). Sometimes this procedure is called 'spectral interpolation'. As a result, the appearance in the frequency domain (DFT) resembles the true spectrum (Fourier integral), thus it is useful for demonstration purposes. However, it does not increase the 'true' resolution, i.e. does not improve the ability to distinguish the closely spaced frequencies. Note that the actual resolvability in the frequency domain depends on the data length T and the window type. Another reason for zero padding is to make the number of sequence a power of two to meet the FFT algorithm. However, this is no longer necessary in many cases such as programming in MATLAB.

Since zero padding may give a wrong impression of the results, it is not used in this book except for some demonstration and special purposes.

---

*Example 4.7:* **Effects of windowing on the modulus of the Fourier transform**

Case 2: Hann window
In this example, we use the same signal as in the previous example.

| Line | MATLAB code | Comments |
|---|---|---|
| 1 | clear all | Same as in the previous example. |
| 2 | f1=10; f2=20; f3=21; fs=60; | |
| 3 | T=0.6; | |
| | % try different values: 0.6, 0.8, 1.0, 1.5, 2, 2.5, 3, 4 | |
| 4 | t=0:1/fs:T-1/fs; | |
| 5 | x=2*sin(2*pi*f1*t)+ 2*sin(2*pi*f2*t)+ 2*sin(2*pi*f3*t); | |
| 6 | N=length(x); | |
| 7 | whan=hanning(N); | Generate the Hann window with |
| 8 | x=x.*whan'; | the same size of vector as x, and |
| 9 | X=fft(x); | multiply by x. Then, perform the |
| 10 | f=fs*(0:N-1)/N; | DFT of the windowed signal. |
| 11 | Xz=fft([x zeros(1,2000-N)]); % zero padding | Same as in the previous example. |
| 12 | Nz=length(Xz); | |
| 13 | fz=fs*(0:Nz-1)/Nz; | |
| 14 | figure(1) | Same as in the previous example, |
| 15 | stem(f(1:N/2+1), sqrt(8/3)*abs(X(1:N/2+1)/fs/T), 'r:') | except that the magnitude |
| 16 | axis([0 30 0 1.2]) | spectrum is multiplied by the |
| 17 | xlabel('Frequency (Hz)'); ylabel('Modulus') | scale factor 'sqrt(8/3)' (see |
| 18 | hold on | Equation (4.114)). |
| 19 | plot(fz(1:Nz/2+1), sqrt(8/3)*abs(Xz(1:Nz/2+1)/fs/T)) | |
| 20 | hold off; grid on | |

## Results of Examples 4.6 and 4.7

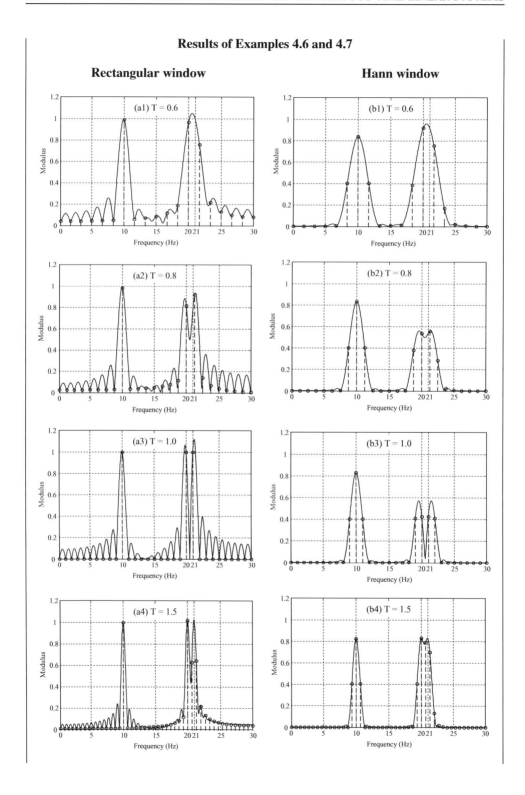

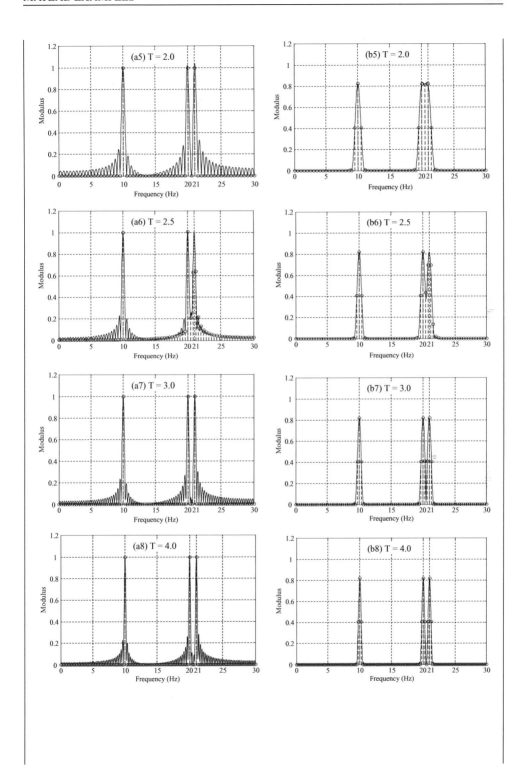

**Comments:**

1. The 10 Hz component is included as a reference, i.e. for the purpose of comparison with the other two peaks.
2. The solid line (DFT with zero padding) is mainly for demonstration purposes, and the dashed stem line with circles is the actual DFT of the windowed sequence. From the results of the DFT (without zero padding), it is shown that the two sinusoidal components (20 Hz and 21 Hz) are separated after T = 2 for the case of a rectangular window. On the other hand, they are not completely separable until T = 4 if the Hann window is used. This is because of its wider main lobe. However, we note that the leakage is greatly reduced by the Hann window.
3. For the case of the Hann window, the magnitudes of peaks are underestimated even if the scale factor is used. (Note that the main lobe contains more frequency lines than in the rectangular window.)
4. However, for the case of the rectangular window, the peaks are estimated correctly when the data length corresponds to exact periods of the signal, i.e. when T = 1, 2, 3 and 4. Note that the peak frequencies are located precisely in this case (see the 21 Hz component). Compare this with the other cases (non-integer T) and with MATLAB Example 3.2 in Chapter 3.

---

*Example 4.8:* **Comparison between the rectangular window and the Hann window: side roll-off characteristics**

Consider the signal $x(t) = A_1 \sin(2\pi f_1 t) + A_2 \sin(2\pi f_2 t)$, where $A_1 \gg A_2$. In this example, we use $A_1 = 1$, $A_2 = 0.001$, $f_1 = 9$, $f_2 = 14$, and the data (window) length 'T = 15.6 seconds'.

| Line | MATLAB code | Comments |
|------|-------------|----------|
| 1 | clear all | Define parameters and the time |
| 2 | f1=9; f2=14; fs=50; T=15.6; | variable. 'T=15.6' is chosen to |
| 3 | t=0:1/fs:T-1/fs; | introduce some windowing effect. The sampling rate is chosen as 50 Hz. |
| 4 | x=1*sin(2*pi*f1*t) + 0.001*sin(2*pi*f2*t); | Expression of the above equation. |
| 5 | N=length(x); | Create the Hann windowed signal xh, |
| 6 | whan=hanning(N); xh=x.*whan'; | and then perform the DFT of both x |
| 7 | X=fft(x); Xh=fft(xh); | and xh. Also, calculate the frequency |
| 8 | f=fs*(0:N-1)/N; | variable. |
| 9 | figure(1) | Plot the results: solid line for the |
| 10 | plot(f(1:N/2+1), 20*log10(abs(X(1:N/2+1)/fs/T))); hold on | rectangular window, and the dashed line for the Hann window. |
| 11 | plot(f(1:N/2+1), 20*log10(sqrt(8/3)* abs(Xh(1:N/2+1)/fs/T)),'r:') | |
| 12 | axis([0 25 -180 0]) | |
| 13 | xlabel('Frequency (Hz)'); ylabel('Modulus (dB)') | |
| 14 | hold off | |

**Results**

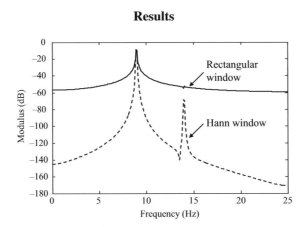

**Comments:** The second frequency component is hardly noticeable with the rectangular window owing to the windowing effect. But, using the Hann window, it becomes possible to see even a very small amplitude component, due to its good side lobe roll-off characteristic.

---

*Example 4.9:* **Comparison between the rectangular window and the Hann window for a transient signal**

Case 1: Response of a single-degree-of-freedom system
Consider the free response of a single-degree-of-freedom system

$$x(t) = \frac{A}{\omega_d} e^{-\zeta \omega_n t} \sin(\omega_d t) \text{ and } F\{x(t)\} = \frac{A}{\omega_n^2 - \omega^2 + j2\zeta \omega_n \omega}$$

where $\omega_d = \omega_n \sqrt{1 - \zeta^2}$. In this example, we use $A = 200$, $\zeta = 0.01$, $\omega_n = 2\pi f_n = 2\pi(20)$.

---

| Line | MATLAB code | Comments |
|------|-------------|----------|
| 1<br>2<br>3 | clear all<br>fs=100; t=[0:1/fs:5-1/fs];<br>A=200; zeta=0.01; wn=2*pi*20;<br>wd=sqrt(1-zeta^2)*wn; | The sampling rate is chosen as 100 Hz. The time variable and other parameters are defined. |
| 4 | x=(A/wd)*exp(-zeta*wn*t).*sin(wd*t); | Expression of the time signal. |
| 5<br>6<br>7<br>8 | N=length(x);<br>whan=hanning(N); xh=x.*whan';<br>X=fft(x); Xh=fft(xh);<br>f=fs*(0:N-1)/N; | Create the Hann windowed signal xh, and then perform the DFT of both x and xh. Also, calculate the frequency variable. |
| 9 | H=A./(wn^2 - (2*pi*f).^2 + i*2*zeta*wn*(2*pi*f)); | Expression of the true Fourier transform, $F\{x(t)\}$. |

| 10 | figure(1) | Plot the results in dB scale: |
|---|---|---|
| 11 | plot(f(1:N/2+1), 20*log10(abs(X(1:N/2+1)/fs))); | Solid line (upper) for the |
|  | hold on | rectangular window, solid line |
| 12 | plot(f(1:N/2+1), 20*log10(sqrt(8/3)* | (lower) for the Hann window, |
|  | abs(Xh(1:N/2+1)/fs)), 'r') | and dashed line for the 'true' |
| 13 | plot(f(1:N/2+1), 20*log10(abs(H(1:N/2+1)))), 'g:') | Fourier transform. |
| 14 | axis([0 50 -150 0]) |  |
| 15 | xlabel('Frequency (Hz)'); ylabel('Modulus (dB)') |  |
| 16 | hold off |  |
|  |  |  |
| 17 | figure(2) | Plot the results in linear scale: |
| 18 | plot(f(1:N/2+1), abs(X(1:N/2+1)/fs)); hold on | underestimation of the |
| 19 | plot(f(1:N/2+1), (sqrt(8/3)*abs(Xh(1:N/2+1)/fs)), 'r') | magnitude spectrum by the |
| 20 | plot(f(1:N/2+1), abs(H(1:N/2+1)), 'g:') | Hann window is more clearly |
| 21 | axis([0 50 0 0.7]) | seen. |
| 22 | xlabel('Frequency (Hz)'); |  |
|  | ylabel('Modulus (linear scale)') |  |
| 23 | hold off |  |

### Results

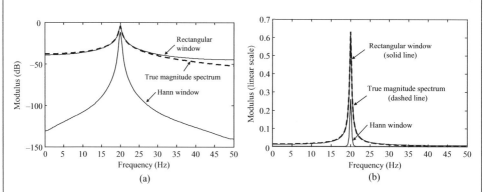

(a)  (b)

**Comments:** Note that the magnitude spectrum is considerably underestimated if the Hann window is used, because a significant amount of energy is lost by windowing. Thus, in general, windowing is not applicable to transient signals.

---

Case 2: Response of a two-degree-of-freedom system, when the contributions of two modes are considerably different. This example is similar to MATLAB Example 4.8.

Consider the free response of a two-degree-of-freedom system, e.g.

$$x(t) = \frac{A}{\omega_{d1}} e^{-\zeta_1 \omega_{n1} t} \sin(\omega_{d1} t) + \frac{B}{\omega_{d2}} e^{-\zeta_2 \omega_{n2} t} \sin(\omega_{d2} t)$$

Then, its Fourier transform is

$$F\{x(t)\} = \frac{A}{\omega_{n1}^2 - \omega^2 + j2\zeta_1 \omega_{n1} \omega} + \frac{B}{\omega_{n2}^2 - \omega^2 + j2\zeta_2 \omega_{n2} \omega}$$

In this example, we use $A = 200$, $B = 0.001A$, $\zeta_1 = \zeta_2 = 0.01$, $\omega_{n1} = 2\pi(20)$ and $\omega_{n2} = 2\pi(30)$. Note that $A \gg B$.

| Line | MATLAB code | Comments |
|---|---|---|
| 1 | clear all | Same as Case 1, except that the |
| 2 | fs=100; t=[0:1/fs:5-1/fs]; | parameters for the second mode |
| 3 | A=200; B=0.001*A; zeta1=0.01; zeta2=0.01; | are also defined. |
| 4 | wn1=2*pi*20; wd1=sqrt(1-zeta1^2)*wn1; | |
| 5 | wn2=2*pi*30; wd2=sqrt(1-zeta2^2)*wn2; | |
| 6 | x=(A/wd1)*exp(-zeta1*wn1*t).*sin(wd1*t) + (B/wd2)*exp(-zeta2*wn2*t).*sin(wd2*t); | Expression of the time signal, $x(t)$. |
| 7 | N=length(x); | Same as Case 1. |
| 8 | whan=hanning(N); xh=x.*whan'; | |
| 9 | X=fft(x); Xh=fft(xh); | |
| 10 | f=fs*(0:N-1)/N; | |
| 11 | H=A./(wn1^2-(2*pi*f).^2+i*2*zeta1*wn1*(2*pi*f)) + B./(wn2^2-(2*pi*f).^2+i*2*zeta2*wn2*(2*pi*f)); | Expression of the true Fourier transform, $F\{x(t)\}$. |
| 12 | figure(1) | Plot the results of the rectangular |
| 13 | plot(f(1:N/2+1), 20*log10(abs(X(1:N/2+1)/fs))); hold on | window in dB scale: solid line for the rectangular window and |
| 14 | plot(f(1:N/2+1), 20*log10(abs(H(1:N/2+1))), 'g:') | dashed line for the 'true' Fourier |
| 15 | axis([0 50 -60 0]) | transform. |
| 16 | xlabel('Frequency (Hz)'); ylabel('Modulus (dB)') | |
| 17 | hold off | |
| 18 | figure(2) | Plot the results of the Hann |
| 19 | plot(f(1:N/2+1), 20*log10(sqrt(8/3)* abs(Xh(1:N/2+1)/fs))) | window in dB scale: solid line for the Hann window, and dashed line |
| 20 | hold on | for the 'true' Fourier transform. |
| 21 | plot(f(1:N/2+1), 20*log10(abs(H(1:N/2+1))), 'g:') | |
| 22 | axis([0 50 -160 0]) | |
| 23 | xlabel('Frequency (Hz)'); ylabel('Modulus (dB)') | |
| 24 | hold off | |

### Results

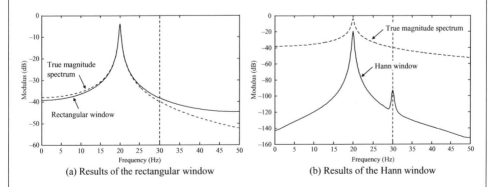

(a) Results of the rectangular window

(b) Results of the Hann window

**Comments:** Similar to MATLAB Example 4.8, the second mode is clearly noticeable when the Hann window is used, although the magnitude spectrum is greatly underestimated. Note that the second mode is almost negligible, i.e. $B \ll A$. So, it is almost impossible to see the second mode in the true magnitude spectrum and even in the phase spectrum as shown in Figure (c).

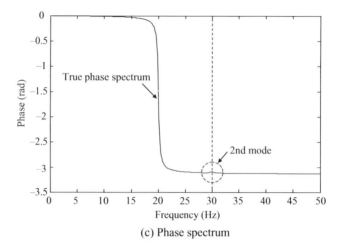

(c) Phase spectrum

The reason for these results is not as clear as in MATLAB Example 4.8 where the two sinusoids are compared. However, it might be argued that the convolution operation in the frequency domain results in magnifying (or sharpening) the resonance region owing to the frequency characteristic of the Hann window.

# 5

# Time Sampling and Aliasing

## Introduction

So far, we have developed the Fourier transform of a *continuous* signal. However, we usually utilize a *digital* computer to perform the transform. Thus, it is necessary to re-examine Fourier methods so as to be able to transform sampled data. We would 'hope' that the discrete version of the Fourier transform resembles (or approximates) the Fourier integral (Equation (4.6)), such that it represents the frequency characteristic (within the range of interest) of the original signal. In fact, from the MATLAB examples given in the previous chapter, we have already seen that the results of the discrete version (DFT) and the continuous version (Fourier integral) appear to be not very different. However, there are fundamental differences between these two versions, and in this chapter we shall consider the effect of sampling, and relate the Fourier transform of a continuous signal and the transform of a discrete signal (or a sequence).

## 5.1 THE FOURIER TRANSFORM OF AN IDEAL SAMPLED SIGNAL

### Impulse Train Modulation

We introduce the Fourier transform of a sequence by using the mathematical notion of 'ideal sampling' of a continuous signal. Consider a 'train' of delta functions $i(t)$ which is expressed as

$$i(t) = \sum_{n=-\infty}^{\infty} \delta(t - n\Delta) \tag{5.1}$$

i.e. delta functions located every $\Delta$ seconds as depicted in Figure 5.1.

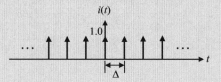

**Figure 5.1**   Train of delta functions

Starting with a continuous signal $x(t)$, an ideal uniformly sampled discrete sequence is $x(n\Delta) = x(t)|_{t=n\Delta}$ evaluated every $\Delta$ seconds of the continuous signal $x(t)$. Since the sequence $x(n\Delta)$ is discrete, we cannot apply the Fourier integral. Instead, the *ideally* sampled signal is often modelled mathematically as the product of the continuous signal $x(t)$ with the train of delta functions $i(t)$, i.e. the sampled signal can be written as

$$x_s(t) = x(t)i(t) \tag{5.2}$$

The reciprocal of the sampling interval, $f_s = 1/\Delta$, is called the sampling rate, which is the number of samples per second. The sampling procedure can be illustrated as in Figure 5.2.

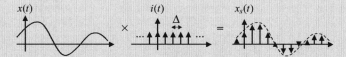

**Figure 5.2**   Impulse train representation of a sampled signal

In this way we see that $x_s(t)$ is an amplitude-modulated train of delta functions. We also note that $x_s(t)$ is not the same as $x(n\Delta)$ since it involves delta functions. However, it is a convenient step to help us form the Fourier transform of the sequence $x(n\Delta)$, as follows. Let $X_s(f)$ denote the Fourier transform of the sampled signal $x_s(t)$. Then, using properties of the delta function,

$$X_s(f) = \int_{-\infty}^{\infty} \left[ x(t) \sum_{n=-\infty}^{\infty} \delta(t - n\Delta) \right] e^{-j2\pi ft} dt = \sum_{n=-\infty}^{\infty} \left[ \int_{-\infty}^{\infty} x(t)e^{-j2\pi ft} \cdot \delta(t - n\Delta)dt \right]$$

$$= \sum_{n=-\infty}^{\infty} x(n\Delta)e^{-j2\pi fn\Delta} \tag{5.3}$$

The summation (5.3) now involves the sequence $x(n\Delta)$ and is (in principle) computable. It is this expression that defines the Fourier transform of a sequence. We are now in a position to note some fundamental differences between the Fourier transform $X(f)$ of the original continuous signal $x(t)$ and $X_s(f)$, the Fourier transform of the uniformly sampled version $x(n\Delta)$.

Note that Equation (5.3) implies that $X_s(f)$ has a *periodic* structure in frequency with period $1/\Delta$. For example, for an integer number $r$, $X_s(f + r/\Delta)$ becomes

$$X_s(f + r/\Delta) = \sum_{n=-\infty}^{\infty} x(n\Delta)e^{-j2\pi(f+r/\Delta)n\Delta} = \sum_{n=-\infty}^{\infty} x(n\Delta)e^{-j2\pi fn\Delta}e^{-j2\pi rn}$$

$$= \sum_{n=-\infty}^{\infty} x(n\Delta)e^{-j2\pi fn\Delta} = X_s(f) \tag{5.4}$$

This periodicity in frequency will be discussed further shortly. The inverse Fourier transform of $X_s(f)$ can be found by multiplying both sides of Equation (5.3) by $e^{j2\pi fr\Delta}$ and integrating with respect to $f$ from $-1/2\Delta$ to $1/2\Delta$ (since $X_s(f)$ is periodic, we need to integrate over only one period), and taking account of the orthogonality of the exponential function. Then

$$\int_{-1/2\Delta}^{1/2\Delta} X_s(f)e^{j2\pi fr\Delta}df = \int_{-1/2\Delta}^{1/2\Delta} \left[\sum_{n=-\infty}^{\infty} x(n\Delta)e^{-j2\pi fn\Delta}\right]e^{j2\pi fr\Delta}df$$

$$= \sum_{n=-\infty}^{\infty} \left[\int_{-1/2\Delta}^{1/2\Delta} x(n\Delta)e^{-j2\pi fn\Delta}e^{j2\pi fr\Delta}df\right]$$

$$= \sum_{n=-\infty}^{\infty} \left[x(n\Delta)\int_{-1/2\Delta}^{1/2\Delta} e^{-j2\pi f(n-r)\Delta}df\right] = x(r\Delta)\frac{1}{\Delta} \tag{5.5}$$

Thus, we summarize the Fourier transform pair for the 'sampled sequence' as below, where we rename $X_s(f)$ as $X(e^{j2\pi f\Delta})$:

$$X(e^{j2\pi f\Delta}) = \sum_{n=-\infty}^{\infty} x(n\Delta)e^{-j2\pi fn\Delta} \tag{5.6}$$

$$x(n\Delta) = \Delta \int_{-1/2\Delta}^{1/2\Delta} X(e^{j2\pi f\Delta})e^{j2\pi fn\Delta}df \tag{5.7}$$

Note that the scaling factor $\Delta$ is present in Equation (5.7).

### The Link Between $X(e^{j2\pi f\Delta})$ and $X(f)$

At this stage, we may ask: 'How is the Fourier transform of a sequence $X(e^{j2\pi f\Delta})$ related to the Fourier transform of a continuous signal $X(f)$?' In order to answer this, we need to examine the periodicity of $X(e^{j2\pi f\Delta})$ as follows.

Note that $i(t)$ in Equation (5.1) is a periodic signal with period $\Delta$, thus it has a Fourier series representation. Since the fundamental period $T_P = \Delta$, we can write the train of delta functions as (see Equation (3.34))

$$i(t) = \sum_{n=-\infty}^{\infty} c_n e^{j2\pi nt/\Delta} \tag{5.8}$$

where the Fourier coefficients are found from Equation (3.35) such that

$$c_n = \frac{1}{\Delta} \int_{-\Delta/2}^{\Delta/2} i(t) e^{-j2\pi nt/\Delta} dt = \frac{1}{\Delta} \tag{5.9}$$

Thus, Equation (5.8) can be rewritten as

$$i(t) = \frac{1}{\Delta} \sum_{n=-\infty}^{\infty} e^{j2\pi nt/\Delta} \tag{5.10}$$

(Recall Equation (3.30) which is equivalent to this.) Using the property of the delta function $\int_{-\infty}^{\infty} e^{\pm j2\pi at} dt = \delta(a)$, the Fourier transform of Equation (5.10) can be calculated as

$$I(f) = F\{i(t)\} = \int_{-\infty}^{\infty} \left[ \frac{1}{\Delta} \sum_{n=-\infty}^{\infty} e^{j2\pi nt/\Delta} \right] e^{-j2\pi ft} dt = \frac{1}{\Delta} \sum_{n=-\infty}^{\infty} \left[ \int_{-\infty}^{\infty} e^{j2\pi nt/\Delta} e^{-j2\pi ft} dt \right]$$

$$= \frac{1}{\Delta} \sum_{n=-\infty}^{\infty} \left[ \int_{-\infty}^{\infty} e^{-j2\pi(f - n/\Delta)t} dt \right] = \frac{1}{\Delta} \sum_{n=-\infty}^{\infty} \delta \left( f - \frac{n}{\Delta} \right) \tag{5.11}$$

Thus, the Fourier transform of the train of delta functions can be drawn in the frequency domain as in Figure 5.3.

Since the Fourier transform of $x_s(t)$ results in the convolution of $X(f)$ with $I(f)$ in the frequency domain, i.e. $X_s(f) = F\{x(t)i(t)\} = X(f) * I(f)$, it follows that

$$X_s(f) = I(f) * X(f) = \int_{-\infty}^{\infty} I(g)X(f - g)dg = \int_{-\infty}^{\infty} \frac{1}{\Delta} \sum_{n=-\infty}^{\infty} \delta \left( g - \frac{n}{\Delta} \right) X(f - g)dg$$

$$= \frac{1}{\Delta} \sum_{n=-\infty}^{\infty} \left[ \int_{-\infty}^{\infty} \delta \left( g - \frac{n}{\Delta} \right) X(f - g)dg \right] = \frac{1}{\Delta} \sum_{n=-\infty}^{\infty} X \left( f - \frac{n}{\Delta} \right) \tag{5.12}$$

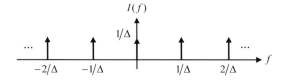

**Figure 5.3**   Fourier transform of the train of delta functions

This gives an alternative form of Equation (5.6), which is

$$X(e^{j2\pi f\Delta}) = \frac{1}{\Delta} \sum_{n=-\infty}^{\infty} X\left(f - \frac{n}{\Delta}\right) \tag{5.13}$$

This important equation describes the relationship between the Fourier transform of a continuous signal and the Fourier transform of a sequence obtained by ideal sampling every $\Delta$ seconds. That is, the Fourier transform of the sequence $x(n\Delta)$ is the sum of shifted versions of the Fourier transform of the underlying continuous signal. This reinforces the periodic nature of $X(e^{j2\pi f\Delta})$. Note also that the 'scaling' effect $1/\Delta$, i.e. the sampling rate $f_s = 1/\Delta$, is a multiplier of the sum in Equation (5.13).

So, the '*sampling* in the time domain' implies a '*periodic* and *continuous* structure in the frequency domain' as illustrated in Figure 5.4. From Equation (5.6), it can be seen that $X_s(f_s - f) = X_s^*(f)$, where $^*$ denotes complex conjugate. This is confirmed (for the modulus) from Figure 5.4. Thus, all the information in $X_s(f)$ lies in the range $0 \le f \le f_s/2$. This figure emphasizes the difference between $X(f)$ and $X(e^{j2\pi f\Delta})$, and leads to the concept of 'aliasing', which arises from the possible overlapping between the replicas of $X(f)$. This will be discussed further in the next section.

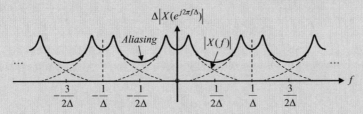

**Figure 5.4**   Fourier transform of the sampled sequence

## An Alternative Route to the Derivation of the Fourier Transform of a Sequence

### *The z-transform*

The expression for the Fourier transform of a sequence, Equation (5.6), can also be obtained via the $z$-transform of a sequence. The $z$-transform is widely used in the solution of difference equations, just as the Laplace transform is used for differential equations. The definition of the $z$-transform $X(z)$ of a sequence of numbers $x(n)$ is

$$X(z) = \sum_{n=-\infty}^{\infty} x(n)z^{-n} \tag{5.14}$$

where $z$ is the complex-valued argument of the transform and $X(z)$ is a function of a complex variable. In Equation (5.14), the notion of time is not explicitly made, i.e. we write $x(n)$ for $x(n\Delta)$. It is convenient here to regard the sampling interval as set to unity. Since $z$ is complex, it can be written in polar form, i.e. using the magnitude and phase such that $z = re^{j\omega}$, and is represented in a complex plane (polar coordinates) as shown in Figure 5.5(a). If this expression

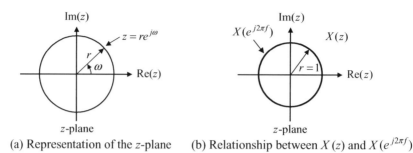

(a) Representation of the $z$-plane    (b) Relationship between $X(z)$ and $X(e^{j2\pi f})$

**Figure 5.5**   Representation of the $z$-plane and the Fourier transform of a sequence

is substituted into Equation (5.14), it gives

$$X(re^{j\omega}) = \sum_{n=-\infty}^{\infty} x(n)(re^{j\omega})^{-n} = \sum_{n=-\infty}^{\infty} x(n)r^{-n}e^{-j\omega n} \tag{5.15}$$

If we further restrict our interest to the unit circle in the $z$-plane, i.e. $r = 1$, so $z = e^{j\omega} = e^{j2\pi f}$, then Equation (5.15) is reduced to

$$X(e^{j2\pi f}) = \sum_{n=-\infty}^{\infty} x(n)e^{-j2\pi fn} \tag{5.16}$$

which is exactly same form for the Fourier transform of a sequence as given in Equation (5.6) for sampling interval $\Delta = 1$.

Thus, it can be argued that the evaluation of the $z$-transform on the unit circle in the $z$-plane yields the Fourier transform of a sequence as shown in Figure 5.5(b). This is analogous to the continuous-time case where the Laplace transform reduces to the Fourier transform if it is evaluated on the imaginary axis, i.e. $s = j\omega$.

### Relationship Between the Laplace Transform and the z-transform

To see the effect of sampling on the $z$-plane, we consider the relationship between the Laplace transform and the $z$-transform. The Laplace transform of $x(t)$, $L\{x(t)\}$, is defined as

$$X(s) = \int_{-\infty}^{\infty} x(t)e^{-st}dt \tag{5.17}$$

where $s = \sigma + j2\pi f$ is a complex variable. Note that if $s = j2\pi f$, then $X(f) = X(s)|_{s=j2\pi f}$. Now, let $\hat{X}(s)$ be the Laplace transform of an (ideally) sampled function; then

$$\hat{X}(s) = L\{x(t)i(t)\} = \int_{-\infty}^{\infty} x(t) \sum_{n=-\infty}^{\infty} \delta(t - n\Delta)e^{-st}dt$$

$$= \sum_{n=-\infty}^{\infty} \left[ \int_{-\infty}^{\infty} x(t)e^{-st}\delta(t - n\Delta)dt \right] = \sum_{n=-\infty}^{\infty} x(n\Delta)e^{-sn\Delta} \tag{5.18}$$

If $z = e^{s\Delta}$, then Equation (5.18) becomes the z-transform, i.e.

$$\hat{X}(s)\Big|_{z=e^{s\Delta}} = \sum_{n=-\infty}^{\infty} x(n\Delta)z^{-n} = X(z) \tag{5.19}$$

Comparing Equation (5.19) with Equation (5.6), it can be shown that if $z = e^{j2\pi f \Delta}$, then

$$X(e^{j2\pi f\Delta}) = X(z)\Big|_{z=e^{j2\pi f\Delta}} \tag{5.20}$$

Also, using the similar argument made in Equation (5.13), i.e. using

$$i(t) = \frac{1}{\Delta} \sum_{n=-\infty}^{\infty} e^{j2\pi nt/\Delta}$$

an alternative form of $\hat{X}(s)$ can be written as

$$\hat{X}(s) = \int_{-\infty}^{\infty} x(t)\frac{1}{\Delta}\sum_{n=-\infty}^{\infty} e^{j2\pi nt/\Delta}e^{-st}dt = \frac{1}{\Delta}\sum_{n=-\infty}^{\infty}\left[\int_{-\infty}^{\infty} x(t)e^{-(s-j2\pi n/\Delta)t}dt\right] \tag{5.21}$$

Thus,

$$\hat{X}(s) = \frac{1}{\Delta}\sum_{n=-\infty}^{\infty} X\left(s - \frac{j2\pi n}{\Delta}\right) \tag{5.22}$$

From Equation (5.22), we can see that $\hat{X}(s)$ is related to $X(s)$ by adding shifted versions of scaled $X(s)$ to produce $\hat{X}(s)$ as depicted in Figure 5.6(b) below, which is similar to the relationship between the Fourier transform of a sampled sequence and the Fourier transform.

Note that, as we can see from Equation (5.19), $X(z)$ is not directly related to $X(s)$, but it is related to $\hat{X}(s)$ via $z = e^{s\Delta}$. Further, if we let $s = j2\pi f$, then we have the following relationship:

$$X_s(f)\left(= X(e^{j2\pi f\Delta})\right) = \hat{X}(s)\Big|_{s=j2\pi f} = X(z)\Big|_{z=e^{j2\pi f\Delta}} \tag{5.23}$$

The relationships between $X(s)$, $\hat{X}(s)$ and $X(z)$ are illustrated in Figure 5.6. In this figure, a pole is included in the s-plane to demonstrate how it is mapped to the z-plane. We can see that a single pole in the $X(s)$-plane results in an infinite number of poles in the $\hat{X}(s)$-plane; then this infinite series of poles all map onto a single pole in the $X(z)$-plane. In effect, the left hand side of the s-plane is mapped to the inside of the unit circle of the z-plane. However, we must realize that, due to the sampling process, what maps onto the z-plane is not $X(s)$, but $\hat{X}(s)$, and each 'strip' in the left hand side of $\hat{X}(s)$ is mapped onto the z-plane plane such that it fills the complete unit circle. This indicates the 'periodic' structure in the frequency domain as well as possible aliasing in the frequency domain.

The above mapping process is sometimes used in designing an IIR (Infinite Impulse Response) digital filter from an existing analogue filter, and is called the impulse-invariant method.

(a) Analogue (continuous-time) domain:

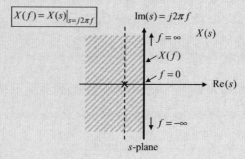

(b) Laplace transform of a sampled function:

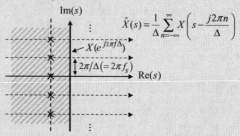

(c) Digital (discrete-time) domain:

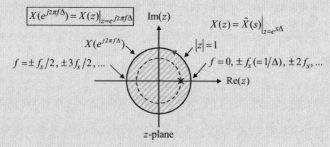

**Figure 5.6**   Relationship between $s$-plane and $z$-plane

## 5.2 ALIASING AND ANTI-ALIASING FILTERS[M5.1–5.3]

As noted in the previous section, Equation (5.13) describes how the frequency components of the sampled signal are related to the Fourier transform of the original continuous signal. A pictorial description of the sampling effect follows. Consider the Fourier transform that has $X(f) = 0$ for $|f| > f_H$, as given in Figure 5.7.

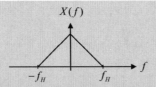

**Figure 5.7** Fourier transform of a continuous signal such that $X(f) = 0$ for $|f| > f_H$

Assuming that the sampling rate $f_s = 1/\Delta$ is such that $f_s > 2f_H$, i.e. $f_H < 1/2\Delta$, then Figure 5.8 shows the corresponding (scaled) Fourier transform of a sampled sequence $\Delta \cdot X_s(f)$ (or $\Delta \cdot X(e^{j2\pi f\Delta})$). Note that the scaling factor $\Delta$ is introduced (see Equation (5.13)), and some commonly used terms are defined in the figure. Thus $\Delta \cdot X_s(f)$ accurately represents $X(f)$ for $|f| < 1/2\Delta$.

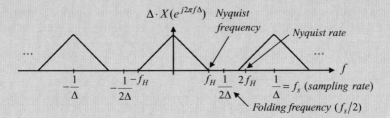

**Figure 5.8** Fourier transform of a sampled sequence $f_s > 2f_H$

Suppose now that $f_s < 2f_H$. Then there is an *overlapping* of the shifted versions of $X(f)$ resulting in a *distortion* of the frequencies for $|f| < 1/2\Delta$ as shown in Figure 5.9.

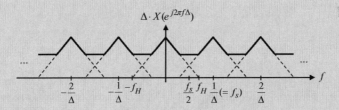

**Figure 5.9** Fourier transform of a sampled sequence $f_s < 2f_H$

This 'distortion' is due to the fact that high-frequency components in the signal are not distinguishable from lower frequencies because the sampling rate $f_s$ is not high enough. Thus, it is clear that to avoid this distortion the highest frequency in the signal $f_H$ should be less than $f_s/2$. This upper frequency limit is often called the Nyquist frequency (see Figure 5.8).

This distortion is referred to as *aliasing*. Consider the particular case of a harmonic wave of frequency $p$ Hz, e.g. $\cos(2\pi pt)$ as in Figure 5.10. We sample this signal every $\Delta$ seconds, i.e. $f_s = 1/\Delta$ (with, say, $p < f_s/2$), to produce the sampled sequence $\cos(2\pi pn\Delta)$. Now, consider another cosine wave of frequency $(p + 1/\Delta)$Hz, i.e. $\cos[2\pi(p + 1/\Delta)t]$; again we sample this every $\Delta$ seconds to give $\cos[2\pi(p + 1/\Delta)n\Delta]$ which can be shown to be $\cos(2\pi pn\Delta)$,

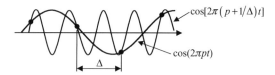

**Figure 5.10**   Illustration of the aliasing phenomenon

identical to the above. So, simply given the sample values, how do we know which cosine wave they come from?

In fact, the same sample values could have arisen from any cosine wave having frequency $\pm p + (k/\Delta)$ ($k = 1, 2, \ldots$), i.e. $\cos(2\pi pn\Delta)$ is indistinguishable from $\cos[2\pi(\pm p + k/\Delta)n\Delta]$. So if a frequency component is detected at $p$ Hz, any one of these higher frequencies can be responsible for this rather than a 'true' component at $p$ Hz. This phenomenon of higher frequencies looking like lower frequencies is called *aliasing*. The values $\pm p + k/\Delta$ are possible aliases of frequency $p$ Hz, and can be seen graphically for some $p$ Hz between 0 and $1/2\Delta$ by 'pleating' the frequency axis as shown in Figure 5.11 (Bendat and Piersol, 2000).

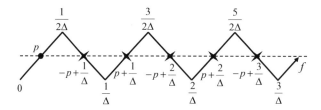

**Figure 5.11**   Possible aliases of frequency $p$ Hz

To avoid aliasing the signal must be band-limited, i.e. it must not have any frequency component above a certain frequency, say $f_H$, and the sampling rate must be chosen to be greater than twice the highest frequency contained in the signal, namely

$$f_s > 2f_H \qquad\qquad (5.24)$$

So, it would appear that we need to know the highest frequency component in the signal. Unfortunately, in many cases the frequency content of a signal will not be known and so the choice of sampling rate is problematic. The way to overcome this difficulty is to *filter* the signal before sampling, i.e. filter the *analogue* signal using an *analogue low-pass filter*. This filter is often referred to as an *anti-aliasing filter*.

### *Anti-aliasing Filters*

In general, the signal $x(t)$ may not be band-limited, thus aliasing will distort the spectral information. Thus, we must eliminate 'undesirable' high-frequency components by applying

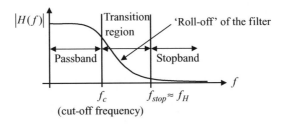

**Figure 5.12**   Typical characteristics of a low-pass filter

an anti-aliasing low-pass filter to the analogue signal prior to digitization. The 'anti-aliasing' filter should have the following properties:

- flat passband;
- sharp cut-off characteristics;
- low distortion (i.e. linear phase characteristic in the passband);
- multi-channel analysers need a set of parallel anti-aliasing filters which must have matched amplitude and phase characteristics.

Filters are characterized by their frequency response functions $H(f)$, e.g. as shown in Figure 5.12.
    Some typical anti-aliasing filters are shown in Figure 5.13.

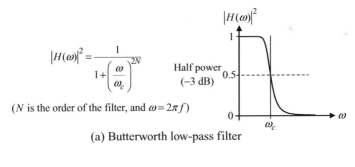

(a) Butterworth low-pass filter

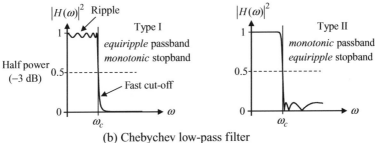

(b) Chebychev low-pass filter

**Figure 5.13**   Some commonly used anti-aliasing low-pass filters

We shall assume that the anti-aliasing filter operates on the signal $x(t)$ to produce a signal to be digitized as illustrated in Figure 5.14.

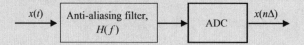

**Figure 5.14** The use of anti-aliasing filter prior to sampling

But we still need to decide what the highest frequency $f_H$ is just prior to the ADC (analogue-to-digital converter). The critical features in deciding this are:

- the 'cut-off' frequency of the filter $f_c$, usually the $f_c$ (Hz) = 3 dB point of the filter;
- the 'roll-off rate' of the filter in dB/octave ($B$ in Figure 5.15);
- the 'dynamic range' of the acquisition system in dB ($A$ in Figure 5.15). (Dynamic range is discussed in the next section.)

These terms and the effect of sampling rate are depicted in Figure 5.15. Note that, in this figure, if $f_s > 2f_{stop}(\approx 2f_H)$ there is no aliasing, and if $f_s > 2f_A$ there is no aliasing up to $f_c$. Also note that it is not the 3 dB point of the filter which should satisfy the Nyquist criterion. But at the Nyquist frequency the filter response should be negligible (e.g. at least 40 dB down on the passband).

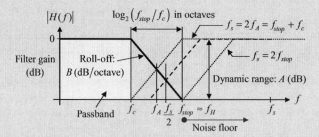

**Figure 5.15** Characteristics of the anti-aliasing filter

If the spectrum is to be used up to $f_c$ Hz, then the figure indicates how $f_s$ is chosen. Using simple trigonometry,

$$-\frac{A}{\log_2(f_{stop}/f_c)} = -B \text{ (dB/octave)} \tag{5.25}$$

Note that if $B$ is dB/decade, then the logarithm is to base 10. Some comments on the octave are: if $f_2 = 2^n f_1$, then $f_2$ is '$n$' octaves; thus, $\log_2 f_2 = n + \log_2 f_1$ and $\log_2 f_2 - \log_2 f_1 = \log_2(f_2/f_1) = n$ (octaves).

From Equation (5.25), it can be shown that $f_{stop} = 2^{A/B} f_c$. Substituting this expression into $f_s > f_{stop} + f_c$, which is the condition for no aliasing up to the cut-off frequency $f_c$

(see Figure 5.15), then we have the following condition for the sampling rate:

$$f_s > f_c(1 + 2^{A/B}) \approx f_c(1 + 10^{0.3A/B}) \qquad (5.26)$$

For example, if $A = 70\,\text{dB}$ and $B = 70\,\text{dB/octave}$, then $f_s > 3 f_c$, and if $A = 70\,\text{dB}$ and $B = 90\,\text{dB/octave}$, then $f_s > f_c(1 + 2^{70/90}) \approx 2.7 f_c$. However, the following practical guide (which is based on twice the frequency at the noise floor) is often used:

$$f_s = 2 f_{stop}(\approx 2 f_H) \approx 2 \times 10^{0.3A/B} f_c \qquad (5.27)$$

For example, if $A = 70\,\text{dB}$ and $B = 90\,\text{dB/octave}$, then $f_s \approx 3.42 f_c$, which gives a more conservative result than Equation (5.26).

In general, the cut-off frequency $f_c$ and the roll-off rate of the anti-aliasing filter should be chosen with the particular application in mind. But, very roughly speaking, if the 3 dB point of the filter is a quarter of the sampling rate $f_s$ and the roll-off rate better than 48 dB/octave, then this gives a 40 to 50 dB reduction in the folding frequency $f_s/2$. This may result in an acceptable level of aliasing (though we note that this may not be adequate for some applications).

Choosing an appropriate sampling rate is important. Although we must avoid aliasing, unnecessarily high sampling rates are not desirable. The 'optimal' sampling rate must be selected according to the specific applications (the bandwidth of interest) and the characteristics of the anti-aliasing filter to be used.

There is another very important aspect to note. If the sampled sequence $x(n\Delta)$ is sampled again (digitally, i.e. downsampled), the resulting sequence can be aliased if an appropriate anti-aliasing 'digital' low-pass filter is not applied before the sampling. This is demonstrated by MATLAB Examples 5.2 and 5.3. Also note that aliasing does occur in most computer simulations. For example, if a numerical integration method (such as the Runge–Kutta method) is applied to solve ordinary differential equations, in this case there is no simple way to avoid the aliasing problem (see comments of MATLAB Example 6.5 in Chapter 6).

## 5.3 ANALOGUE-TO-DIGITAL CONVERSION AND DYNAMIC RANGE

An ADC is a device that takes a continuous (analogue) time signal as an input and produces a sequence of numbers (digital) as an output that are sample values of the input. It may be convenient to consider the ADC process as consisting of two phases, namely *sampling* and *quantization*, as shown in Figure 5.16.

Note that actual ADCs do not consist of two separate stages (as in the conceptual figure), and various different types are available. In Figure 5.16, $x(n\Delta)$ is the exact value of time signal $x(t)$ at time $t = n\Delta$, i.e. it is the *ideally* sampled sequence with sample interval $\Delta$. $\tilde{x}(n\Delta)$ is

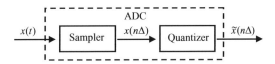

**Figure 5.16** Conceptual model of the analogue-to-digital conversion

the representation of $x(n\Delta)$ on a computer, and is different from $x(n\Delta)$ since a 'finite number of bits' are used to represent each number. Thus, we can expect that some errors are produced in the quantization process.

Now, consider the problem of quantization, in Figure 5.17.

Figure 5.17    Quantization process

Suppose the ADC represents a number using 3 bits (and a sign bit), i.e. a 4 bit ADC as given in Figure 5.18.

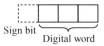

Figure 5.18    A digital representation of a 4 bit ADC

Each bit is either 0 or 1, i.e. two states, so there are $2^3 = 8$ possible states to represent a number. If the input voltage range is $\pm 10$ volts then the 10 volts range must be allocated to the eight possible states in some way, as shown in Figure 5.19.

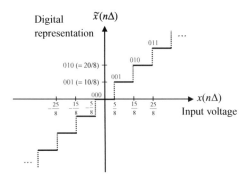

Figure 5.19    Digital representation of an analogue signal using a 4 bit ADC

In Figure 5.19, any input voltage between $-5/8$ and $5/8$ volts will be represented by the bit pattern [000], and from $5/8$ to $15/8$ volts by [010], etc. The rule for assigning the bit pattern to the input range depends on the ADC. In the above example the steps are *uniform* and the 'error' can be expressed as

$$e(n\Delta) = \tilde{x}(n\Delta) - x(n\Delta) \tag{5.28}$$

Not that, for the particular quantization process given in Figure 5.19, the error $e(n\Delta)$ has values between $-5/8$ and $5/8$. This error is called the *quantization noise* (or *quantization error*). From this it is clear that 'small' signals will be poorly represented, e.g. within the input voltage range of $\pm 10$ volts, a sine wave of amplitude $\pm 1.5$ volts, say, will be represented by the 4 bit ADC as shown in Figure 5.20.

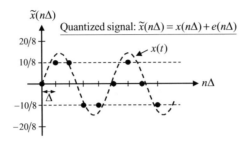

**Figure 5.20**   Example of poor digital representation

What will happen for a sine wave of amplitude $\pm 10$ volts and another sine wave of amplitude $\pm 11$ volts? The former corresponds to the *maximum dynamic range*, and the latter signal will be *clipped*.

Details of quantization error can be found in various references (Oppenheim and Schafer, 1975; Rabiner and Gold, 1975; Childers and Durling, 1975; Otnes and Enochson, 1978). A brief summary is given below. The error $e(n\Delta)$ is often treated as random 'noise'. The probability distributions of $e(n\Delta)$ depend on the particular way in which the quantization occurs. Often it is assumed that this error has a uniform distribution (with zero mean) over one quantization step, and is stationary and 'white'. The probability density function of $e(n\Delta)$ is shown in Figure 5.21, where $\delta = X/2^b$ for a $b$ bit word length (excluding the sign bit), and $X$ (volts) corresponds to the full range of the ADC. Note that $\delta = 10/2^b = 10/2^3 = 10/8$ in our example above. The variance of $e(n\Delta)$ is then

$$\text{Var}(e) = \sigma_e^2 = \int_{-\infty}^{\infty} (e - \mu_e)^2 p(e) de = \frac{1}{\delta} \int_{-\delta/2}^{\delta/2} e^2 de$$

$$= \frac{\delta^2}{12} = \frac{(X/2^b)^2}{12} \tag{5.29}$$

where $\mu_e$ is the mean value of $e(n\Delta)$. (See Chapter 7 for details of statistical quantities.)

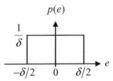

**Figure 5.21**   Probability density function of $e(n\Delta)$

Now, if we assume that the signal $x(t)$ is random and $\sigma_x^2$ is the variance of $x(n\Delta)$, then a measure of signal-to-noise ratio (SNR) is defined as

$$\frac{S}{N} = 10 \log_{10} \left( \frac{\text{signal power}}{\text{error power}} \right) = 10 \log_{10} \left( \frac{\sigma_x^2}{\sigma_e^2} \right) \quad \text{(for zero mean)} \tag{5.30}$$

where 'signal power' is $Average[x^2(n\Delta)]$ and 'error power' or the quantization noise is $Average[e^2(n\Delta)]$. This describes the *dynamic range* (or quantization signal-to-noise ratio) of the ADC. Since we assume that the error is random and has a uniform probability density function, for the full use of the dynamic range of ADC with $b$ bit word length, e.g. $\sigma_x = X$, Equation (5.30) becomes

$$\frac{S}{N} = 10\log_{10}\left(\sigma_x^2/\sigma_e^2\right) = 10\log_{10}[12X^2/(X/2^b)^2]$$

$$= 10\log_{10}(12 \times 2^{2b}) \approx 10.8 + 6b \text{ dB} \qquad (5.31)$$

For example, a 12 bit ADC (11 bit word length) has a maximum dynamic range of about 77 dB. However, we note that this would undoubtedly result in clipping. So, if we choose $\sigma_x = X/4$ to 'avoid' clipping, then the dynamic rage is reduced to

$$\frac{S}{N} = 10\log_{10}(\sigma_x^2/\sigma_e^2) \approx 6b - 1.25 \text{ dB} \qquad (5.32)$$

In this case, a 12 bit ADC gives a dynamic range of about 65 dB. This may be reduced further by practical considerations of the quality of the acquisition system (Otnes and Enochson, 1978). For example, the sampler in Figure 5.16 cannot be realized with a train of delta functions (thus producing aperture error and jitter). Nevertheless, it is emphasized that we must *avoid clipping* but always try to *use the maximum dynamic range*.

## 5.4 SOME OTHER CONSIDERATIONS IN SIGNAL ACQUISITION

### *Signal Conditioning*

We have already noted that signals should use as much of the ADC range as possible — but *without overloading* — or clipping of the signal will occur. 'Signal conditioning' refers to the procedures used to ensure that 'good data' are delivered to the ADC. This includes the correct choice of transducer and its operation and subsequent manipulation of the data before the ADC.

Specifically, transducer outputs must be 'conditioned' to accommodate cabling, environmental considerations and features of the recording instrumentation. Conditioning includes amplification and filtering, with due account taken of power supplies and cabling. For example, some transducers, such as strain gauges, require power supplies. Considerations in this case include: stability of power supply with little ripple, low noise, temperature stability, low background noise pick-up, low interchannel interference, etc.

**Amplifiers:** Amplifiers are used to increase (or attenuate) magnitudes in a calibrated fashion; transform signals from one physical variable to another, e.g. charge to voltage; remove d.c. biases; provide impedance matching, etc. The most common types are voltage amplifier, charge amplifier, differential amplifier, preamplifier, etc. In each case, care should be taken to ensure linearity, satisfactory frequency response and satisfactory 'slew rate' (i.e. response to maximum rate of rise of a signal). In any case, the result of amplification should not cause 'overload' which exceeds the limit of input (or output) range of a device.

**Filters:** Filters are used to limit signal bandwidth. Typically these are low-pass filters (anti-aliasing filters), high-pass filters, band-pass filters and band-stop filters. (Note that high-pass and band-stop filters would need additional low-pass filtering before sampling.) Most filters here are 'analogue' electronic filters. Sometimes natural 'mechanical filtering' is very helpful.

**Cabling:** Cabling must be suited to the application. Considerations are cable length, impedance of cable and electronics, magnetic and capacitive background noise, environment, interferences, transducer type, etc.

*Triboelectric noise* (static electricity) is generated when a coaxial cable is used to connect a high-impedance piezoelectric transducer to a charge amplifier, and undergoes mechanical distortion. *Grounding* must be considered. Suitable common earthing must be established to minimize electromagnetic interference manifesting itself as background noise. *Shielding* confines radiated electromagnetic energy.

Note that none of the considerations listed above is 'less important' to obtain (and generate) good data. A couple of practical examples are demonstrated below. First, consider generating a signal using a computer to excite a shaker. The signal must pass through a digital-to-analogue converter (DAC), a low-pass filter (or reconstruction filter) and the power amplifier before being fed into the shaker. Note that, in this case, it is not only the reconstruction filter, but also the power amplifier that is a filter in some sense. Thus, each device may distort the original signal, and consequently the signal which the shaker receives may not properly represent the original (or intended) signal. The frequency response of the power amplifier in particular should be noted carefully. Most power amplifiers have a band-limited frequency response with a reasonably high enough upper frequency limit suitable for general sound and vibration problems. However, some have a lower frequency limit (as well as the upper limit), which acts as a band-pass filter. This type of power amplifier can distort the signal significantly if the signal contains frequency components outside the frequency band of the amplifier. For example, if a transient signal such as a half-sine pulse is fed to the power amplifier, the output will be considerably distorted owing to the loss of energy in the low-frequency region. This effect is shown in Figure 5.22, where a half-sine wave is generated by a computer and measured before and after the power amplifier which has a lower frequency limit.

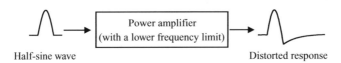

Half-sine wave        Distorted response

**Figure 5.22** Example of distortion due to the power amplifier

As another practical example, consider the beam experimental setup in Chapter 1 (Figure 1.11). In Figure 1.11, all the cables are secured adequately to minimize additional dynamic effects. Note that the beam is very light and flexible, so any excessive movement and interference of the cables can affect the dynamics of the beam. Now, suppose the cable connected to the accelerometer is loosely laid down on the table as shown in Figure 5.23. Then, the movement of the beam causes the cable to slide over the table. This results in additional friction damping to the structure (and also possibly additional stiffness). The system frequency response functions for each case are shown in Figure 5.24, where the effects of this cable interference are clearly seen.

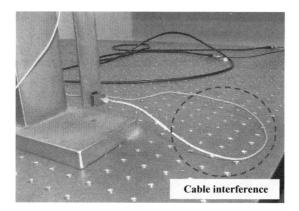

**Figure 5.23**    Experiment with cable interference

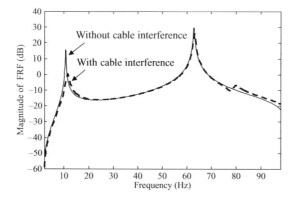

**Figure 5.24**    FRF of the system with/without cable interference

## Data Validation

As demonstrated in the above experimental results, every possible effort should be made *early* in any experiment to ensure good data are captured. Data validation refers generally to the many and varied checks and tests one may perform prior to 'serious' signal processing. This will occur at both analogue and digital stages. Obviously it would be best always to process only 'perfect' signals. This ideal is impossible and a very clear understanding of any shortcomings in the data is vital.

A long list of items for consideration can be compiled, some of which are as follows:

- Most signals will be recorded, even if some real-time processing is carried out. Identify any physical events for correlation with data.
- Inspect time histories critically, e.g. if periodic signals are expected, check for other signals such as noise, transients.
- Ensure non-stationary signals are adequately captured and note any changing 'physics' that might account for the non-stationarity.

- Check for signal clipping.
- Check for adequate signal levels (dynamic range).
- Check for excessive background noise, sustained or intermittent (spikes or bursts).
- Check for power line pick-up.
- Check for spurious trends, i.e. drifts, d.c. offsets.
- Check for signal drop-outs.
- Check for ADC operation.
- Check for aliasing.
- Always carry out some sample analyses (e.g. moments, spectra and probability densities, etc; these statistical quantities are discussed in Part II of this book).

## 5.5 SHANNON'S SAMPLING THEOREM (SIGNAL RECONSTRUCTION)

This chapter concludes with a look at digital-to-analogue conversion and essentially starts from the fact that, to avoid aliasing, the sampling rate $f_s$ should be greater than twice the highest frequency contained in the signal. This begs a fundamental question: is it possible to reconstruct the original analogue signal exactly from the sample values or has the information carried by the original analogue signal been lost? As long as there is no aliasing, we can indeed reconstruct the signal exactly and this introduces the concept of an ideal digital-to-analogue conversion. This is simple to understand using the following argument.

Recall the pictorial representation of the Fourier transforms of a continuous signal $x(t)$ and its sampled equivalent $x(n\Delta)$, i.e. $X(f)$ and $X(e^{j2\pi f\Delta})$ respectively, as shown in Figure 5.25. The figure shows the situation when no aliasing occurs. Also, note the scale factor.

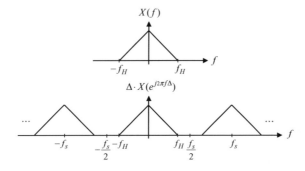

**Figure 5.25** Fourier transforms: $X(f)$ and $X(e^{j2\pi f\Delta})$

In digital-to-analogue conversion, we want to operate on $x(n\Delta)$ (equivalently $X(e^{j2\pi f\Delta})$) to recover $x(t)$ (equivalently $X(f)$). It is clear that to achieve this we simply need to multiply $X(e^{j2\pi f\Delta})$ by a frequency window function $H(f)$, where

$$H(f) = \Delta(= 1/f_s) \quad -f_s/2 < f < f_s/2$$
$$= 0 \quad \text{elsewhere}$$

(5.33)

Then

$$X(f) = H(f)X(e^{j2\pi f\Delta})$$

(5.34)

Taking the inverse Fourier transform of this gives

$$x(t) = h(t) * x(n\Delta) \tag{5.35}$$

where

$$h(t) = \frac{\sin \pi f_s t}{\pi f_s t} \tag{5.36}$$

Note that Equation (5.35) is not a mathematically correct expression. Thus, using the expression for $x(n\Delta)$ as $x_s(t) = x(t)i(t)$ where $i(t) = \sum_{n=-\infty}^{\infty} \delta(t - n\Delta)$, then Equation (5.35) becomes

$$x(t) = h(t) * x_s(t) = \int_{-\infty}^{\infty} \left[ \frac{\sin \pi f_s \tau}{\pi f_s \tau} \sum_{n=-\infty}^{\infty} x(t - \tau)\delta(t - n\Delta - \tau) \right] d\tau$$

$$= \sum_{n=-\infty}^{\infty} \left[ \int_{-\infty}^{\infty} \frac{\sin \pi f_s \tau}{\pi f_s \tau} x(t - \tau)\delta(t - n\Delta - \tau) d\tau \right]$$

$$= \sum_{n=-\infty}^{\infty} x(n\Delta) \frac{\sin \pi f_s(t - n\Delta)}{\pi f_s(t - n\Delta)} \tag{5.37}$$

i.e. the 'ideal' interpolating function is the sinc function of the form $\sin x / x$. Equation (5.37) can be depicted as in Figure 5.26 which shows how to reconstruct $x(t)$ at time $t$ that requires the infinite sum of scaled sinc functions.

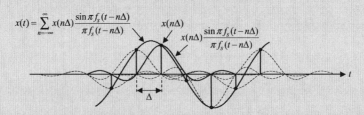

**Figure 5.26** Graphical representation of Equation (5.37)

Note that, with reference to Figure 5.25, if the highest frequency component of the signal is $f_H$ then the window function $H(f)$ need only be $\Delta$ for $|f| \leq f_H$ and zero elsewhere. Using this condition and applying the arguments above, the reconstruction algorithm can be expressed as

$$x(t) = \sum_{n=-\infty}^{\infty} x(n\Delta) \frac{2 f_H}{f_s} \frac{\sin 2\pi f_H(t - n\Delta)}{2\pi f_H(t - n\Delta)} \tag{5.38}$$

This result is called *Shannon's sampling theorem*.

This ideal reconstruction algorithm is not fully realizable owing to the infinite summation, and practical digital-to-analogue converters (DACs) are much simpler — notably the zero-order hold converter. Typical digital-to-analogue conversion using the zero-order hold is shown in

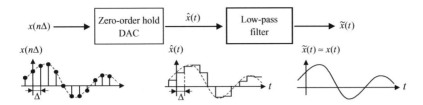

**Figure 5.27** Reconstruction of a signal using a zero-order hold DAC

Figure 5.27. The zero-order hold DAC generates a sequence of rectangular pulses by holding each sample for $\Delta$ seconds. The output of the zero-order hold DAC, however, inevitably contains a large amount of unwanted high-frequency components. Thus, in general, we need a low-pass filter to eliminate these high frequencies following the DAC. This low-pass filter is often called the reconstruction filter (or anti-imaging filter), and has a similar (or identical) design to the anti-aliasing low-pass filter. The cut-off frequency of the reconstruction filter is usually set to half the sampling rate, i.e. $f_s/2$.

Note that not only does the zero-order hold DAC produce undesirable high frequencies, but also its frequency response is no longer flat in both magnitude and phase (it has the shape of a sinc function). Thus the output signal $\hat{x}(t)$ has reduced amplitude and phase change in its passband (frequency band of the original (or desired) signal $x(t)$). To compensate for this effect, a pre-equalization digital filter (before the DAC) or post-equalization analogue filter (after the reconstruction filter) is often used. Another method of reducing this effect is by 'increasing the update rate' of the DAC. Similar to the sampling rate, the update rate is the rate at which the DAC updates its value.

For example, if we can generate a sequence $x(n\Delta)$ in Figure 5.27 such that $1/\Delta$ is much higher than $f_H$ (the highest frequency of the desired signal $x(t)$), and if the DAC is capable of generating the signal accordingly, then we have a much smoother analogue signal $\hat{x}(t)$, i.e. $\hat{x}(t) \approx \tilde{x}(t)$. In this case, we may not need to use the reconstruction filter. In effect, for a given band-limited signal, by representing $\hat{x}(t)$ using much narrower rectangular pulses, we have the frequency response of the DAC with flatter passband and negligible high-frequency side roll-off of the sinc function (note that the rectangular pulse (or a sinc function in the frequency domain) can be considered as a crude low-pass filter). Since many modern DAC devices have an update rate of 1MHz or above, in many situations in sound and vibration applications, we may reasonably approximate the desired signal simply by using the maximum capability of the DAC device.

## 5.6 BRIEF SUMMARY

1. The Fourier transform pair for a sampled sequence is given by

$$x(n\Delta) = \Delta \int_{-1/2\Delta}^{1/2\Delta} X(e^{j2\pi f\Delta})e^{j2\pi fn\Delta}\, df \quad \text{and} \quad X(e^{j2\pi f\Delta}) = \sum_{n=-\infty}^{\infty} x(n\Delta)e^{-j2\pi fn\Delta}$$

In this case, the scaling factor $\Delta$ is introduced.

2. The relationship between the Fourier transform of a continuous signal and the Fourier transform of the corresponding sampled sequence is

$$X(e^{j2\pi f\Delta}) = \frac{1}{\Delta} \sum_{n=-\infty}^{\infty} X\left(f - \frac{n}{\Delta}\right)$$

i.e. $X(e^{j2\pi f\Delta})$ is a continuous function consisting of replicas of scaled $X(f)$, and is periodic with period $1/\Delta$. This introduces possible aliasing.

3. To avoid aliasing, an 'analogue' low-pass filter (anti-aliasing filter) must be used before the analogue-to-digital conversion, and the sampling rate of the ADC must be high enough. In practice, for a given anti-aliasing filter with a roll-off rate of $B$ dB/octave and an ADC with a dynamic range of $A$ dB, the sampling rate is chosen as

$$f_s \approx 2 \times 10^{0.3A/B} f_c$$

4. To obtain 'good' data, we need to use the maximum dynamic range of the ADC (but must avoid clipping). Also, care must be taken with any signal conditioning, filters, amplifiers, cabling, etc.

5. When generating an analogue signal, for some applications, we may not need a reconstruction filter if the update rate of the DAC is high.

## 5.7 MATLAB EXAMPLES

***Example 5.1:*** **Demonstration of aliasing**

Case A: This example demonstrates that the values $\pm p + k/\Delta$ Hz become aliases of frequency $p$ Hz. (see Figure 5.11).

Consider that we want to sample a sinusoidal signal $x(t) = \sin 2\pi pt$ with the sampling rate $f_s = 100$ Hz. We examine three cases: $x_1(t) = \sin 2\pi p_1 t$, $x_2(t) = \sin 2\pi p_2 t$ and $x_3(t) = \sin 2\pi p_3 t$ where $p_1 = 20$ Hz, $p_2 = 80$ Hz and $p_3 = 120$ Hz. Note that all the frequencies will appear at the same frequency of 20 Hz.

| Line | MATLAB code | Comments |
|------|-------------|----------|
| 1 | clear all | Define the sampling rate fs = 100 Hz, total record |
| 2 | fs=100; T=10; | time T = 10 seconds, and the time variable t from 0 |
| 3 | t=0:1/fs:T-1/fs; | to 'T-1/fs' seconds. Also define the frequencies for |
| 4 | p1=20; p2=80; p3=120; | each sinusoid. |
| 5 | x1=sin(2*pi*p1*t); | Generate the signals $x_1(t)$, $x_2(t)$ and $x_3(t)$. Note that |
| 6 | x2=sin(2*pi*p2*t); | all these signals use the same time variable 't', thus it |
| 7 | x3=sin(2*pi*p3*t); | has the same sampling rate. |
| 8 | N=length(t); | Perform the DFT of each signal, and calculate the frequency variable f. |

```
 9  X1=fft(x1); X2=fft(x2);
    X3=fft(x3);
10  f=fs*(0:N-1)/N;
11  figure(1); plot(f, abs(X1)/fs/T)
12  xlabel('Frequency (Hz)');
    ylabel('Modulus')
13  axis([0 100 0 0.55])

14  figure(2); plot(f, abs(X2)/fs/T)
15  xlabel('Frequency (Hz)');
    ylabel('Modulus')
16  axis([0 100 0 0.55])

17  figure(3); plot(f, abs(X3)/fs/T)
18  xlabel('Frequency (Hz)');
    ylabel('Modulus')
19  axis([0 100 0 0.55])
```

Plot the modulus of the DFT of $x_1(t) = \sin 2\pi(20)t$ for the frequency range 0 Hz to 100 Hz (i.e. up to the sampling frequency). Note that the right half of the graph is the mirror image of the left half (except the 0 Hz component).

Plot the modulus of the DFT of $x_2(t) = \sin 2\pi(80)t$.

Plot the modulus of the DFT of $x_3(t) = \sin 2\pi(120)t$.

## Results

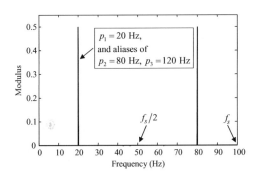

**Comments:** Note that all the frequencies $p_1 = 20\,\text{Hz}$, $p_2 = 80\,\text{Hz}$ and $p_3 = 120\,\text{Hz}$ appear at the same frequency 20 Hz.

---

*Example 5.2:* **Demonstration of aliasing**

Case B: This example demonstrates the aliasing problem on the 'digital' sampling of a sampled sequence $x(n\Delta)$.

Consider a sampled sinusoidal sequence $x(n\Delta) = \sin 2\pi pn\Delta$ where $p = 40\,\text{Hz}$, and the sampling rate is $f_s = 500\,\text{Hz}$ ($f_s = 1/\Delta$). Now, sample this sequence digitally again, i.e. generate a new sequence $x_1(k\Delta) = x[(5k)\Delta]$, $k = 0, 1, 2, \ldots$, by taking every five sample values of $x(n\Delta)$ (this has the effect of reducing the sampling rate to 100 Hz). Also generate a sequence $x_2(k\Delta) = x[(10k)\Delta]$ by taking every 10 sample values of $x(n\Delta)$, which reduces the sampling rate to 50 Hz. Thus, aliasing occurs, i.e. $p = 40\,\text{Hz}$ will appear at 10 Hz.

| Line | MATLAB code | Comments |
|------|-------------|----------|
| 1 | clear all | Define the sampling rate fs = 500 Hz, total record time |
| 2 | fs=500; T=10; | T = 10 seconds, and the time variable t from 0 to |
| 3 | t=0:1/fs:T-1/fs; | 'T-1/fs' seconds. Also generate the sampled sinusoidal |
| 4 | p=40; x=sin(2*pi*p*t); | signal whose frequency is 40 Hz. |
| 5 | x1=x(1:5:end); | Perform digital sampling, i.e. generate new sequences |
| 6 | x2=x(1:10:end); | $x_1(k\Delta)$ and $x_2(k\Delta)$ as described above. |
| 7 | N=length(x); N1=length(x1); N2=length(x2); | Perform the DFT of each signal $x(n\Delta)$, $x_1(k\Delta)$ and $x_2(k\Delta)$, and calculate the frequency variables f, f1 and |
| 8 | X=fft(x); X1=fft(x1); X2=fft(x2); | f2 accordingly. |
| 9 | f=fs*(0:N-1)/N; f1=100*(0:N1-1)/N1; f2=50*(0:N2-1)/N2; | |
| 10 | figure(1); plot(f, abs(X)/fs/T) | Plot the modulus of the DFT of $x(n\Delta) = \sin 2\pi(40)n\Delta$ |
| 11 | xlabel('Frequency (Hz)'); ylabel('Modulus') | for the frequency range 0 Hz to 500 Hz (up to the sampling rate). |
| 12 | axis([0 500 0 0.55]) | |
| 13 | figure(2); plot(f1, abs(X1)/100/T) | Plot the modulus of the DFT of $x_1(k\Delta)$ for the frequency range 0 Hz to 100 Hz (sampling rate of |
| 14 | xlabel('Frequency (Hz)'); ylabel('Modulus') | $x_1(k\Delta)$). |
| 15 | axis([0 100 0 0.55]) | |
| 16 | figure(3); plot(f2, abs(X2)/50/T) | Plot the modulus of the DFT of $x_2(k\Delta)$ for the |
| 17 | xlabel('Frequency (Hz)'); ylabel('Modulus') | frequency range 0 Hz to 50 Hz (sampling rate of $x_2(k\Delta)$). |
| 18 | axis([0 50 0 0.55]) | |

## Results

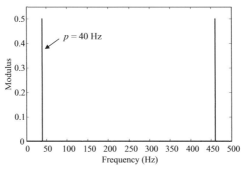

(a) DFT of $x(n\Delta) = \sin 2\pi(40)n\Delta$ with $f_s (= 1/\Delta) = 500$ Hz

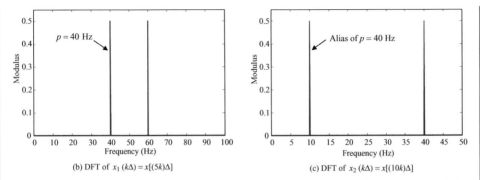

(b) DFT of $x_1$ $(k\Delta) = x[(5k)\Delta]$

(c) DFT of $x_2$ $(k\Delta) = x[(10k)\Delta]$

**Comments:** Note that aliasing occurs in the third case, i.e. $p = 40$ Hz appears at $10$ Hz because the sampling rate is $50$ Hz in this case.

---

*Example 5.3:* **Demonstration of 'digital' anti-aliasing filtering**

This example demonstrates a method to overcome the problem addressed in the previous MATLAB example.

We use the MATLAB function 'resample' to avoid the aliasing problem. The 'resample' function applies the digital anti-aliasing filter to the sequence before the sampling.

Consider a sampled sinusoidal sequence $x(n\Delta) = \sin 2\pi p_1 n\Delta + \sin 2\pi p_2 n\Delta$ where $p_1 = 10$ Hz and $p_2 = 40$ Hz and the sampling rate is $f_s = 500$ Hz ($f_s = 1/\Delta$). Generate new sequences $x_1(k\Delta_1)$ and $x_2(k\Delta_2)$ from $x(n\Delta)$ such that $\Delta_1/\Delta = 5$ and $\Delta_2/\Delta = 10$ without causing aliasing using the 'resample' function.

| Line | MATLAB code | Comments |
|---|---|---|
| 1<br>2<br>3<br>4 | clear all<br>fs=500; T=10;<br>t=0:1/fs:T-1/fs; p1=10; p2=40;<br>x=sin(2\*pi\*p1\*t) + sin(2\*pi\*p2\*t); | Define the sampling rate fs = 500 Hz, total record time T = 10 seconds, and the time variable t from 0 to 'T-1/fs' seconds. Also generate the sampled signal whose frequency components are 10 Hz and 40 Hz. |
| 5<br>6 | x1=resample(x,100,500);<br>x2=resample(x,50,500); | Perform the 'resampling' as described above. For example, the function 'resample(x,100,500)' takes the sequence 'x', applies a low-pass filter appropriately to the sequence, and returns the resampled sequence, where '100' is the new sampling rate and '500' is the original sampling rate. |
| 7 | N=length(x); N1=length(x1);<br>N2=length(x2); | |

```
 8  X=fft(x); X1=fft(x1); X2=fft(x2);
 9  f=fs*(0:N-1)/N; f1=100*(0:N1-1)/N1;
    f2=50*(0:N2-1)/N2;
10  figure(1); plot(f, abs(X)/fs/T)
11  xlabel('Frequency (Hz)');
    ylabel('Modulus')
12  axis([0 500 0 0.55])
13  figure(2); plot(f1, abs(X1)/100/T)
14  xlabel('Frequency (Hz)');
    ylabel('Modulus')
15  axis([0 100 0 0.55])
16  figure(3); plot(f2, abs(X2)/50/T)
17  xlabel('Frequency (Hz)');
    ylabel('Modulus')
18  axis([0 50 0 0.55])
```

Exactly the same code as in the previous example.

Exactly the same code as in the previous example.
Note that, due to the low-pass filtering, the 40 Hz component disappears on this graph.

## Results

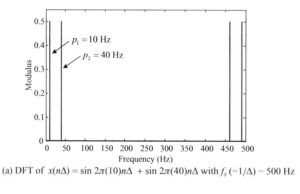

(a) DFT of $x(n\Delta) = \sin 2\pi(10)n\Delta + \sin 2\pi(40)n\Delta$ with $f_s\ (=1/\Delta) = 500$ Hz

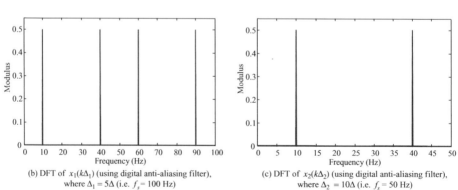

(b) DFT of $x_1(k\Delta_1)$ (using digital anti-aliasing filter), where $\Delta_1 = 5\Delta$ (i.e. $f_s = 100$ Hz)

(c) DFT of $x_2(k\Delta_2)$ (using digital anti-aliasing filter), where $\Delta_2 = 10\Delta$ (i.e. $f_s = 50$ Hz)

**Comments:** Note that, in Figure (c), only the 10 Hz component is shown, and the 40 Hz component disappears owing to the inherent low-pass (anti-aliasing) filtering process in the 'resample' function.

# 6

# The Discrete Fourier Transform

## Introduction

In this chapter we develop the properties of a fundamental tool of digital signal analysis – the discrete Fourier transform (DFT). This will include aspects of linear filtering, and relating the DFT to other Fourier representations. The chapter concludes with an introduction to the fast Fourier transform (FFT).

## 6.1 SEQUENCES AND LINEAR FILTERS

### Sequences

A sequence (or digital signal) is a function which is defined at a discrete set of points. A sequence results from: (i) a process which is naturally discrete such as a daily posted currency exchange rate, and (ii) sampling (at $\Delta$ second intervals (say)) an analogue signal as in Chapter 5. We shall denote a sequences as $x(n)$. This is an ordered set of numbers as shown in Figure 6.1.

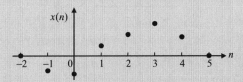

**Figure 6.1** Example of a sequence

Some examples are listed below:

(a) The unit impulse sequence or the Kronecker delta function, $\delta(n)$, is defined as

$$\left.\begin{array}{ll} \delta(n) = 1 & \text{if } n = 0 \\ = 0 & \text{if } n \neq 0 \end{array}\right\} \tag{6.1}$$

It can be depicted as in Figure 6.2

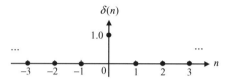

**Figure 6.2**  The unit impulse sequence, $\delta(n)$

This is the *digital* impulse or unit sample, i.e. it is the digital equivalent of the Dirac delta $\delta(t)$. If the unit impulse sequence is *delayed* (or shifted) by $k$, then

$$\left.\begin{array}{ll} \delta(n - k) = 1 & \text{if } n = k \\ = 0 & \text{if } n \neq k \end{array}\right\} \tag{6.2}$$

If $k$ is positive the shift is $k$ steps to the right. For example, Figure 6.3 shows the case for $k = 2$.

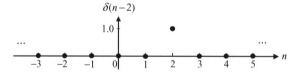

**Figure 6.3**  The delayed unit impulse sequence, $\delta(n-2)$

(b) The unit step sequence, $u(n)$, is defined as

$$\left.\begin{array}{ll} u(n) = 1 & \text{if } n \geq 0 \\ = 0 & \text{if } n < 0 \end{array}\right\} \tag{6.3}$$

The unit sample can be expressed by the difference of the unit step sequences, i.e. $\delta(n) = u(n) - u(n-1)$. Conversely, the unit step can be expressed by the *running sum* of the unit sample, i.e. $u(n) = \sum_{k=-\infty}^{n} \delta(k)$.

Starting with the unit sample, an arbitrary sequence can be expressed as the *sum of scaled, delayed unit impulses*. For example, consider the sequence $x(n)$ shown in Figure 6.4, where the values of the sequence are denoted as $a_n$.

This sequence can be written as $x(n) = a_{-3}\delta(n + 3) + a_1\delta(n - 1) + a_2\delta(n - 2) + a_5\delta(n - 5)$, i.e. in general form any sequence can be represented as

$$x(n) = \sum_{k=-\infty}^{\infty} x(k)\delta(n - k) \tag{6.4}$$

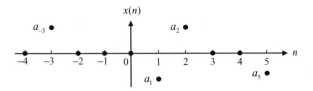

**Figure 6.4** An arbitrary sequence, $x(n)$

## Linear Filters

### Discrete Linear Time (Shift) Invariant Systems[M6.1]

The input–output relationship for a discrete LTI system (a digital filter) is shown in Figure 6.5.

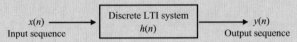

**Figure 6.5** A discrete LTI system

Similar to the continuous LTI system, we define the impulse response sequence of the discrete LTI system as $h(n)$. If the input to the system is a scaled and delayed impulse at $k$, i.e. $x(n) = a_k \delta(n - k)$, then the response of the system at $n$ is $y(n) = a_k h(n - k)$. So, for a general input sequence, the response at $n$ due to input $x(k)$ is $h(n - k)x(k)$. Since any input can be expressed as the sum of scaled, delayed unit impulses as described in Equation (6.4), the total response $y(n)$ to the input sequence $x(n)$ is

$$y(n) = \sum_{k=-\infty}^{n} h(n - k)x(k) \quad \text{if the system is } \textit{causal} \tag{6.5a}$$

or

$$y(n) = \sum_{k=-\infty}^{\infty} h(n - k)x(k) \quad \text{if the system is } \textit{non-causal} \tag{6.5b}$$

We shall use the latter notation (6.5b) which includes the former (6.5a) as a special case when $h(n) = 0$, if $n < 0$. This expression is called the *convolution sum*, which describes the relationship between the input and the output. That is, the input–output relationship of the digital LTI system is expressed by the *convolution of two sequences* $x(n)$ and $h(n)$:

$$y(n) = x(n) * h(n) = \sum_{k=-\infty}^{\infty} h(n - k)x(k) \tag{6.6}$$

Note that the convolution sum satisfies the property of commutativity, i.e.

$$y(n) = \sum_{k=-\infty}^{\infty} h(n - k)x(k) = \sum_{r=-\infty}^{\infty} h(r)x(n - r) \tag{6.7a}$$

or simply

$$y(n) = x(n) * h(n) = h(n) * x(n) \tag{6.7b}$$

The above expressions for the convolution sum are analogous to the convolution integral for a continuous system (see Equations (4.51), (4.53)). An example of the convolution sum is demonstrated graphically in Figure 6.6. In this figure, note that the number of non-zero elements of sequence $y(n)$ is '12' which is one element shorter than the sum of the lengths of non-zero elements of sequences $x(n)$ and $h(n)$.

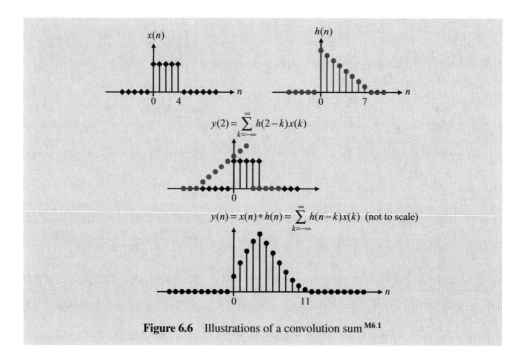

**Figure 6.6**   Illustrations of a convolution sum [M6.1]

### Relationship to Continuous Systems

Starting from $y(t) = h(t) * x(t) = \int_{-\infty}^{\infty} h(\tau)x(t - \tau)d\tau$, consider that the signals are sampled such that $y(n\Delta) = h(n\Delta) * x(n\Delta)$. Then the approximation to the convolution integral becomes

$$y(n\Delta) \approx \sum_{r=-\infty}^{\infty} h(r\Delta)x((n - r)\Delta) \cdot \Delta \qquad (6.8)$$

Note the scaling factor $\Delta$, i.e. if the discrete LTI system $h(n)$ results from the sampling of the corresponding continuous system $h(t)$ with sampling rate $1/\Delta$ and the input $x(n)$ is also the sampled version of $x(t)$, then it follows that

$$y(n\Delta) \approx y(n) \cdot \Delta \qquad (6.9)$$

where $y(n) = h(n) * x(n)$.

The concept of creating a digital filter $h(n)$ by simply sampling the impulse response of an analogue filter $h(t)$ is called 'impulse-invariant' filter design (see Figure 5.6 in Section 5.1).

## Stability and Description of a Digital LTI System

Many digital systems are characterized by difference equations (analogous to differential equations used for continuous systems). The input–output relationship for a digital system (Figure 6.5) can be expressed by

$$y(n) = -\sum_{k=1}^{N} a_k y(n-k) + \sum_{r=0}^{M} b_r x(n-r) \qquad (6.10)$$

Taking the $z$-transform of Equation (6.10) gives

$$Z\{y(n)\} = Y(z) = -Y(z) \sum_{k=1}^{N} a_k z^{-k} + X(z) \sum_{r=0}^{M} b_r z^{-r} \qquad (6.11)$$

Note that we use the time shifting property of the $z$-transform, i.e. $Z\{x(n-r)\} = z^{-r} X(z)$, to obtain Equation (6.11). Rearranging Equation (6.11) gives the transfer function of the digital system as

$$H(z) = \frac{Y(z)}{X(z)} = \frac{\sum\limits_{r=0}^{M} b_r z^{-r}}{1 + \sum\limits_{k=1}^{N} a_k z^{-k}} \qquad (6.12)$$

which is the $z$-transform of impulse response $h(n)$. Since Equation (6.12) is a rational function, i.e. the ratio of two polynomials, it can be written as

$$H(z) = b_0 z^{N-M} \frac{(z - z_1)(z - z_2) \ldots (z - z_M)}{(z - p_1)(z - p_2) \ldots (z - p_N)} \qquad (6.13)$$

Note that $H(z)$ has $M$ zeros (roots of the numerator) and $N$ poles (roots of the denominator). From Equation (6.13), the zeros and poles characterize the system. A causal system is BIBO (Bounded Input/Bounded Output) stable if all its poles lie within the unit circle $|z| = 1$. Or equivalently, the digital LTI system is BIBO stable if $\sum_{n=-\infty}^{\infty} |h(n)| < \infty$, i.e. output sequence $y(n)$ is bounded for every bounded input sequence $x(n)$ (Oppenheim et al., 1997).

The system described in the form of Equation (6.10) or (6.12) is called an auto-regressive moving average (ARMA) system (or model) which is characterized by an output that depends on past and current inputs and past outputs. The numbers $N$, $M$ are the orders of the auto-regressive and moving average components, and characterize the order with the notation $(N, M)$. This ARMA model is widely used for general filter design problems (e.g. Rabiner and Gold, 1975; Proakis and Manolakis, 1988) and for 'parametric' spectral estimation (Marple, 1987).

If all the coefficients of the denominator are zero, i.e. $a_k = 0$ for all $k$, the system is called a moving average (MA) system, and has only zeros (except the stack of *trivial* poles at the origin, $z = 0$). Note that this system is always stable since it does not have a pole. MA systems always have a finite duration impulse response. If all the coefficients of the numerator are zero except $b_0$, i.e. $b_r = 0$ for $k > 0$, the system is called an auto-regressive (AR) system, and has only poles (except the stack of *trivial* zeros at the origin, $z = 0$). The AR systems have a feedback nature and generally have an infinite duration impulse response. In general, ARMA systems also have an infinite duration impulse response.

Sometimes, the ARMA representation of the system can be very useful, especially for real-time processing. For example, if the estimated impulse response sequence $h(n)$ based on the methods in Chapter 9, which can be considered as an MA system, is very large, one can fit the corresponding FRF data to a reduced order ARMA model. This may be useful for some real-time digital signal processing (DSP). (See Comments 2 in MATLAB Example 9.4, Chapter 9.)

## 6.2  FREQUENCY DOMAIN REPRESENTATION OF DISCRETE SYSTEMS AND SIGNALS

Consider the response of a digital filter to a harmonic signal, i.e. $x(n) = e^{j2\pi fn}$. Then the output is

$$
\begin{aligned}
y(n) &= \sum_{k=-\infty}^{\infty} h(k)x(n-k) = \sum_{k=-\infty}^{\infty} h(k)e^{j2\pi f(n-k)} \\
&= e^{j2\pi fn} \sum_{k=-\infty}^{\infty} h(k)e^{-j2\pi fk}
\end{aligned}
\tag{6.14}
$$

We define $H(e^{j2\pi f}) = \sum_{k=-\infty}^{\infty} h(k)e^{-j2\pi fk}$. Then

$$
y(n) = e^{j2\pi fn} H(e^{j2\pi f}) = x(n)H(e^{j2\pi f})
\tag{6.15}
$$

$H(e^{j2\pi f})$ is called the *frequency response function* (FRF) of the system (compare this with Equation (4.57)).

Consider an example. Suppose we have a discrete system whose impulse response is $h(n) = a^n u(n)$, $|a| < 1$, as shown for example in Figure 6.7(a). Then the FRF of the system is

$$
H(e^{j2\pi f}) = \sum_{n=0}^{\infty} a^n e^{-j2\pi fn} = \sum_{n=0}^{\infty} (ae^{-j2\pi f})^n
\tag{6.16}
$$

This is a geometric series, and using the property of a geometric series, i.e.

$$
\sum_{n=0}^{\infty} r^n = \frac{1}{1-r}, \quad |r| < 1
$$

Equation (6.16) can be written as

$$
H(e^{j2\pi f}) = \frac{1}{1-ae^{-j2\pi f}}
\tag{6.17}
$$

The modulus and phase of Equation (6.17) are shown in Figures 6.7(b) and (c), respectively.

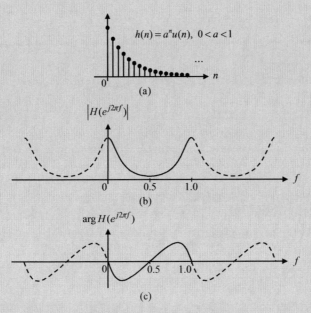

$$h(n) = a^n u(n), \quad 0 < a < 1$$

(a)

$|H(e^{j2\pi f})|$

(b)

$\arg H(e^{j2\pi f})$

(c)

**Figure 6.7**   Example of discrete impulse response and corresponding FRF

Note that, *unlike* the FRF of a continuous system, $H(e^{j2\pi f})$ is *periodic* (with period 1, or $2\pi$ if $\omega$ is used instead of $f$), i.e.

$$H(e^{j2\pi f}) = H(e^{j2\pi(f+k)}) = H(e^{j2\pi f}e^{j2\pi k}) = H(e^{j2\pi f}) \tag{6.18}$$

where $k$ is integer. Note also that this is a periodic *continuous* function, whereas its corresponding impulse response $h(n)$ is discrete in nature. Why should the FRF be periodic? The answer is that the system input is $x(n) = e^{j2\pi fn}$ which is indistinguishable from $x(n) = e^{j(2\pi f + 2\pi k)n}$ and so the system reacts in the same way to both inputs. This phenomenon is very similar to the case of sampled sequences discussed in Chapter 5, and we shall discuss their relation shortly.

Since $H(e^{j2\pi f})$ is periodic it has a 'Fourier series' representation. From Equation (6.14), we already have

$$H(e^{j2\pi f}) = \sum_{n=-\infty}^{\infty} h(n)e^{-j2\pi fn} \tag{6.19}$$

The values $h(n)$ are the Fourier coefficients and this expression can be inverted to give

$$h(n) = \int_{-1/2}^{1/2} H(e^{j2\pi f})e^{j2\pi fn} df \tag{6.20}$$

Equation (6.20) is easily justified by considering the Fourier series pair given in Chapter 3, i.e.

$$x(t) = \sum_{n=-\infty}^{\infty} c_n e^{j2\pi nt/T_P}$$

and

$$c_n = \frac{1}{T_P} \int_0^{T_P} x(t) e^{-j2\pi nt/T_P} dt$$

The two expressions (6.19) and (6.20) are the basis of the Fourier representation of discrete signals and apply to any sequence provided that Equation (6.19) converges. Equation (6.19) is the *Fourier transform of a sequence*, and is often called the *discrete-time Fourier transform* (Oppenheim *et al.*, 1997). However, this should not be confused with *discrete Fourier transform* (DFT) for *finite length* signals that will be discussed in the next section. Alternatives to Equations (6.19) and (6.20) are

$$H(e^{j\omega}) = \sum_{n=-\infty}^{\infty} h(n) e^{-j\omega n} \tag{6.21}$$

$$h(n) = \frac{1}{2\pi} \int_{-\pi}^{\pi} H(e^{j\omega}) e^{j\omega n} d\omega \tag{6.22}$$

Note that, similar to the Fourier integral, if $h(n)$ is real, $\left| H(e^{j2\pi f}) \right|$ is an even and arg $H(e^{j2\pi f})$ is an odd function of '$f$'.

### *The Fourier Transform of the Convolution of Two Sequences*

Let us consider an output sequence of a discrete LTI system, which is the convolution of two sequences $h(n)$ and $x(n)$, i.e. $y(n) = h(n) * x(n) = \sum_{k=-\infty}^{\infty} h(k) x(n-k)$. Since the sequence $x(n)$ has a Fourier representation, i.e. $x(n) = \int_{-1/2}^{1/2} X(e^{j2\pi f}) e^{j2\pi fn} df$, substituting this into the convolution expression gives

$$y(n) = \sum_{k=-\infty}^{\infty} h(k) x(n-k) = \sum_{k=-\infty}^{\infty} h(k) \int_{-1/2}^{1/2} X(e^{j2\pi f}) e^{j2\pi f(n-k)} df$$

$$= \int_{-1/2}^{1/2} X(e^{j2\pi f}) \underbrace{\sum_{k=-\infty}^{\infty} h(k) e^{-j2\pi fk}}_{H(e^{j2\pi f})} e^{j2\pi fn} df = \int_{-1/2}^{1/2} X(e^{j2\pi f}) H(e^{j2\pi f}) e^{j2\pi fn} df$$

$$\tag{6.23}$$

Thus,

$$\boxed{Y(e^{j2\pi f}) = X(e^{j2\pi f}) H(e^{j2\pi f})} \tag{6.24}$$

i.e. the Fourier transform of the convolution of two sequences is the product of their transforms.

### Relation to Sampled Sequences, $x(n\Delta)$

If time is involved, i.e. a sequence results from *sampling* a continuous signal, then Equations (6.19) and (6.20) must be modified appropriately. For a sample sequence $x(n\Delta)$, the Fourier representations are

$$X(e^{j2\pi f\Delta}) = \sum_{n=-\infty}^{\infty} x(n\Delta)e^{-j2\pi f n\Delta} \tag{6.25}$$

$$x(n\Delta) = \Delta \int_{-1/2\Delta}^{1/2\Delta} X(e^{j2\pi f\Delta})e^{j2\pi f n\Delta} df \tag{6.26}$$

which correspond to Equations (6.19) and (6.20), with $\Delta = 1$. Note that we have already seen these equations in Chapter 5, i.e. they are the same as Equations (5.6) and (5.7) which are the Fourier transform pair for a 'sampled sequence'.

## 6.3 THE DISCRETE FOURIER TRANSFORM

So far we have considered sequences that run over the range $-\infty < n < \infty$ ($n$ integer). For the special case where the sequence is of *finite* length (i.e. non-zero for a finite number of values) an alternative Fourier representation is possible called the *discrete Fourier transform* (DFT).

It turns out that the DFT is a Fourier representation of a finite length sequence and *is itself a sequence* rather than a continuous function of frequency, and it corresponds to samples, equally spaced in frequency, of the Fourier transform of the signal. The DFT is fundamental to many digital signal processing algorithms (following the discovery of the *fast Fourier transform* (FFT), which is the name given to *an efficient algorithm for the computation of the DFT*).

We start by considering the Fourier transform of a (sampled) sequence given by Equation (6.25). Suppose $x(n\Delta)$ takes some values for $n = 0, 1, \ldots, N-1$, i.e. $N$ points only, and is zero elsewhere. Then this can be written as

$$X(e^{j2\pi f\Delta}) = \sum_{n=0}^{N-1} x(n\Delta)e^{-j2\pi f n\Delta} \tag{6.27}$$

Note that this is still *continuous* in frequency. Now, let us evaluate this at frequencies $f = k/N\Delta$ where $k$ is integer. Then, the right hand side of Equation (6.27) becomes $\sum_{n=0}^{N-1} x(n\Delta)e^{-j(2\pi/N)nk}$, and we write this as

$$X(k) = \sum_{n=0}^{N-1} x(n\Delta)e^{-j(2\pi/N)nk} \tag{6.28}$$

This is the DFT of a finite (sampled) sequence $x(n\Delta)$. For more usual notation, omitting $\Delta$, the DFT of $x(n)$ is defined as

$$X(k) = \sum_{n=0}^{N-1} x(n)e^{-j(2\pi/N)nk} \tag{6.29}$$

As a result, the relationship between the Fourier transform of a sequence and the DFT of a finite length sequence can be expressed as

$$X(k) = \left[ X(e^{j2\pi f\Delta}) \text{ evaluated at } f = \frac{k}{N\Delta} \text{Hz} \right] (k \text{ integer}) \tag{6.30}$$

i.e. $X(k)$ may be regarded as the *sampled version* of $X(e^{j2\pi f\Delta})$ *in the frequency domain*. Note that, since $X(e^{j2\pi f\Delta})$ is periodic with $1/\Delta$, we may need to evaluate for $k = 0, 1, \ldots, N-1$, i.e. $N$ points only.

The inverse DFT can be found by multiplying both sides of Equation (6.29) by $e^{j(2\pi/N)rk}$ and summing over $k$. Then

$$\sum_{k=0}^{N-1} X(k)e^{j(2\pi/N)rk} = \sum_{k=0}^{N-1}\sum_{n=0}^{N-1} x(n)e^{-j(2\pi/N)nk}e^{j(2\pi/N)rk} = \sum_{k=0}^{N-1}\sum_{n=0}^{N-1} x(n)e^{-j(2\pi/N)k(n-r)} \tag{6.31}$$

Interchanging the summation order on the right hand side of Equation (6.31) and noting that

$$\sum_{k=0}^{N-1} e^{-j(2\pi/N)k(n-r)} = N \quad \text{if } n = r \tag{6.32}$$
$$= 0 \quad \text{otherwise}$$

gives $\sum_{k=0}^{N-1} X(k)e^{j(2\pi/N)rk} = N \cdot x(r)$. Thus, the inverse DFT is given by

$$x(n) = \frac{1}{N} \sum_{k=0}^{N-1} X(k)e^{j(2\pi/N)nk} \tag{6.33}$$

Note that in Equation (6.33), since $e^{j(2\pi/N)(n+N)k} = e^{j(2\pi/N)nk}$, we see that both $X(k)$ and $x(n)$ are periodic with period $N$. It is important to realize that whilst the original sequence $x(n)$ is zero for $n < 0$ and $n \geq N$, the act of 'sampling in frequency' has imposed a *periodic structure* on the sequence. In other words, the DFT of a finite length $x(n)$ implies that $x(n)$ is one period of a periodic sequence $x_p(n)$, where $x(n) = x_p(n)$ for $0 \leq n \leq N - 1$ and $x_p(n) = x_p(n + rN)$ ($r$ integer).

As an example, the DFT of a finite length sequence is shown in Figure 6.8 where the corresponding Fourier transform of a sequence is also shown for comparison. Suppose $x(n)$ has the form shown in Figure 6.8(a); then Figures 6.8(b) and (c) indicate the (continuous) amplitude and phase of $X(e^{j2\pi f\Delta})$. Figures 6.8(e) and (f) are the corresponding $|X(k)|$ and arg $X(k)$ – the DFT of $x(n)$ (equivalently the DFT of $x_p(n)$ in Figure 6.8(d)). These correspond to evaluating Figures 6.8(b) and (c) at frequencies $f = k/N\Delta$. Note that the periodicity is present in all figures except Figure 6.8(a).

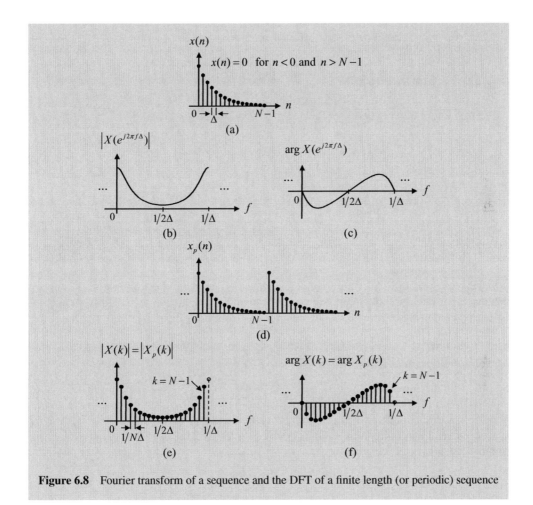

**Figure 6.8** Fourier transform of a sequence and the DFT of a finite length (or periodic) sequence

## Data Truncation[M6.2]

We assumed above that the sequence $x(n)$ was zero for $n$ outside values 0 to $N - 1$. In general, however, signals may not be finite in duration. So, we now consider the truncated sampled data $x_T(n\Delta)$. For example, consider a finite ($N$ points) sequence (*sampled* and *truncated*) as shown in Figure 6.9.

As we would expect from the windowing effect discussed in Chapter 4, there will be some *distortion* in the frequency domain. Let $x_p(n)$ and $w_p(n)$ be the equivalent periodic sequences of $x(n\Delta)$ and $w(n\Delta)$ for $0 \le n \le N - 1$ (omitting $\Delta$ for convenience). Then the DFT of the

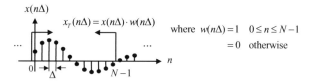

**Figure 6.9** Sampled and truncated sequence $x_T(n\Delta)$

truncated signal, $X_T(k)$, becomes

$$X_T(k) = \text{DFT}\left[x_p(n)w_p(n)\right]$$

$$= \frac{1}{N^2} \sum_{n=0}^{N-1}\sum_{k_1=0}^{N-1} X_p(k_1)e^{j(2\pi/N)nk_1} \sum_{k_2=0}^{N-1} W_p(k_2)e^{j(2\pi/N)nk_2}e^{-j(2\pi/N)nk}$$

$$= \frac{1}{N^2} \sum_{k_1=0}^{N-1} X_p(k_1) \sum_{k_2=0}^{N-1} W_p(k_2) \sum_{n=0}^{N-1} e^{-j(2\pi/N)n(k-k_1-k_2)} = \frac{1}{N}\sum_{k_1=0}^{N-1} X_p(k_1)W_p(k-k_1)$$

$$= \frac{1}{N} X_p(k) \circledast W_p(k) \tag{6.34}$$

It is the convolution of the two periodic sequences — hence the distortion in the frequency domain, where the symbol $\circledast$ denotes circular convolution (this will be explained in Section 6.5). The windowing effect will be demonstrated in MATLAB Example 6.2.

### *Alternative Representation of the DFT*

Starting with the z-transform of $x(n)$, i.e. $X(z)$, then when $z = e^{j2\pi f\Delta}$ a circle is picked out of unit radius, and $X(e^{j2\pi f\Delta})$ is the value of $X(z)$ evaluated at points on the unit circle. When $f = k/N\Delta$, this amounts to evaluating $X(z)$ at specific points on the unit circle, i.e. $N$ evenly spaced points around the unit circle. This gives the DFT expression $X(k)$ as illustrated in Figure 6.10.

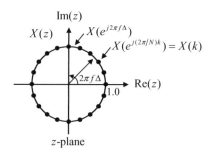

**Figure 6.10** Representation of the DFT in the z-plane

### Frequency Resolution and Zero Padding

As we have seen earlier in Chapter 4, the frequency resolution of Fourier transform $X_T(f)$ depends on the data length (or window length) $T$. Note that the data length of the truncated sampled sequence $x_T(n\Delta)$ is $T = N\Delta$, and the frequency spacing in $X_T(k)$ is $1/N\Delta = 1/T$ Hz. Thus, we may have an arbitrary fine frequency spacing when $T \to \infty$.

If the sequence $x(n\Delta)$ is finite in nature, then the Fourier transform of a sequence $X(e^{j2\pi f\Delta})$ is fully representative of the original sequence without introducing truncation, because

$$X(e^{j2\pi f\Delta}) = \sum_{n=-\infty}^{\infty} x(n\Delta)e^{-j2\pi fn\Delta} = \sum_{n=0}^{N-1} x(n\Delta)e^{-j2\pi fn\Delta}$$

Then, the DFT $X(k) = X(e^{j2\pi f\Delta})\big|_{f=k/N\Delta}$ gives the frequency spacing $1/N\Delta$ Hz. This spacing may be considered sufficient because we do not lose any information, i.e. we can completely recover $x(n\Delta)$ from $X(k)$.

However, we often want to *see* more detail in the frequency domain, such as finer frequency spacing. A convenient procedure is simply to 'add zeros' to $x(n)$, i.e. define

$$\hat{x}(n) = x(n) \quad 0 \le n \le N - 1$$
$$= 0 \quad N \le n \le L - 1 \tag{6.35}$$

Then the $L$-point DFT of $\hat{x}(n)$ is

$$\hat{X}(k) = \sum_{n=0}^{L-1} \hat{x}(n)e^{-j(2\pi/L)nk} = \sum_{n=0}^{N-1} x(n)e^{-j(2\pi/L)nk} \tag{6.36}$$

Thus, we see that $\hat{X}(k) = X(e^{j(2\pi/L)k})$, $k = 0, 1, \ldots, L - 1$, i.e. 'finer' spacing round the unit circle in the $z$-plane (see Figure 6.10), in other words, zero padding in the time domain results in the interpolation in the frequency domain (Smith, 2003). In vibration problems, this can be used to obtain the fine detail near resonances. However, care must be taken with this artificially made finer structure — the zero padding does not increase the 'true' resolution (see MATLAB Example 4.6 in Chapter 4), i.e. the fundamental resolution is fixed and it is only the frequency spacing that is reduced.

An interesting feature is that, with zero padding in the frequency domain, performing the inverse DFT results in interpolation in the time domain, i.e. an increased sampling rate in the time domain (note that zeros are padded symmetrically with respect to $N/2$, and it is assumed that $X(N/2) = 0$ for an even number of $N$). So zero padding in one domain results in a finer structure in the other domain.

Zero padding is sometimes useful for analysing a transient signal that dies away quickly. For example, if we estimate the FRF of a system using the impact testing method, the measured signal (from the force sensor of an impact hammer) quickly falls into the noise level. In this case, we can *artificially* improve the quality of the measured signal by replacing the data in the noise region with zeros (see MATLAB Example 6.7); note that the measurement time may also be increased (in effect) by adding more zeros. This approach can also be applied to the free vibration signal of a highly damped system (see MATLAB Example 6.5).

## *Scaling Effects*[M6.2]

If the sequence $x(n\Delta)$ results from sampling a continuous signal $x(t)$ we must consider the scaling effect on $X(k)$ as compared with $X(f)$. We need to consider the scaling effect differently for transient signals and periodic signals. For a transient signal, the energy of the signal is finite. Assuming that the data window is large enough so that the truncation of data does not introduce a loss of energy, the only scaling factor is the sampling interval $\Delta$. However, if the original signal is periodic the energy is infinite, so in addition to the scaling effect introduced by sampling, the DFT coefficients will have different amplitudes depending on the length of the data window. This effect can be easily justified by comparing Parseval's theorems for a periodic signal (Equation (3.39)) and for a transient signal (Equation (4.17)). The following example shows the relationship between the Fourier integral and the DFT, together with the scaling effect for a periodic signal.

Consider a periodic continuous signal $x(t) = A \cos 2\pi pt$, $p = 1/T_P$, and its Fourier integral, as shown in Figure 6.11(a). Suppose we use the data length $T$ seconds; then its effect is applying the rectangular window as shown in Figure 6.11(b). Note that the magnitude spectrum of $W(f)$ depends on the window length $T$. The Fourier integral of the truncated signal is shown in Figure 6.11(c), and the Fourier transform of a truncated and sampled signal is

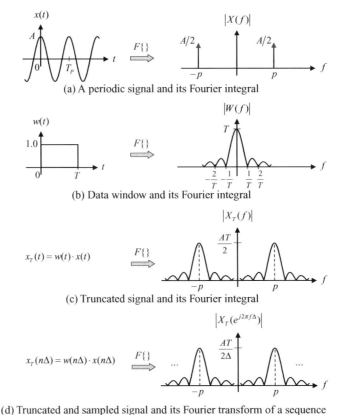

(a) A periodic signal and its Fourier integral

(b) Data window and its Fourier integral

(c) Truncated signal and its Fourier integral

(d) Truncated and sampled signal and its Fourier transform of a sequence

**Figure 6.11** Various Fourier transforms of a sinusoidal signal

shown in Figure 6.11(d). Note the periodicity in this figure. Note also the windowing effects and the amplitude differences for each transform (especially the scaling factor in Figure 6.11(d)).

Now consider the DFT of the truncated and sampled sequence. The DFT results in frequencies at $f_k = k/N\Delta, k = 0, 1, \ldots, N-1$, i.e. the frequency range covers from 0 Hz to $(f_s - f_s/N)$ Hz. Thus, if we want frequency $p$ to be picked out exactly, we need $k/N\Delta = p$ for some $k$. Suppose we sample at every $\Delta = T_P/10$ and take one period (10-point DFT) *exactly*, i.e. $T(=N\Delta) = T_P(= 1/p)$. As shown in Figure 6.12, the frequency separation is $1/N\Delta = 1/T_P = p$ (Hz), thus $p = f_1 = 1/N\Delta$ which is the second line on the *discrete* frequency axis ($f_k = k/N\Delta$). Note that the first line is $f_0 = 0$ (Hz), i.e. the d.c. component. All other frequencies ($f_k$ except $f_1$ and $f_9$) are 'zeros', since these frequencies correspond to the zero points of the side lobes that are separated by $1/T = 1/T_P$. Thus, the resulting DFT is *one single* spike (up to $k = N/2$).

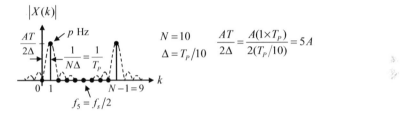

**Figure 6.12** The 10-point DFT of the truncated and sampled sinusoidal signal, $T = T_P$

Since the DFT has a periodic structure, $X(10)$ (if it is evaluated) will be equal to $X(0)$. Also, due to the symmetry property of the magnitude spectrum of $X(k)$, the right half of the figure is the mirror image of the left half such that $|X(1)| = |X(9)|, |X(2)| = |X(8)|, \ldots$, $|X(4)| = |X(6)|$. Note that the magnitude of $X(1)$ is $5A$. Also note that $X(5)$ is the value at the folding frequency $f_s/2$. From the fact that we have taken an 'even'-numbered DFT, we have the DFT coefficient at the folding frequency. However, if we take an 'odd'-numbered DFT, then it cannot be evaluated at the folding frequency. For example, if we take the nine-point DFT, the symmetric structure will become $|X(1)| = |X(8)|, |X(2)| = |X(7)|, \ldots, |X(4)| = |X(5)|$ (see Section 6.4 and MATLAB Example 6.4).

For the same sampling interval, if we take five periods *exactly*, i.e. $T(=N\Delta) = 5T_P$ (50-point DFT), then the frequency separation is $1/N\Delta = 1/(50 \cdot T_P/10) = 1/5T_P = p/5$ (Hz) as shown in Figure 6.13. Thus, $p = f_5 = 5/N\Delta$ which is the sixth line on the discrete frequency axis. Again, all other frequencies $f_k$ (except $f_5$ and $f_{45}$) are 'zeros', since these frequencies also correspond to the zero points of the side lobes that are now separated by

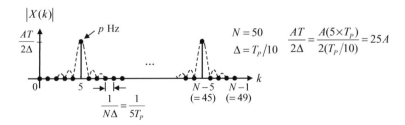

**Figure 6.13** The 50-point DFT of the truncated and sampled sinusoidal signal, $T = 5T_P$

$1/T = 1/5T_P$. Note the magnitude change at the peak frequency, which is now $25A$ (compare this with the previous case, the 10-point DFT).

If a non-integer number of periods are taken, this will produce *all* non-zero frequency components (as we have seen in MATLAB Example 4.6 in Chapter 4; see also MATLAB Example 6.2b). The 'scaling' effect is due to both 'sampling' and 'windowing', and so different window types may produce different scaling effects (see MATLAB Examples 4.6 and 4.7 in Chapter 4). Since the DFT evaluates values at frequencies $f_k = k/N\Delta$, the frequency resolution can only be improved by increasing $N\Delta$ ($=$ window length, $T$). Thus, if the sampling rate is increased (i.e. smaller $\Delta$ is used), then we need more data (larger $N$) in order to maintain the same resolution (see Comments in MATLAB Example 6.3).

## 6.4 PROPERTIES OF THE DFT

The properties of the DFT are fundamental to signal processing. We summarize a few here:

(a) The DFT of the Kronecker delta function $\delta(n)$ is

$$\text{DFT}\,[\delta(n)] = \sum_{n=0}^{N-1} \delta(n) e^{-j(2\pi/N)nk} = e^{-j(2\pi/N)0 \cdot k} = 1 \qquad (6.37)$$

(Note that the Kronecker delta function $\delta(n)$ is analogous to its continuous counterpart, the Dirac delta function $\delta(t)$, but it cannot be related as the sampling of $\delta(t)$.)

(b) *Linearity:* If DFT $[x(n)] = X(k)$ and DFT $[y(n)] = Y(k)$, then

$$\text{DFT}\,[ax(n) + by(n)] = aX(k) + bY(k) \qquad (6.38)$$

(c) *Shifting property:* If DFT $[x(n)] = X(k)$, then

$$\text{DFT}\,[x(n - n_0)] = e^{-j(2\pi/N)n_0 k} X(k) \qquad (6.39)$$

Special attention must be given to the meaning of a time shift of a finite duration sequence. Shown in Figure 6.14 is the finite sequence $x(n)$ of duration $N$ samples (marked ●). The $N$-point DFT of $x(n)$ is $X(k)$. Also shown are the samples of the 'equivalent' periodic sequence $x_p(n)$ with the same DFT as $x(n)$.

If we want the DFT of $x(n - n_0)$, $n_0 < N$, we must consider a shift of the periodic sequence $x_p(n - n_0)$ and the equivalent finite duration sequence with DFT $e^{-j(2\pi/N)n_0 k} X(k)$ is that part of $x_p(n - n_0)$ in the interval $0 \leq n \leq N - 1$, as shown in Figure 6.15 for $n_0 = 2$ (for example), i.e. shift to *right*.

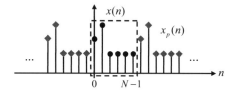

**Figure 6.14**   Finite sequence $x(n)$ and equivalent periodic sequence $x_p(n)$

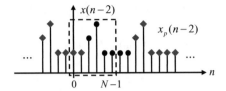

**Figure 6.15** Shifted finite sequence $x(n - n_0)$ and equivalent shifted periodic sequence $x_p(n - n_0)$

Examining Figures 6.14 and 6.15, we might imagine the sequence $x(n)$ as displayed around the circumference of a cylinder in such a way that the cylinder has $N$ points on it. As the cylinder revolves we see $x_p(n)$, i.e. we can talk of a 'circular' shift.

(d) *Symmetry properties*[M6.4]: For real data $x(n)$, we have the following symmetry properties. An example is shown in Figure 6.16 (compare the symmetric structures for even and odd numbers of $N$). Note that, at $N/2$, the *imaginary part* must be 'zero', and the phase can be either 'zero or $\pi$' depending on the sign of real part:

$$\mathrm{Re}\,[X(k)] = \mathrm{Re}\,[X(N - k)] \tag{6.40a}$$

$$\mathrm{Im}\,[X(k)] = -\mathrm{Im}\,[X(N - k)] \tag{6.40b}$$

$$|X(k)| = |X(N - k)| \tag{6.41a}$$

$$\mathrm{arg}\,X(k) = -\,\mathrm{arg}\,X(N - k) \tag{6.41b}$$

Or, we may express the above results as (*denotes complex conjugate)

$$\boxed{X(N - k) = X^*(k)} \tag{6.42}$$

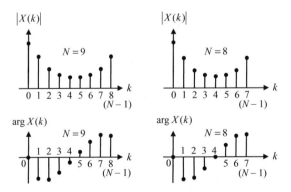

**Figure 6.16** Symmetry properties of the DFT

## 6.5 CONVOLUTION OF PERIODIC SEQUENCES[M6.6]

Consider two periodic sequences with the *same* length of period, $x_p(n)$ and $h_p(n)$, and their DFTs as follows:

$$X_p(k) = \sum_{n=0}^{N-1} x_p(n)e^{-j(2\pi/N)nk} \tag{6.43a}$$

$$H_p(k) = \sum_{n=0}^{N-1} h_p(n)e^{-j(2\pi/N)nk} \tag{6.43b}$$

Then, similar to the property of Fourier transforms, the DFT of the convolution of two *periodic* sequences is the product of their DFTs, i.e. DFT $\left[y_p(n) = x_p(n) * h_p(n)\right]$ is

$$Y_p(k) = X_p(k)H_p(k) \tag{6.44}$$

The proof of this is given below:

$$Y_p(k) = \text{DFT}\left[x_p(n) * h_p(n)\right] = \text{DFT}\left[\sum_{r=0}^{N-1} x_p(r)h_p(n-r)\right]$$

$$= \sum_{n=0}^{N-1}\sum_{r=0}^{N-1} x_p(r)h_p(n-r)e^{-j(2\pi/N)nk}$$

$$= \sum_{r=0}^{N-1} x_p(r)\sum_{n=0}^{N-1} h_p(n-r)e^{-j(2\pi/N)(n-r)k}e^{-j(2\pi/N)rk}$$

$$= \sum_{r=0}^{N-1} x_p(r)e^{-j(2\pi/N)rk} \cdot \sum_{n=0}^{N-1} h_p(n-r)e^{-j(2\pi/N)(n-r)k} = X_p(k) \cdot H_p(k) \tag{6.45}$$

This is important – so we consider its interpretation carefully. $y_p(n)$ is called a *circular convolution*, or sometimes a *periodic convolution*. Let us look at the result of convolving two periodic sequences in Figure 6.17.

Now, from $y_p(n) = x_p(n) * h_p(n) = \sum_{r=0}^{N-1} x_p(r)h_p(n-r)$, we draw the sequences in question as functions of $r$. To draw $h_p(n-r)$, we first draw $h_p(-r)$, i.e. we 'reverse' the sequence $h_p(r)$ and then move it $n$ places to the *right*. For example, $h_p(0-r)$, $h_p(2-r)$ and $x_p(r)$ are as shown in Figure 6.18.

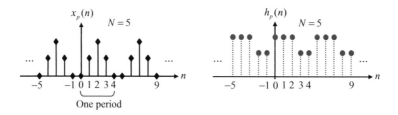

**Figure 6.17**   Two periodic sequences $x_p(n)$ and $h_p(n)$

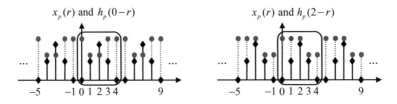

$x_p(r)$ and $h_p(0-r)$          $x_p(r)$ and $h_p(2-r)$

**Figure 6.18**  Illustration of the circular convolution process

As $n$ varies, $h_p(n-r)$ slides over $x_p(r)$ and it may be seen that the result of the convolution is the same for $n = 0$ as it is for $n = N$ and so on, i.e. $y_p(r)$ is periodic – hence the term *circular* or *periodic* convolution. The resulting convolution is shown in Figure 6.19.

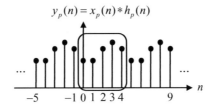

$$y_p(n) = x_p(n) * h_p(n)$$

**Figure 6.19**  Resulting sequence of the convolution of $x_p(n)$ and $h_p(n)$

Often the symbol ⊛ is used to denote circular convolution to distinguish it from linear convolution. Let us consider another simple example of circular convolution. Suppose we have two finite sequences $x(n) = [1, \ 3, \ 4]$ and $h(n) = [1, \ 2, \ 3]$. Then the values of the circular convolution $y(n) = x(n) \circledast h(n)$ are

$$y(0) = \sum_{r=0}^{2} x(r)h(0-r) = 18, \quad \text{where } h(0-r) = [1, \ 3, \ 2]$$

$$y(1) = \sum_{r=0}^{2} x(r)h(1-r) = 17, \quad \text{where } h(1-r) = [2, \ 1, \ 3] \tag{6.46}$$

$$y(2) = \sum_{r=0}^{2} x(r)h(2-r) = 13, \quad \text{where } h(2-r) = [3, \ 2, \ 1]$$

Note that $y(3) = y(0)$ and $h(3-r) = h(0-r)$ if they are to be evaluated.

If we are working with finite duration sequences, say $x(n)$ and $h(n)$, and then take DFTs of these, there are then 'equivalent' periodic sequences with the same DFTs, i.e. $X_p(k) = X(k)$ and $H_p(k) = H(k)$. If we form the inverse DFT (IDFT) of the product of these, i.e. IDFT $\left[H_p(k)X_p(k)\right]$ or IDFT $[H(k)X(k)]$, then the result will be circular convolution of the two finite sequences:

$$\boxed{x(n) \circledast h(n) = \text{IDFT}\,[X(k)H(k)]} \tag{6.47}$$

Sometimes, we may wish to form the linear convolution of the two sequences as discussed in Section 6.1. Consider two finite sequences $x(n)$ and $h(n)$, where $n = 0, 1, \ldots, N-1$ as shown in Figures 6.20(a) and (b). Note that these are the same sequences as in Figure 6.17, but their lengths are now only one period. The linear convolution of these two sequences, $y(n) = x(n) * h(n)$, results in a sequence with nine points as shown in Figure 6.20(c).

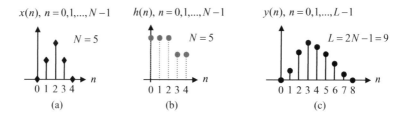

**Figure 6.20** Linear convolution of two finite sequences

The question is: can we do it using DFTs? (We might wish to do this because the FFT offers a procedure that could be quicker than direct convolution.)

We can do this using DFTs once we recognize that the $y(n)$ may be regarded as one period of a periodic sequence of period 9. To get this periodic sequence we add *zeros* to $x(n)$ and $h(n)$ to make $x(n)$ and $h(n)$ of length 9 (as shown in Figures 6.21(a) and (b)), and form the nine-point DFT of each. Then we take the IDFT of the product to get the required convolution, i.e. $x(n) \circledast h(n) = \text{IDFT}[X(k)H(k)]$. The result of this approach is shown in Figure 6.21(c) which is the same as Figure 6.20(c).

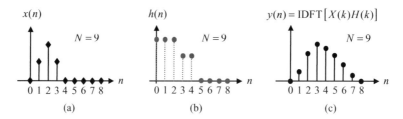

**Figure 6.21** Linear convolution of two finite sequences using the DFT

More generally, suppose we wish to convolve two sequences $x(n)$ and $h(n)$ of length $N_1$ and $N_2$, respectively. The linear convolution of these two sequences is a sequence $y(n)$ of length $N_1 + N_2 - 1$. To obtain this sequence from a circular convolution we require $x(n)$ and $h(n)$ to be sequences of $N_1 + N_2 - 1$ points, which is achieved by simply adding zeros to $x(n)$ and $h(n)$ appropriately. Then we take the DFTs of these augmented sequences, multiply them together and take the IDFT of the product. A single period of the resulting sequence is the required convolution. (The extra zeros on $x(n)$ and $h(n)$ eliminate the 'wrap-around' effect.) This process is called *fast convolution*. Note that the number of zeros added must ensure that $x(n)$ and $h(n)$ are of length greater than or equal to $N_1 + N_2 - 1$ and both the same length.

## 6.6 THE FAST FOURIER TRANSFORM

A set of algorithms known as the fast Fourier transform (FFT) has been developed to reduce the computation time required to evaluate the DFT coefficients. The FFT algorithm was rediscovered by Cooley and Tukey (1965) – the same algorithm had been used by the German mathematician Karl Friedrich Gauss around 1805 to interpolate the trajectories of asteroids. Owing to the high computational efficiency of the FFT, so-called *real-time* signal processing became possible. This section briefly introduces the basic 'decimation in time' method for a radix 2 FFT. For more details of FFT algorithms, see various references (Oppenheim and Schafer, 1975; Rabiner and Gold, 1975; Duhamel and Vetterli, 1990).

### *The Radix 2 FFT*

Since the DFT of a sequence is defined by $X(k) = \sum_{n=0}^{N-1} x(n)e^{-j(2\pi/N)nk}$, $k = 0, 1, \ldots,$ $N - 1$, by defining $W_N = e^{-j(2\pi/N)}$ the DFT can be rewritten as

$$X(k) = \sum_{n=0}^{N-1} x(n)W_N^{nk} \tag{6.48}$$

It is this expression that we shall consider. Note that $W_N^{nk}$ is periodic with period $N$ (in both $k$ and $n$), and the subscript $N$ denotes the periodicity. The number of multiply and add operations to calculate the DFT directly is approximately $N^2$, so we need more efficient algorithms to accomplish this. The FFT algorithms use the periodicity and symmetry property of $W_N^{nk}$, and reduce the number of operations $N^2$ to approximately $N \log_2 N$ (e.g. if $N = 1024$ the number of operations is reduced by a factor of about 100).

In particular, we shall consider the case of $N$ to be the power of two, i.e. $N = 2^\nu$. This leads to the base 2 or radix 2 algorithm. The basic principle of the algorithm is that of decomposing the computation of a DFT of length $N$ into successively smaller DFTs. This may be done in many ways, but we shall look at the *decimation in time* (DIT) method.

The name indicates that the sequence $x(n)$ is successively decomposed into smaller subsequences. We take a general sequence $x(n)$ and define $x_1(n)$, $x_2(n)$ as sequences with half the number of points and with

$$x_1(n) = x(2n), \quad n = 0, 1, \ldots, \frac{N}{2} - 1, \quad \text{i.e. even number of } x(n) \tag{6.49a}$$

$$x_2(n) = x(2n + 1), \quad n = 0, 1, \ldots, \frac{N}{2} - 1, \quad \text{i.e. odd number of } x(n) \tag{6.49b}$$

Then

$$X(k) = \sum_{n=0}^{N-1} x(n)W_N^{nk} = \sum_{\substack{n=0 \\ (\text{even})}}^{N-1} x(n)W_N^{nk} + \sum_{\substack{n=1 \\ (\text{odd})}}^{N-1} x(n)W_N^{nk}$$

$$= \sum_{n=0}^{N/2-1} x(2n)W_N^{2nk} + \sum_{n=0}^{N/2-1} x(2n + 1)W_N^{(2n+1)k} \tag{6.50}$$

Noting that $W_N^2 = [e^{-j(2\pi/N)}]^2 = e^{-j[2\pi/(N/2)]} = W_{N/2}$, Equation (6.50) can be written as

$$X(k) = \sum_{n=0}^{N/2-1} x_1(n)W_{N/2}^{nk} + W_N^k \sum_{n=0}^{N/2-1} x_2(n)W_{N/2}^{nk} \tag{6.51}$$

i.e.

$$X(k) = X_1(k) + W_N^k X_2(k) \tag{6.52}$$

where $X_1(k)$ and $X_2(k)$ are $N/2$-point DFTs of $x_1(n)$ and $x_2(n)$. Note that, since $X(k)$ is defined for $0 \le k \le N - 1$ and $X_1(k)$, $X_2(k)$ are periodic with period $N/2$, then

$$X(k) = X_1 \left( k - \frac{N}{2} \right) + W_N^k X_2 \left( k - \frac{N}{2} \right) \quad \frac{N}{2} \le k \le N - 1 \tag{6.53}$$

The above Equations (6.52) and (6.53) can be used to develop the computational procedure. For example, if $N = 8$ it can be shown that two four-pont DFTs are needed to make up the full eight-point DFT. Now we do the same to the four-point DFT, i.e. divide $x_1(n)$ and $x_2(n)$ each into two sequences of even and odd numbers, e.g.

$$X_1(k) = A(k) + W_{N/2}^k B(k) = A(k) + W_N^{2k} B(k) \quad \text{for } 0 \le k \le \frac{N}{2} - 1 \tag{6.54}$$

where $A(k)$ is a two-point DFT of even numbers of $x_1(n)$, and $B(k)$ is a two-point DFT of odd numbers of $x_1(n)$. This results in four two-pont DFTs in total. Thus, finally, we only need to compute two-point DFTs.

In general, the total number of multiply and add operations is $N \log_2 N$. Finally, we compare the number of operations $N^2$(DFT) versus $N \log_2 N$ (FFT) in Table 6.1.

**Table 6.1**  Number of multiply and add operations, FFT versus DFT

| $N$ | $N^2$ (DFT) | $N \log_2 N$ (FFT) | $N^2/(N \log_2 N)$ |
|---|---|---|---|
| 16 | 256 | 64 | 4.0 |
| 512 | 262 144 | 4608 | 56.9 |
| 2048 | 4 194 304 | 22 528 | 186.2 |

## 6.7  BRIEF SUMMARY

1. The input–output relationship of a digital LTI system is expressed by the convolution of two sequences of $h(n)$ and $x(n)$, i.e.

$$y(n) = \sum_{k=-\infty}^{\infty} h(n - k)x(k) = \sum_{r=-\infty}^{\infty} h(r)x(n - r) \quad \text{or}$$

$$y(n) = x(n) * h(n) = h(n) * x(n)$$

The Fourier transform of the sequence $h(n)$, $H(e^{j2\pi f})$, is called the system frequency response function (FRF), where

$$H(e^{j2\pi f}) = \sum_{n=-\infty}^{\infty} h(n)e^{-j2\pi f n} \quad \text{and} \quad h(n) = \int_{-1/2}^{1/2} H(e^{j2\pi f})e^{j2\pi f n} df$$

Note that $H(e^{j2\pi f})$ is continuous and periodic in frequency.

2. The Fourier transform of the convolution of two sequences is the product of their transforms, i.e.

$$F\{y(n) = x(n) * h(n)\} = Y(e^{j2\pi f}) = X(e^{j2\pi f})H(e^{j2\pi f})$$

3. The DFT pair for a *finite* (or periodic) sequence is

$$x(n) = \frac{1}{N}\sum_{k=0}^{N-1} X(k)e^{j(2\pi/N)nk} \quad \text{and} \quad X(k) = \sum_{n=0}^{N-1} x(n)e^{-j(2\pi/N)nk}$$

Note that the $N$-point DFT of a finite length sequence $x(n)$ imposes a *periodic structure* on the sequence.

4. Frequency spacing in $X(k)$ can be increased by adding zeros to the end of sequence $x(n)$. However, care must be taken since this is not a 'true' improvement in resolution (ability to distinguish closely spaced frequency components).

5. The relationship between the DFT $X(k)$ and the Fourier transform of a (sampled) sequence $X(e^{j2\pi f\Delta})$ is

$$X(k) = \left[X(e^{j2\pi f\Delta}) \text{ evaluated at } f = \frac{k}{N\Delta}\text{Hz}\right]$$

Note that this sampling in frequency imposes the periodicity in the time domain (as does the sampling in the time domain which results in periodicity in the frequency domain).

6. If a signal is *sampled* and *truncated*, we must consider the windowing effect (distortion in the frequency domain) and the scaling factor as compared with the Fourier transform of the original signal.

7. Symmetry properties of the DFT are given by

$$X(N - k) = X^*(k)$$

8. The *circular convolution* of two finite sequences can be obtained by the inverse DFT of the product of their DFTs, i.e.

$$x(n) \circledast h(n) = \text{IDFT}[X(k)H(k)]$$

The linear convolution of these two sequences, $y(n) = x(n) * h(n)$, can also be obtained via the DFT by adding zeros to $x(n)$ and $h(n)$ appropriately.

9. The fast Fourier transform (FFT) is an efficient algorithm for the computation of the DFT (the same algorithm can be used to compute the inverse DFT). There are many FFT algorithms. There used to be a restriction of data length $N$ to be a power of two, but there are algorithms available that do not have this restriction these days (see FFTW, http://www.fftw.org).

10. Finally, we summarize the various Fourier transforms in Figure 6.22 (we follow the display method given by Randall, 1987) and the pictorial interpretation of the DFT of a sampled and truncated signal is given in Figure 6.23 (see Brigham, 1988).

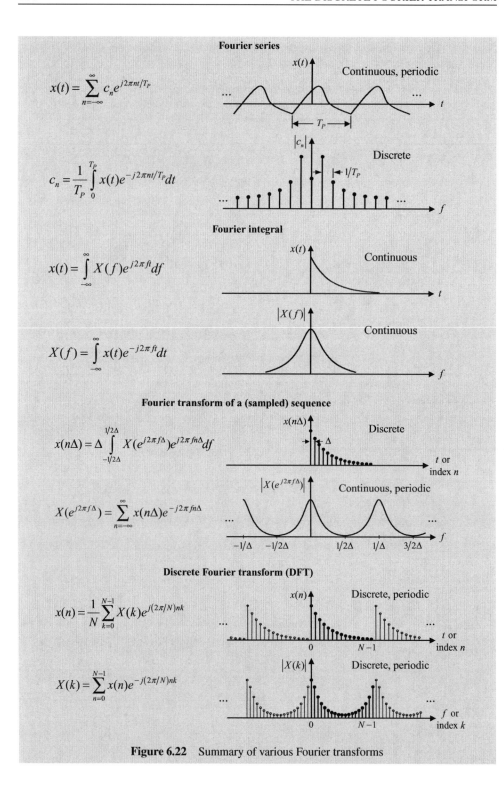

**Figure 6.22**  Summary of various Fourier transforms

**Fourier transform of original signal**

Data $x(t)$

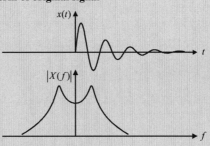

$X(f) = F\{x(t)\}$

**Fourier transform of truncated signal**

$x_T(t) = x(t)w(t)$

$w(t)$ is a data window

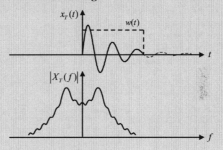

$X_T(f) = F\{x_T(t)\}$

**Fourier transform of sampled, truncated signal**

$x_T(t)$ is sampled
every $\Delta$ seconds

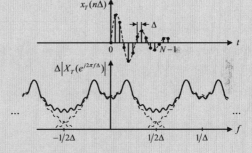

$X_T(e^{j2\pi f\Delta}) = F\{x_T(n\Delta)\}$

**DFT of sampled, truncated signal**

DFT imposes periodicity
in the time domain

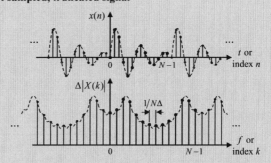

$X(k) = \text{DFT}[x(n)]$,

$X_T(e^{j2\pi f\Delta})$ is sampled
every $1/N\Delta\,\text{Hz}$

**Figure 6.23**  Pictorial interpretations (from the Fourier integral to the DFT)

## 6.8 MATLAB EXAMPLES

*Example 6.1:* **Example of convolution** (see Figure 6.6)

In this example, we demonstrate the convolution sum, $y(n) = x(n) * h(n)$, and its commutative property, i.e. $x(n) * h(n) = h(n) * x(n)$.

| Line | MATLAB code | Comments |
|---|---|---|
| 1<br>2<br>3<br>4<br>5 | clear all<br>x=[1 1 1 1 1 0 0 0 0];<br>h=[8 7 6 5 4 3 2 1 0 0 0];<br>nx=[0:length(x)-1];<br>nh=[0:length(h)-1]; | Define a sequence $x(n)$ whose total length is 9, but the length of non-zero elements is 5. Also define a sequence $h(n)$ whose total length is 11, but the length of non-zero elements is 8. And define indices for $x(n)$ and $h(n)$.<br>Note that MATLAB uses the index from 1, whereas we define the sequence from $n = 0$. |
| 6<br>7<br>8 | y1=conv(h,x);<br>y2=conv(x,h);<br>ny=[0:length(y1)-1]; | Perform the convolution sum using the MATLAB function 'conv', where $y_1(n) = h(n) * x(n)$ and $y_2(n) = x(n) * h(n)$.<br>Both will give the same results. Note that the length of 'conv(h,x)' is 'length(h) + length(x) −1'. And define the index for both $y_1(n)$ and $y_2(n)$. |
| 9<br>10<br>11<br>12<br>13<br>14<br>15<br>16 | figure(1); stem(nx,x, 'd', 'filled')<br>xlabel('\itn'); ylabel('\itx\rm(\itn\rm)')<br>figure(2); stem(nh,h, 'filled')<br>xlabel('\itn'); ylabel('\ith\rm(\itn\rm)')<br>figure(3); stem(ny,y1, 'filled')<br>xlabel('\itn'); ylabel('\ity_1\rm(\itn\rm)')<br>figure(4); stem(ny,y2, 'filled')<br>xlabel('\itn'); ylabel('\ity_2\rm(\itn\rm)') | Plot the sequences $x(n)$, $h(n)$, $y_1(n)$ and $y_2(n)$.<br>Note that $y_1(n)$ and $y_2(n)$ are the same, the total length of $y_1(n)$ is 19, which is '11 + 9 −1', and the length of the non-zero elements is 12, which is '8+ 5 − 1'. |

**Results**

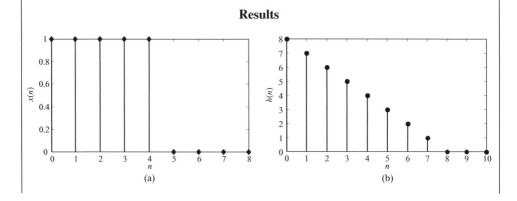

(a)                    (b)

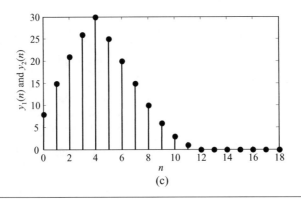

(c)

---

***Example 6.2a:* DFT of a sinusoidal signal**

Case A: Truncated exact number of periods. (see Figures 6.12 and 6.13).

Consider a sinusoidal signal $x(t) = A \sin 2\pi pt$, $p = 1/T_P$ Hz. Sample this signal at the sampling rate $f_s = 10/T_P$ Hz. We examine two cases: (i) data are truncated at exactly one period (10-point DFT), (ii) data are truncated at exactly five periods (50-point DFT). For this example, we use $A = 2$ and $p = 1$ Hz. Note that the Fourier integral gives the value $A/2 = 1$ at $p$ Hz.

| Line | MATLAB code | Comments |
|---|---|---|
| 1 | clear all | Define parameters and the sampling rate fs |
| 2 | A=2; p=1; Tp=1/p; fs=10/Tp; | such that 10 samples per period Tp. Truncate |
| 3 | T1=1*Tp; T2=5*Tp; | the data exactly one period (T1) and five |
| 4 | t1=[0:1/fs:T1-1/fs]; | periods (T2). Define time variables t1 and t2 |
| 5 | t2=[0:1/fs:T2-1/fs]; | for each case. |
| 6 | x1=A*cos(2*pi*p*t1); | Generate the sampled and truncated signals x1 |
| 7 | x2=A*cos(2*pi*p*t2); | (one period) and x2 (five periods). Perform the |
| 8 | X1=fft(x1); X2=fft(x2); | DFT of each signal. |
| 9 | N1=length(x1); N2=length(x2); | Calculate the frequency variables f1 and f2 for |
| 10 | f1=fs*(0:N1-1)/N1;<br>f2=fs*(0:N2-1)/N2; | each case. |
| 11 | figure(1) | Plot the results (modulus) of 10-point DFT. |
| 12 | stem(f1, abs(X1), 'fill') | Note the frequency range 0 to 9 Hz |
| 13 | xlabel('Frequency (Hz)') | ($f_s - f_s/N$) and the peak amplitude |
| 14 | ylabel('Modulus of \itX\rm(\itk\rm)');<br>axis ([0 9.9 0 10]) | $AT/2\Delta = 5A = 10$ (see Figure 6.12). Since exact number of period is taken for DFT, all the frequency components except $p = 1$ Hz (and 9 Hz, which is the mirror image of $p$ Hz) are zero. |

| 15 | figure(2) | This plots the same results, but now the |
| 16 | stem(f1, abs(X1)/fs/T1, 'fill'); % this is the | DFT coefficients are scaled appropriately. |
|    | same as stem(f1, abs(X1)/N1, 'fill') | Note that the modulus of $X(k)$ is divided |
| 17 | xlabel('Frequency (Hz)') | by the sampling rate (fs) and window |
| 18 | ylabel('Modulus (scaled)'); | length (T1). Note that it also gives |
|    | axis ([0 9.9 0 1]) | the same scaling effect if $X(k)$ is divided |

This plots the same results, but now the DFT coefficients are scaled appropriately. Note that the modulus of $X(k)$ is divided by the sampling rate (fs) and window length (T1). Note that it also gives the same scaling effect if $X(k)$ is divided by the number of points N1. The result corresponds to the Fourier integral, i.e. the peak amplitude is now $A/2 = 1$ at $p$ Hz.

| 19 | figure(3) |
| 20 | stem(f2, abs(X2), 'fill') |
| 21 | xlabel('Frequency (Hz)') |
| 22 | ylabel('Modulus of \itX\rm(\itk\rm)') |

Plot the results (modulus) of 50-point DFT. Note that the peak amplitude is $AT/2\Delta = 25A = 50$. In this case, we used the data five times longer in 'time' than in the previous case. This results in an increase of frequency resolution, i.e. the resolution is increased five times that in the previous case.

| 23 | figure(4) |
| 24 | stem(f2, abs(X2)/fs/T2, 'fill'); % this is the |
|    | same as stem(f2, abs(X2)/N2, 'fill') |
| 25 | xlabel('Frequency (Hz)'); |
|    | ylabel('Modulus (scaled)') |

This plots the same results, but, as before, the DFT coefficients are scaled appropriately, thus $A/2 = 1$ at $p$ Hz.

## Results

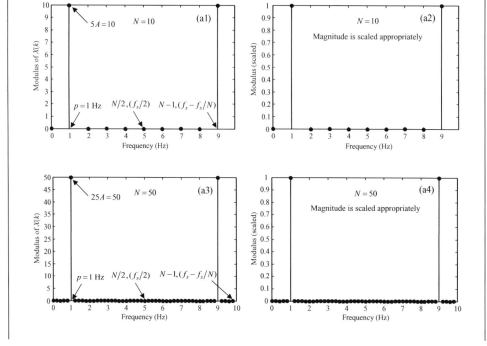

**Comment:** In this example, we applied the scaling factor $1/(f_s T) = 1/N$ to $X(k)$ to relate its amplitude to the corresponding Fourier integral $X(f)$. However, this is only true for periodic signals which have discrete spectra in the frequency domain. In fact, using the DFT, we have computed the *Fourier coefficients* (amplitudes of specific frequency components) for a periodic signal, i.e. $c_k \approx X_k/N$ (see Equation (3.45) in Chapter 3).

For transient signals, since we compute the amplitude *density* rather than amplitude at a specific frequency, the correct scaling factor is $1/f_s$ or $\Delta$ (assuming that the rectangular window is used), although there is some distortion in the frequency domain due to the windowing effect. The only exception of this scaling factor may be the delta function. Note that $\delta(n)$ is *not* the result of sampling the Dirac delta function $\delta(t)$ which is a mathematical idealization.

---

*Example 6.2b:* **DFT of a sinusoidal signal**

Case B: Truncated with a non-integer number of periods. (See also the windowing effect in Sections 4.11 and 3.6.)

We use the same signal as in MATLAB Example 6.2a, i.e. $x(t) = A \sin 2\pi pt$, $p = 1/T_P$ Hz, $f_s = 10/T_P$ Hz, $A = 2$, and $p = 1$ Hz. However, we truncate the data in two cases: (i) data are truncated one and a half periods (15-point DFT), (ii) data are truncated three and a half periods (35-point DFT). Note that we use an odd number for the DFT.

| Line | MATLAB code | Comments |
|---|---|---|
| 1 | clear all | Exactly the same as previous example |
| 2 | A=2; p=1; Tp=1/p; fs=10/Tp; | (Case A), except T1 is one and a half |
| 3 | T1=1.5*Tp; T2=3.5*Tp; | periods of the signal and T2 is three and |
| 4 | t1=[0:1/fs:T1-1/fs];<br>t2=[0:1/fs: T2-1/fs]; | a half periods of the signal. |
| 5 | x1=A*cos(2*pi*p*t1); x2=A*cos(2*pi*p*t2); | |
| 6 | X1=fft(x1); X2=fft(x2); | |
| 7 | N1=length(x1); N2=length(x2); | |
| 8 | f1=fs*(0:N1-1)/N1;<br>f2=fs*(0:N2-1)/N2; | |
| 9 | X1z=fft([x1 zeros(1,5000-N1)]); % zero padding | Perform 5000-point DFT by adding |
| 10 | X2z=fft([x2 zeros(1,5000-N2)]); % zero padding | zeros at the end of each sequence x1 |
| 11 | Nz=length(X1z); | and x2, i.e. 'zero padding' is applied for |
| 12 | fz=fs*(0:Nz-1)/Nz; | demonstration purpose. Calculate new frequency variable accordingly. |
| 13 | figure(1) | Plot the results (modulus) of 15-point |
| 14 | stem(f1, abs(X1)/fs/T1, 'fill');<br>hold on | DFT (stem plot) and DFT with zero padding (dashed line). Magnitudes of |
| 15 | plot(fz, abs(X1z)/fs/T1, 'r:'); hold off | DFT coefficients are scaled |
| 16 | xlabel('Frequency (Hz)');<br>ylabel('Modulus (scaled)') | appropriately. Examine the effect of windowing in this figure. Note the |
| 17 | axis([0 10 0 1.02]) | change of magnitude at the peak (compare this with the previous example). Also, note that we do not have the value at the frequency $p = 1$ Hz. |

| Line | MATLAB code | Comments |
|---|---|---|
| 18 | figure(2) | Plot the results (modulus) of 35-point |
| 19 | stem(f2, abs(X2)/fs/T2, 'fill'); hold on | DFT (stem plot) and DFT with zero |
| 20 | plot(fz, abs(X2z)/fs/T2, 'r:'); hold off | padding (dashed line). Note that the |
| 21 | xlabel('Frequency (Hz)'); | resolution is improved, but there is still |
|  | ylabel('Modulus (scaled)') | a significant amount of smearing and |
| 22 | axis([0 10 0 1.02]) | leakage due to windowing. Again, we |
|  |  | do not have the DFT coefficient at the |
|  |  | frequency $p = 1$ Hz. |

## Results

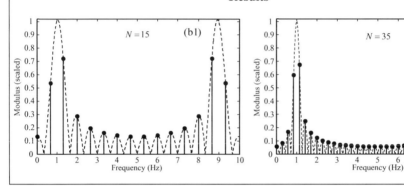

*Example 6.3:* **DFT of a sinusoidal signal**

Increase of sampling rate does not improve the frequency resolution; it only increases the frequency range to be computed (with a possible benefit of avoiding aliasing, see aliasing in Chapter 5).

We use the same signal as in the previous MATLAB example, i.e. $x(t) = A \sin 2\pi pt$, $p = 1/T_P$ Hz, $A = 2$ and $p = 1$ Hz. However, the sampling rate is increased twice, i.e. $f_s = 20/T_P$ Hz. We examine two cases: (a) data length $T = T_P$ (20-point DFT; this corresponds to the first case of MATLAB Example 6.2a), (b) data length $T = 1.5T_P$ (30-point DFT; this corresponds to the first case of MATLAB Example 6.2b).

| Line | MATLAB code | Comments |
|---|---|---|
| 1 | clear all | Exactly the same as previous examples |
| 2 | A=2; p=1; Tp=1/p; fs=20/Tp; | (MATLAB Examples 6.2a and 6.2b), |
| 3 | T1=1*Tp; T2=1.5*Tp; | except that the sampling rate fs is now |
| 4 | t1=[0:1/fs:T1-1/fs]; t2=[0:1/fs:T2-1/fs]; | doubled. |
| 5 | x1=A*cos(2*pi*p*t1); | |
|  | x2=A*cos(2*pi*p*t2); | |
| 6 | X1=fft(x1); X2=fft(x2); | |
| 7 | N1=length(x1); N2=length(x2); | |
| 8 | f1=fs*(0:N1-1)/N1; f2=fs*(0:N2-1)/N2; | |

| 9 | figure(1) | Plot the results (modulus) of 20-point DFT (i.e. for the case of $T = T_P$). Note that the frequency spacing is 1 Hz which is exactly the same as MATLAB Example 6.2a (when $N = 10$), and the folding frequency is now 10 Hz (5 Hz in the previous example). |
| 10 | stem(f1, abs(X1)/fs/T1, 'fill') | |
| 11 | xlabel('Frequency (Hz)'); ylabel('Modulus (scaled)') | |
| 12 | axis([0 20 0 1]) | |
| 13 | figure(2) | Plot the results (modulus) of 30-point DFT (i.e. for the case of $T = 1.5T_P$). Again, the result is the same as MATLAB Example 6.2b (when $N = 15$), within the frequency range 0 to 5 Hz. |
| 14 | stem(f2, abs(X2)/fs/T2, 'fill') | |
| 15 | xlabel('Frequency (Hz)'); ylabel('Modulus (scaled)') | |
| 16 | axis([0 20 0 1]) | |

## Results

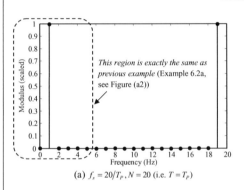

(a) $f_s = 20/T_P$, $N = 20$ (i.e. $T = T_P$)

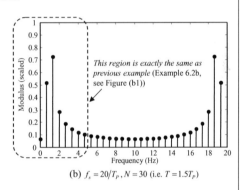

(b) $f_s = 20/T_P$, $N = 30$ (i.e. $T = 1.5T_P$)

**Comments:** Compare these results with the previous examples (MATLAB Example 6.2a, 6.2b). Recall that the only way of increasing frequency resolution is by increasing data length (in time). Note that, since the sampling rate is doubled, double the amount of data is needed over the previous example in order to get the same frequency resolution.

---

*Example 6.4:* **Symmetry properties of DFT** (see Section 6.4)

Consider a discrete sequence $x(n) = a^n u(n)$, $0 < a < 1$, $n = 0, 1, \ldots, N - 1$. In this example, we use $a = 0.3$ and examine the symmetry properties of the DFT for two cases: (a) $N$ is an odd number ($N = 9$), and (b) $N$ is an even number ($N = 10$).

| Line | MATLAB code | Comments |
|---|---|---|
| 1 | clear all | Define the parameter $a$, and variables n1 (for the odd-numbered sequence) and n2 (for the even-numbered sequence). |
| 2 | a=0.3; | |
| 3 | n1=0:8; % 9-point sequence | |
| 4 | n2=0:9; % 10-point sequence | |

| | | |
|---|---|---|
| 5 | `x1=a.^n1; x2=a.^n2;` | Create two sequences x1 and x2 according to the above equation, i.e. $x(n) = a^n u(n)$. |
| 6 | `X1=fft(x1); X2=fft(x2);` | Perform the DFT of each sequence, i.e. $X(k) = \text{DFT}[x(n)]$. |
| 7 | `figure(1)` | Plot the real part of the DFT of the first |
| 8 | `subplot(2,2,1);` | sequence x1. The MATLAB command |
| | `stem(n1, real(X1), 'fill')` | 'subplot(2,2,1)' divides the figure(1) into four |
| 9 | `axis([-0.5 8.5 0 1.6])` | sections (2×2) and allocates the subsequent |
| 10 | `xlabel('\itk');` | graph to the first section. |
| | `ylabel('Re[\itX\rm(\itk\rm)]')` | |
| 11 | `subplot(2,2,2);` | Plot the imaginary part of the DFT of the first |
| | `stem(n1, imag(X1), 'fill')` | sequence x1. Note that, since $N/2$ is not an |
| 12 | `axis([-0.5 8.5 -0.4 0.4])` | integer number, we cannot evaluate the DFT |
| 13 | `xlabel('\itk');` | coefficient for this number. Thus, the |
| | `ylabel('Im[\itX\rm(\itk\rm)]')` | zero-crossing point cannot be shown in the figure. |
| 14 | `subplot(2,2,3);` | Plot the modulus of the DFT of the first |
| | `stem(n1, abs(X1), 'fill')` | sequence x1. |
| 15 | `axis([-0.5 8.5 0 1.6])` | |
| 16 | `xlabel('\itk');` | |
| | `ylabel('|\itX\rm(\itk\rm)|')` | |
| 17 | `subplot(2,2,4);` | Plot the phase of the DFT of the first sequence |
| | `stem(n1, angle(X1), 'fill')` | x1. Similar to the imaginary part of the DFT, |
| 18 | `axis([-0.5 8.5 -0.4 0.4])` | there is no zero-crossing point (or $\pi$) in the |
| 19 | `xlabel('\itk');` | figure. |
| | `ylabel('arg\itX\rm(\itk\rm)')` | |
| 20 | `figure(2)` | Plot the real part of the DFT of the second |
| 21 | `subplot(2,2,1);` | sequence x2. |
| | `stem(n2, real(X2), 'fill')` | |
| 22 | `axis([-0.5 9.5 0 1.6])` | |
| 23 | `xlabel('\itk');` | |
| | `ylabel('Re[\itX\rm(\itk\rm)]')` | |
| 24 | `subplot(2,2,2);` | Plot the imaginary part of the DFT of the |
| | `stem(n2, imag(X2), 'fill')` | second sequence x2. Since $N/2$ is an integer |
| 25 | `axis([-0.5 9.5 -0.4 0.4])` | number, we can evaluate the DFT coefficient |
| 26 | `xlabel('\itk');` | for this number. Note that the value is zero at |
| | `ylabel('Im[\itX\rm(\itk\rm)]')` | $n = N/2$. |
| 27 | `subplot(2,2,3);` | |
| | `stem(n2, abs(X2), 'fill')` | Plot the modulus of the DFT of the second |
| 28 | `axis([-0.5 9.5 0 1.6])` | sequence x2. |
| 29 | `xlabel('\itk');` | |
| | `ylabel('|\itX\rm(\itk\rm)|')` | |
| 30 | `subplot(2,2,4);` | Plot the phase of the DFT of the second |
| | `stem(n2, angle(X2), 'fill')` | sequence x2. Similar to the imaginary part of |
| 31 | `axis([-0.5 9.5 -0.4 0.4])` | the DFT, there is a zero-crossing point at |
| 32 | `xlabel('\itk');` | $n = N/2$. (The value is zero because the real |
| | `ylabel('arg\itX\rm(\itk\rm)')` | part is positive. If the real part is negative the |
| | | value will be $\pi$.) |

**Results**

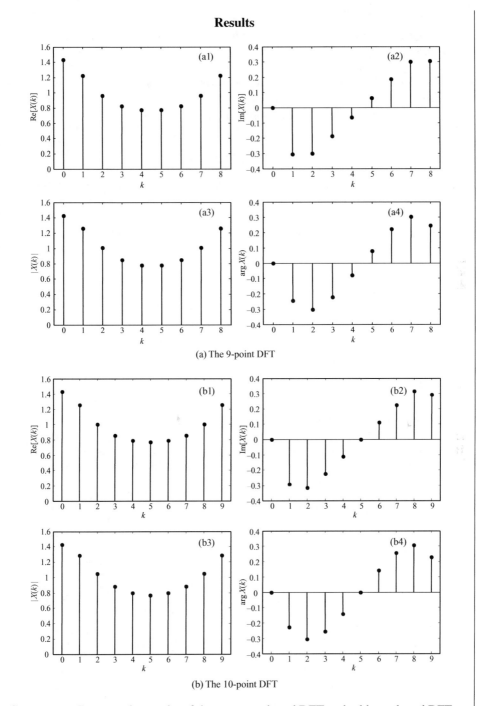

(a) The 9-point DFT

(b) The 10-point DFT

**Comments:** Compare the results of the even-numbered DFT and odd-numbered DFT.

***Example 6.5:* Zero-padding approach to improve (*artificially*) the quality of a measured signal**

Consider the free response of a single-degree-of-freedom system

$$x(t) = \frac{A}{\omega_d} e^{-\zeta \omega_n t} \sin(\omega_d t) \quad \text{and} \quad F\{x(t)\} = \frac{A}{\omega_n^2 - \omega^2 + j2\zeta\omega_n\omega}$$

where $A = 200$, $\omega_n = 2\pi f_n = 2\pi(10)$ and $\omega_d = \omega_n\sqrt{1 - \zeta^2}$. In order to simulate a practical situation, a small amount of noise (Gaussian white) is added to the signal. Suppose the system is heavily damped, e.g. $\zeta = 0.3$; then the signal $x(t)$ falls into the noise level quickly.

Now, there are two possibilities of performing the DFT. One is to use only the beginning of the signal where the signal-to-noise ratio is high, but this will give a poor frequency resolution. The other is to use longer data (including the noise-dominated part) to improve the frequency resolution. However, it is significantly affected by noise in the frequency domain.

The above problem may be resolved by truncating the beginning part of signal and adding zeros to it (this increases the measurement time *artificially*).

| Line | MATLAB code | Comments |
|---|---|---|
| 1 | clear all | Define the sampling rate fs = 100 Hz, |
| 2 | fs–100; T–5; | total record time T = 5 seconds, and the |
| 3 | t=[0:1/fs:T-1/fs]; | time variable t from 0 to 'T-1/fs' seconds. |
| 4 | A=200; zeta=0.3; wn=2*pi*10; wd=sqrt(1-zeta^2)*wn; | Also generate the sampled signal according to the equation above. |
| 5 | x=(A/wd)*exp(-zeta*wn*t).*sin(wd*t); | |
| 6 | var_x=sum((x-mean(x)).^2)/(length(x)-1); % var_x=var(x) | Calculate the variance of the signal (note that the MATLAB function 'var(x)' can also be used). |
| 7 | randn('state',0); | MATLAB function 'randn(size(x))' generates the normally distributed random numbers with the same size as x, and 'randn('state', 0)' initializes the random number generator. |
| 8 | noise=0.05*sqrt(var_x)*randn(size(x)); | Generate the noise sequence whose power |
| 9 | xn=x+noise; | is 0.25 % of the signal power that gives the SNR of approximately 26 dB (see Equation (5.30)). Then, add this noise to the original signal. |
| 10 | figure(1) | Plot the noisy signal. It can be easily |
| 11 | plot(t, xn) | observed that the signal falls into the noise |
| 12 | axis([0 2 -0.8 2.2]) | level at about 0.4 seconds. Note that 0.4 |
| 13 | xlabel('\itt\rm (seconds)'); ylabel('\itx\rm(\itt\rm)') | seconds corresponds to the 40 data points. Thus, for the DFT, we may use the signal up to 0.4 seconds (40-point DFT) at the expense of the frequency resolution, or use the whole *noisy* signal (500-point DFT) to improve the resolution. |

| 14 | Xn1=fft(xn,40); % 40 corresponds to 0.4  seconds in time | First, perform the DFT using only the first 40 data points of the signal. The MATLAB function 'fft(xn, 40)' performs the DFT of xn using the first 40 elements of xn. Next, perform the DFT using the whole noisy signal (500-point DFT). Calculate the corresponding frequency variables. |
|----|----|----|
| 15 | N1=length(Xn1); f1=fs*(0:N1-1)/N1; | |
| 16 | Xn2=fft(xn); | |
| 17 | N2=length(xn); f2=fs*(0:N2-1)/N2; | |
| | | |
| 18 | Xa=A./(wn^2 - (2*pi*f2).^2 + i*2*zeta*wn*(2*pi*f2)); | Calculate the Fourier integral according to the formula above. This will be used for the purpose of comparison. |
| | | |
| 19 | figure(2) | Plot the modulus of the 40-point DFT (solid line), and plot the true magnitude spectrum of the Fourier transform (dashed line). Note the poor frequency resolution in the case of the 40-point DFT. |
| 20 | plot(f1(1:N1/2+1), 20*log10(abs(Xn1(1:N1/2 +1)/fs))) | |
| 21 | hold on | |
| 22 | plot(f2(1:N2/2+1), 20*log10(abs(Xa(1:N2/2+1))), 'r:') | |
| 23 | xlabel('Frequency (Hz)'); ylabel('Modulus (dB)'); hold off | |
| | | |
| 24 | figure(3) | Plot the modulus of the DFT of the whole noisy signal (solid line), and plot the true magnitude spectrum of the Fourier transform (dashed line). Note the effect of noise in the frequency domain. |
| 25 | plot(f2(1:N2/2+1), 20*log10(abs(Xn2(1:N2/2+1)/fs))) | |
| 26 | hold on | |
| 27 | plot(f2(1:N2/2+1), 20*log10(abs(Xa(1:N2/2+1))), 'r:') | |
| 28 | xlabel('Frequency (Hz)'); ylabel('Modulus (dB)'); hold off | |
| | | |
| 29 | Xnz=fft(xn(1:40),N2); | Now, perform the DFT of the truncated and zero-padded signal. The MATLAB function 'fft(xn(1:40),N2)' takes only the first 40 data elements of xn, then adds zeros up to the number N2. |
| | | |
| 30 | figure(4) | Plot the modulus of the DFT of the zero-padded signal (solid line), and plot the true magnitude spectrum of the Fourier transform (dashed line). Note the improvement in the frequency domain. |
| 31 | plot(f2(1:N2/2+1), 20*log10(abs(Xnz(1:N2/2+1)/fs))) | |
| 32 | hold on | |
| 33 | plot(f2(1:N2/2+1), 20*log10(abs(Xa(1:N2/2+1))), 'r:') | |
| 34 | xlabel('Frequency (Hz)'); ylabel('Modulus (dB)'); hold off | |

**Results**

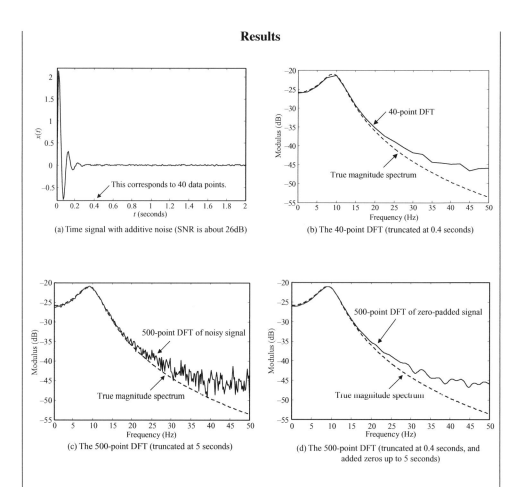

(a) Time signal with additive noise (SNR is about 26dB)

(b) The 40-point DFT (truncated at 0.4 seconds)

(c) The 500-point DFT (truncated at 5 seconds)

(d) The 500-point DFT (truncated at 0.4 seconds, and added zeros up to 5 seconds)

**Comments:** In this example, apart from the zero-padding feature, there is another aspect to consider. Consider the DFT of the noise-free signal (i.e. noise is not added), and compare it with the Fourier integral. To do this, add the following lines at the end of the above MATLAB code:

```
X=fft(x);
figure(5)
plot(f2(1:N2/2+1), 20*log10(abs(X(1:N2/2+1)/fs))); hold on
plot(f2(1:N2/2+1), 20*log10(abs(Xa(1:N2/2+1)))), 'r:')
xlabel('Frequency (Hz)'); ylabel('Modulus (dB)'); hold off
```

The results are shown in Figure (e). Note the occurrence of *aliasing* in the DFT result. In computer simulations, we have evaluated the values of $x(t)$ at $t = 0, 1/f_s, 2/f_s, \ldots, T - 1/f_s$ simply inserting the time variable in the equation *without* doing any preprocessing. In the MATLAB code, the act of defining the time variable 't=[0:1/fs:T-1/fs];' is the 'sampling' of the analogue signal $x(t)$. Since we cannot (in a simple way in computer programming) apply the low-passfilter before the sampling, we always have to face the aliasing problem in computer simulations. Note that aliasing does occur even if the signal is obtained by solving the corresponding ordinary differential equation using a numerical integration method such as the Runge–Kutta method. Thus, we may use a much higher sampling rate to minimize the aliasing problem, but we cannot avoid it completely.

Note also that aliasing occurs over the 'entire' frequency range, since the original analogue signal is not band-limited. It is also interesting to compare the effect of aliasing in the low-frequency region (compared with the natural frequency, $f_n = 10\,\text{Hz}$) and in the high-frequency region, i.e. the magnitude spectrum is increased at high frequencies, but decreased at low frequencies. This is due to the phase structure of the Fourier transform of the original signal, i.e. $\arg X(f)$. Note further that there is a phase shift at the natural frequency (see Fahy and Walker, 1998). Thus the phase difference between $X(f)$ and its mirror image is approximately $2\pi$ at the folding frequency and is approximately $\pi$ at zero frequency. In other words, $X(f)$ and the aliased part are in phase at high frequencies (increase the magnitude) and out of phase at low frequencies (decrease the magnitude), as can be seen from Figures (f) and (g).

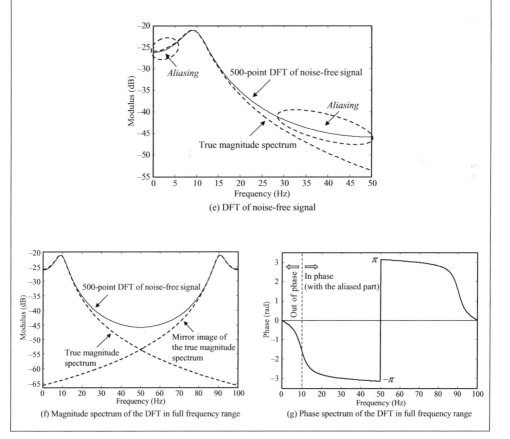

(e) DFT of noise-free signal

(f) Magnitude spectrum of the DFT in full frequency range

(g) Phase spectrum of the DFT in full frequency range

---

***Example 6.6:*** **Circular (periodic) and linear convolutions using the DFT**

Consider the following two finite sequences of length $N = 5$:

$$x(n) = [1\,3\,5\,3\,1] \quad \text{and} \quad h(n) = [9\,7\,5\,3\,1]$$

Perform the circular convolution and the linear convolution using the DFT.

| Line | MATLAB code | Comments |
|------|-------------|----------|
| 1 | clear all | Define the sequences $x(n)$ and $h(n)$. |
| 2 | x=[1 3 5 3 1]; h=[9 7 5 3 1]; | |
| 3 | X=fft(x); H=fft(h); | Perform the DFT of each sequence. Take the |
| 4 | yp=ifft(X.*H); | inverse DFT of the product $X(k)$ and $H(k)$ to |
| 5 | np=0:4; | obtain the circular convolution result. Define the variable for the x-axis. |
| 6 | figure(1) | Plot the sequences x and h, and the results of |
| 7 | subplot(3,1,1); stem(np, x, 'd', 'fill') | circular convolution. Note that the sequences |
| 8 | axis([-0.4 4.4 0 6]) | x and h are *periodic* in effect. |
| 9 | xlabel('\itn'); | |
|   | ylabel('\itx_p\rm(\itn\rm)') | |
| 10 | subplot(3,1,2); stem(np, h, 'fill') | |
| 11 | axis([-0.4 4.4 0 10]) | |
| 12 | xlabel('\itn'); | |
|   | ylabel('\ith_p\rm(\itn\rm)') | |
| 13 | subplot(3,1,3); stem(np, yp, 'fill') | |
| 14 | axis([-0.4 4.4 0 90]) | |
| 15 | xlabel('\itn'); | |
|   | ylabel('\ity_p\rm(\itn\rm)') | |
| 16 | Xz=fft([x zeros(1,length(h)-1)]); | Perform the linear convolution using the |
| 17 | Hz=fft([h zeros(1,length(x)-1)]); | DFT. Note that zeros are added appropriately |
| 18 | yz=ifft(Xz.*Hz); | when calculating DFT coefficients. |
| 19 | nz=0:8; | Also, note that the MATLAB function 'conv(x, h)' will give the same result (in fact, this function uses the same algorithm). |
| 20 | figure(2) | Plot the zero-padded sequences, and the |
| 21 | subplot(3,1,1); | results of linear convolution using the DFT. |
|   | stem(nz, [x 0 0 0 0], 'd', 'fill') | |
| 22 | axis([-0.4 8.4 0 6]) | |
| 23 | xlabel('\itn'); | |
|   | ylabel('\itx\rm(\itn\rm)') | |
| 24 | subplot(3,1,2); | |
|   | stem(nz, [h 0 0 0 0], 'fill') | |
| 25 | axis([-0.4 8.4 0 10]) | |
| 26 | xlabel('\itn'); | |
|   | ylabel('\ith\rm(\itn\rm)') | |
| 27 | subplot(3,1,3); stem(nz, yz, 'fill') | |
| 28 | axis([-0.4 8.4 0 90]) | |
| 29 | xlabel('\itn'); | |
|   | ylabel('\ity\rm(\itn\rm)') | |

**Results**

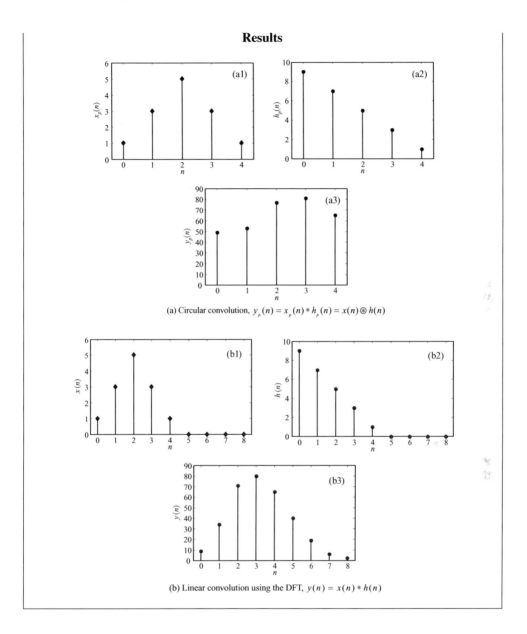

(a) Circular convolution, $y_p(n) = x_p(n) * h_p(n) = x(n) \circledast h(n)$

(b) Linear convolution using the DFT, $y(n) = x(n) * h(n)$

---

*Example 6.7:* **System identification (impact testing of a structure)**

Consider the experimental setup shown in Figure (a) (see also Figure 1.11 in Chapter 1), and suppose we want to identify the system (FRF between **A** and **B**) by the impact testing method. Note that many modern signal analysers are equipped with built-in signal conditioning modules.

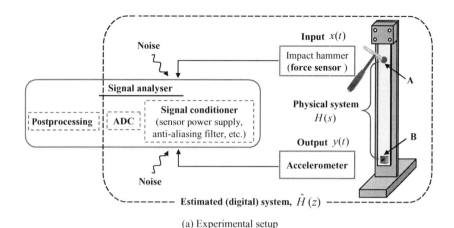

(a) Experimental setup

If the measurement noise is ignored, both input and output are deterministic and transient. Thus, provided that the input $x(t)$ is sufficiently narrow in time (broad in frequency), we can obtain the FRF between **A** and **B** over a desired frequency range from the relationship

$$Y(f) = H(f)X(f) \quad \rightarrow \quad H(f) = \frac{Y(f)}{X(f)} \tag{6.55}$$

However, as illustrated in Figure (a), the actual signals are contaminated with noise. Also, the system we are identifying is not the actual physical system $H$ (between **A** and **B**) but the $\hat{H}$ that includes the individual frequency responses of sensors and filters, the effects of quantization noise, measurement (external) noise and the experimental rig. Nevertheless, for convenience we shall use the notation $H$ rather than $\hat{H}$.

Measurement noise makes it difficult to use Equation (6.55). Thus, we usually perform the same experiment several times and average the results to estimate $H(f)$. The details of various estimation methods are discussed in Part II of this book. Roughly speaking, one estimation method of FRF may be expressed as

$$H_1(f) \approx \frac{\dfrac{1}{N}\displaystyle\sum_{n=1}^{N} X_n^*(f)Y_n(f)}{\dfrac{1}{N}\displaystyle\sum_{n=1}^{N} X_n^*(f)X_n(f)} \tag{6.56}$$

where $N$ is the number of times the experiment is replicated (equivalently it is the number of averages). Note that different values of $X_n(f)$ and $Y_n(f)$ are produced in each experiment, and if $N = 1$ Equations (6.55) and (6.56) are the same. In this MATLAB example, we shall estimate the FRF based on both Equations (6.55) and (6.56), and compare the results.

The experiment is performed 10 times, and the measured data are stored in the file 'impact_data_raw.mat',[1] where the sampling rate is chosen as $f_s = 256$ Hz, and each signal is recorded for 8 seconds (which results in the frequency resolution of $\Delta f = 1/8 = 0.125$ Hz, and each signal is 2048 elements long). The variables in the file are 'in1, in2, ..., in10' (input signals) and 'out1, out2, ..., out10' (output signals). The anti-aliasing filter is automatically controlled by the signal analyser according to the sampling rate (in this case, the cut-off frequency is about 100 Hz). Also, the signal analyser is configured to remove the d.c. component of the measured signal (i.e. high-pass filtering with cut-on at about 5 Hz).

Before performing the DFT of each signal, let us investigate the measured signals. If we type the following script in the MATLAB command window:

```
load impact_data_raw
fs=256; N=length(in1); f=fs*(0:N-1)/N;
T=N/fs; t=0:1/fs:T-1/fs;
figure(1); plot(t, in1); axis([-0.1 8 -1.5 2.5])
xlabel('\itt\rm (seconds)'); ylabel('\itx\rm(\itt\rm)')
figure(2); plot(t, out1); axis([-0.1 8 -4 4])
xlabel('\itt\rm (seconds)'); ylabel('\ity\rm(\itt\rm)')
```

The results will be as shown in Figure (b1) and (b2).

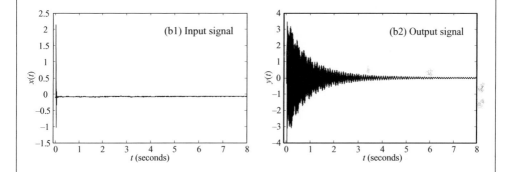

Note that the output signal is truncated before the signal dies away completely. However, the input signal dies away quickly and noise dominates later. If we type in the following script we can see the effect of noise on the input signal, i.e. the DFT of the input signal shows a noisy spectrum as in Figure (c):

```
In1=fft(in1);
figure (3); plot(f(1:N/2+1), 20*log10(abs(In1(1:N/2+1))))
xlabel('Frequency (Hz)'); ylabel('Modulus (dB)')
axis([0 128 -70 30])
```

---

[1] The data files can be downloaded from the Companion Website (www.wiley.com/go/shin_hammond)

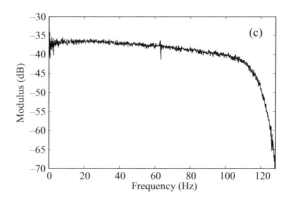

Now let us look at the input signal in more detail by typing

plot(in1(1:50)); grid on

As shown in Figure (d1), the input signal after the 20th data point and before the 4th data point is dominated by noise. Thus, similar to MATLAB Example 6.5, the data in this region are replaced by the noise level (note that they are not replaced by zeros due to the offset of the signal). The following MATLAB script replaces the noise region with constant values and compensates the offset (note that the output signal is not offset, so it is replaced with zeros below the 4th data point):

```
in1(1:4)=in1(20); in1(20:end)=in1(20); in1=in1-in1(20);
out1(1:4)=0;
```

The result is shown in Figure (d2).

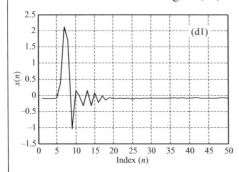
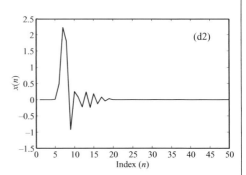

Now we type the script below to see the effect of this preprocessing, which is a much cleaner spectrum as in Figure (e). Note that each signal has a different transient characteristic, so it is preprocessed individually and differently. The preprocessed data set is stored in the file 'impact_data_pre_processed.mat'.

```
In1=fft(in1);
plot(f(1:N/2+1), 20*log10(abs(In1(1:N/2+1))))
xlabel('Frequency (Hz)'); ylabel('Modulus (dB)')
axis([0 128 -70 30])
```

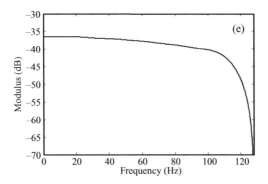

Now, using these two data sets, we shall estimate the FRF based on both Equations (6.55) and (6.56).

---

Case A: FRF estimate by Equation (6.55), i.e.

$$H(f) = \frac{Y(f)}{X(f)}$$

| Line | MATLAB code | Comments |
|------|-------------|----------|
| 1 | clear all | Load the data set which is not preprocessed. |
| 2 | load impact_data_raw | Define frequency and time variables. |
| 3 | fs = 256; N = length(in1);<br>f=fs*(0:N-1)/N; | |
| 4 | T=N/fs; t=0:1/fs:T-1/fs; | |
| 5 | In1=fft(in1); Out1=fft(out1); | Perform the DFT of input signal and output |
| 6 | H=Out1./In1; | signal (only one set of input–output records).<br>Then, calculate the FRF according to<br>Equation (6.55). |
| 7 | figure(1) | Plot the magnitude and phase spectra of the |
| 8 | plot(f(41:761),<br>20*log10(abs(H(41:761)))) | FRF (for the frequency range from 5 Hz to<br>95 Hz). |
| 9 | axis([5 95 -30 50]) | |
| 10 | xlabel('Frequency (Hz)');<br>ylabel('FRF (Modulus, dB)') | |
| 11 | figure(2) | |
| 12 | plot(f(41:761),<br>unwrap(angle(H(41:761)))) | |
| 13 | axis([5 95 -3.5 3.5]) | |
| 14 | xlabel('Frequency (Hz)');<br>ylabel('FRF (Phase, rad)') | |
| 15 | load impact_data_pre_processed | Load the preprocessed data set, and perform |
| 16 | In1=fft(in1); Out1=fft(out1); | the DFT. Then, calculate the FRF according |
| 17 | H=Out1./In1; | to Equation (6.55). |

| 18 | figure(3) | Plot the magnitude and phase spectra of the FRF. |
|----|-----------|---|
| 19 | plot(f(41:761), | |
|    | 20*log10(abs(H(41:761)))) | |
| 20 | axis([5 95 -30 50]) | |
| 21 | xlabel('Frequency (Hz)'); | |
|    | ylabel('FRF (Modulus, dB)') | |
| 22 | figure(4) | |
| 23 | plot(f(41:761), | |
|    | unwrap(angle(H(41:761)))) | |
| 24 | axis([5 95 -3.5 3.5]) | |
| 25 | xlabel('Frequency (Hz)'); | |
|    | ylabel('FRF (Phase, rad)') | |

## Results

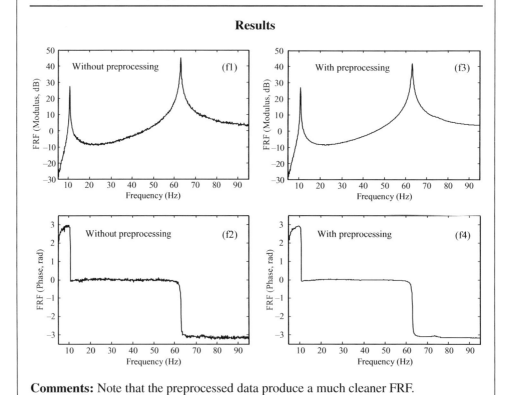

**Comments:** Note that the preprocessed data produce a much cleaner FRF.

Case B: FRF estimate by Equation (6.56), i.e.

$$H_1(f) \approx \frac{\dfrac{1}{N}\displaystyle\sum_{n=1}^{N} X_n^*(f)Y_n(f)}{\dfrac{1}{N}\displaystyle\sum_{n=1}^{N} X_n^*(f)X_n(f)}$$

| Line | MATLAB code | Comments |
|------|-------------|----------|
| 1 | clear all | Load the data set which is not preprocessed |
| 2 | load impact_data_raw | (Line 2). |
| 3 | % load impact_data_pre_processed | Later, comment out this line (with %), and |
| 4 | fs = 256; N = length(in1); | uncomment Line 3 to load the preprocessed |
|   | f=fs*(0:N-1)/N; | data set. |
| 5 | T=N/fs; t=0:1/fs:T-1/fs; | Define frequency and time variables. |
| 6 | Navg=10; | Define the number of averages $N = 10$ (see |
|   | % Navg=3 for preprocessed data set | Equation (6.56)). Later, use $N = 3$ for the |
|   |   | preprocessed data set. |
| 7 | for n=1:Navg | This 'for' loop produces variables: In1, |
| 8 | In = ['In', int2str(n), '= fft(in', | In2, ..., In10; Out1, Out2, ..., Out10; Sxx1, |
|   | int2str(n), ');']; | Sxx2, ..., Sxx10; Sxy1, Sxy2, ..., Sxy10. |
| 9 | eval(In); | They are the DFTs of input and output signals, |
| 10 | Out = ['Out', int2str(n), '= fft(out', | and the elements of the numerator and |
|   | int2str(n), ');']; | denominator of Equation (6.56), such that, for |
| 11 | eval(Out); | example, In1 $= X_1$, Out1 $= Y_1$, |
| 12 | Sxx = ['Sxx', int2str(n), '=conj(In', | Sxx1 $= X_1^* X_1$ and Sxy1 $= X_1^* Y_1$. |
|   | int2str(n), ')''.*In', int2str(n), ';']; | For more details of the 'eval' function see the |
| 13 | eval(Sxx); | MATLAB help window. |
| 14 | Sxy = ['Sxy', int2str(n), '= conj(In', |   |
|   | int2str(n), ')''.*Out', int2str(n), ';']; |   |
| 15 | eval(Sxy); |   |
| 16 | end |   |
| 17 | Sxx=[]; Sxy=[]; | Define empty matrices which will be used in |
| 18 | for n=1:Navg | the 'for' loop. |
| 19 | tmp1= ['Sxx', int2str(n), ';']; | The 'for' loop produces two matrices Sxx and |
| 20 | Sxx=[Sxx; eval(tmp1)]; | Sxy, where the $n$th row of the matrices is |
| 21 | tmp2= ['Sxy', int2str(n), ';']; | $X_n^* X_n$ and $X_n^* Y_n$, respectively. |
| 22 | Sxy=[Sxy; eval(tmp2)]; |   |
| 23 | end |   |
| 24 | Sxx=mean(Sxx); Sxy=mean(Sxy); | First calculate the numerator and denominator |
| 25 | H1=Sxy./Sxx; | of Equation (6.56), and then $H_1$ is obtained. |
| 26 | figure(1) | Plot the magnitude and phase spectra of the |
| 27 | plot(f(41:761), | FRF. |
|   | 20*log10(abs(H(41:761)))) | Run this MATLAB program again using the |
| 28 | axis([5 95 -30 50]) | preprocessed data set, and compare the results. |
| 29 | xlabel('Frequency (Hz)'); |   |
|   | ylabel('FRF (Modulus, dB)') |   |
| 30 | figure(2) |   |
| 31 | plot(f(41:761), |   |
|   | unwrap(angle(H1(41:761)))) |   |
| 32 | axis([5 95 -3.5 3.5]) |   |
| 33 | xlabel('Frequency (Hz)'); |   |
|   | ylabel('FRF (Phase, rad)') |   |

## Results

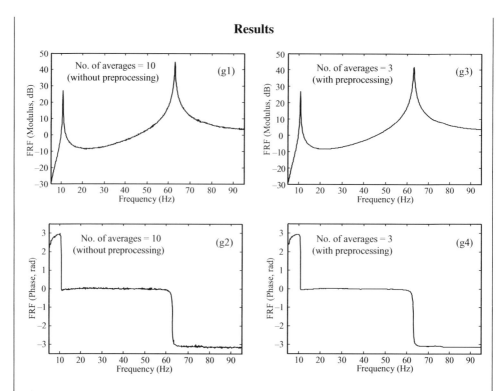

**Comments:** Comparing Figures (g1), (g2) with (f1), (f2) in Case A, it can be seen that averaging improves the FRF estimate. The effect of averaging is to remove the noises which are 'uncorrelated' with the signals $x(t)$ and $y(t)$, as will be seen later in Part II of this book. Note that preprocessing results in a much better FRF estimate using far fewer averages, as can be seen from Figures (g3) and (g4).

# Part II

## Introduction to Random Processes

# 7

# Random Processes

## Introduction

In Part I, we discussed Fourier methods for analysing deterministic signals. In Part II, our interest moves to the treatment of non-deterministic signals. There are many ways in which a signal may be characterized as non-deterministic. At this point we shall say that the time history of the signal cannot be predicted exactly. We may consider the signal shown in Figure 7.1 as a sample of a non-deterministic signal.

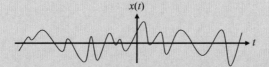

**Figure 7.1**   A sample of a non-deterministic signal

An example of such a signal might be the time history measured from an accelerometer mounted on a ventilation duct. In order to be able to describe the characteristics of such a time history we need some basic ideas of probability and statistics. So we shall now introduce relevant concepts and return to showing how we can use them for time histories in Chapter 8.

## 7.1  BASIC PROBABILITY THEORY

The mathematical theory of describing *uncertain* (or *random*) phenomena is probability theory. It may be best explained by examples – games of chance such as tossing coins, rolling dice,

*Fundamentals of Signal Processing for Sound and Vibration Engineers*
K. Shin and J. K. Hammond.    © 2008 John Wiley & Sons, Ltd

etc. First, we define a few terms:

(a) *An experiment of chance* is an experiment whose outcome is not predictable.
(b) *The sample space* is the collection (set) of all possible outcomes of an experiment, and is denoted by $\Omega$. For example, if an experiment is tossing a coin, then its sample space is $\Omega = (H, T)$, where $H$ and $T$ denote head and tail respectively, and if an experiment is rolling a die, then its sample space is $\Omega = (1, 2, \ldots, 6)$.
(c) *An event* is the outcome of an experiment and is the collection (subset) of points in the sample space, and denoted by $E$. For example, 'the event that a number $\leq 4$ occurs when a die is rolled' is indicated in the Venn diagram shown in Figure 7.2. Individual events in the sample space are called *elementary events*, thus *events* are collections of elementary events.

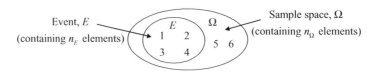

**Figure 7.2**   Sample space ($\Omega$) and event ($E$)

The sample space $\Omega$ is the set of all possible outcomes, containing $n_\Omega$ elements. The event $E$ is a subset of $\Omega$, containing $n_E$ elementary events.

(d) *Probability:* To each event $E$ in a sample space $\Omega$, we may assign a number which measures our belief that $E$ will occur. This is the *probability* of occurrence of event $E$, which is written as $\text{Prob}[E] = P(E)$. In the case where each elementary event is *equally likely*, then it is 'logical' that

$$P(E) = \frac{n_E}{n_\Omega} \tag{7.1}$$

This is a measure of the 'likelihood of occurrence' of an 'event' in an 'experiment of chance', and the probability of event $E$ in the above example is $P(E) = n_E/n_\Omega = 2/3$. Note that $P(E)$ is a 'number' such that

$$0 \leq P(E) \leq 1 \tag{7.2}$$

From this, we conclude that the probability of occurrence of a 'certain' event is one and the probability of occurrence of an 'impossible' event is zero.

### Algebra of Events

Simple 'set operations' visualized with reference to Venn diagrams are useful in setting up the basic axioms of probability. Given events $A, B, C, \ldots$ in a sample space $\Omega$, we can define certain operations on them which lead to other events in $\Omega$. These may be represented by Venn diagrams. If event $A$ is a subset of $\Omega$ (but *not* equal to $\Omega$) we can draw and write $A$ and $\Omega$ as in Figure 7.3(a), and the complement of $A$ is $A'$ (i.e. not $A$) denoted by the shaded area as shown in Figure 7.3(b).

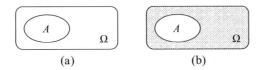

**Figure 7.3**   Event $A$ as a subset of $\Omega$ and its complement

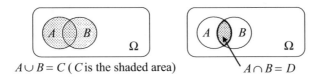

$A \cup B = C$ ($C$ is the shaded area)                  $A \cap B = D$

**Figure 7.4**   Union and intersection of two sets $A$ and $B$

The *union* (sum) of two sets $A$ and $B$ is the set of elements belonging to $A$ or $B$ or both, and is denoted by $A \cup B$. The *intersection* (or product) of two sets $A$ and $B$ is the set of elements common to *both* $A$ and $B$, denoted by $A \cap B$. In Venn diagram terms, they are shown as in Figure 7.4.

If two sets have no elements in common, we write $A \cap B = \Phi$ (the null set). Such events are said to be *mutually exclusive*. For example, in rolling a die, if $A$ is the event that a number $\le 2$ occurs, and $B$ is the event that a number $\ge 5$ occurs, then $A \cap B = \Phi$, i.e. $\Phi$ corresponds to an impossible event.

Some properties of set operations are:

(a) $A \cup (B \cap C) = (A \cup B) \cap (A \cup C)$                                                   (7.3)
(b) $A \cap (B \cup C) = (A \cap B) \cup (A \cap C)$                                                   (7.4)
(c) $(A \cup B)' = A' \cap B'$                                                                                       (7.5)
(d) For any set $A$, let $n(A)$ denote the number of elements in $A$; then

$$n(A \cup B) = n(A) + n(B) - n(A \cap B) \qquad (7.6)$$

Two different cases are shown in Figure 7.5 to demonstrate the use of Equation (7.6).

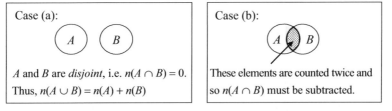

**Figure 7.5**   Demonstration of $n(A \cup B)$ for two different cases

### Algebra of Probabilities

The above intuitive ideas are formalized into the axioms of probability as follows. To each event $E_i$ (in a sample space $\Omega$), we assign a number called the probability of $E_i$ (denoted $P(E_i)$) such that

(a) $0 \leq P(E_i) \leq 1$                                                                 (7.7)

(b) If $E_i$ and $E_j$ are mutually exclusive, then

$$P(E_i \cup E_j) = P(E_i) + P(E_j) \tag{7.8}$$

(c) If $\bigcup E_i = \Omega$, then

$$P\left(\bigcup E_i\right) = 1 \tag{7.9}$$

(d) $P(\Phi) = 0$                                                                           (7.10)

(e) For any events $E_1$, $E_2$, not necessarily mutually exclusive,

$$P(E_1 \cup E_2) = P(E_1) + P(E_2) - P(E_1 \cap E_2) \tag{7.11}$$

### Equally Likely Events

If $n$ events, $E_1, E_2, \ldots, E_n$, are judged to be equally likely, then

$$P(E_i) = \frac{1}{n} \tag{7.12}$$

As an example of this, throw two dice and record the number on each face. What is the probability of the event that the total score is 5? The answer is $P(E_5) = n_{E_5}/n_\Omega = 4/36 = 1/9$.

### Joint Probability

The probability of occurrence of events $A$ and $B$ jointly is called a joint probability and is denoted $P(A \cap B)$ or $P(A, B)$. With reference to Figure 7.6, this is the occurrence of the shaded area, i.e.

$$P(A \cap B) = \frac{n_{A \cap B}}{n_\Omega} \tag{7.13}$$

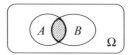

**Figure 7.6** The intersection of two sets $A$ and $B$ in a sample space $\Omega$

### Conditional Probability

The probability of occurrence of an event $A$ given that event $B$ has occurred is written as $P(A|B)$, and is called a *conditional probability*. To explain this, consider the intersection of two sets $A$ and $B$ in a sample space $\Omega$ as shown in Figure 7.6. To compute $P(A|B)$, in effect we are computing a probability with respect to a 'reduced sample space', i.e. it is the ratio of the number of elements in the shaded area relative to the number of elements in $B$, namely $n_{A\cap B}/n_B$, which may be written $(n_{A\cap B}/n_\Omega)/(n_B/n_\Omega)$, or

$$P(A|B) = \frac{P(A \cap B)}{P(B)} = \frac{P(A, B)}{P(B)} \tag{7.14}$$

### Statistical Independence

If $P(A|B) = P(A)$, we say event $A$ and $B$ are *statistically independent*. Note that this is so if $P(A \cap B) = P(A)P(B)$. As an example of this, toss a coin and roll a die. The probability that a coin lands head and a die scores 3 is $P(A \cap B) = 1/2 \cdot 1/6 = 1/12$ since the events are independent, i.e. knowing the result of the first event (a coin lands head or tail) does not give us any information on the second event (score on the die).

### Relative Frequencies[M7.1]

As defined in Equations (7.1) and (7.2), the *probability* of event $E$ in a sample space $\Omega$, $P(E)$ is a theoretical concept which can be computed without conducting an experiment. In the simple example above this has worked based on the assumption of equal likelihood of occurrence of the elementary events. When this is not the case we resort to measurements to 'estimate' the probability of occurrence of events. We approach this via the notion of *relative frequency* (or proportion) of times that $E$ occurs in a long series of trials. Thus, if event $E$ occurs $n_E$ times in $N$ trials, then the relative frequency of $E$ is given by

$$f_E = \frac{n_E}{N} \tag{7.15}$$

Obviously, as $N$ changes, so does $f_E$. For example, toss a coin and note $f_H$ (the relative frequency of a head occurring) as $N$ increases. This is shown in Figure 7.7.

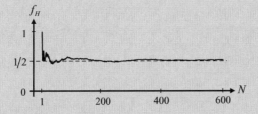

**Figure 7.7**   Illustration of $f_H$ as $N$ increases

This graph suggests that the *error* is 'probably' reduced as $N$ gets larger.

The above notion of relative frequency is not useful as a definition of probability because its values are not unique, but it is intuitively appealing and is used to *estimate* probabilities where applicable, i.e. $f_E$ is often taken as an *estimate* of $P(E)$. The relative frequency is sometimes referred to as the 'empirical' probability since it is deduced from observed data. This estimate has the following properties:

(a) For all events $A$,

$$f_A \geq 0 \text{ (non-negativity)} \tag{7.16}$$

(b) For all mutually exclusive events,

$$f_{A \cup B} = \frac{n_A + n_B}{N} = f_A + f_B \text{ (additivity)} \tag{7.17}$$

(c) For any set of collectively exhaustive events, $A_1, A_2, \ldots$, i.e. $A_1 \cup A_2 \cup \cdots = \bigcup A_i$,

$$f_{\bigcup A_i} = \frac{N}{N} = 1 \text{ (certainty)} \tag{7.18}$$

i.e. a 'certain' event has a relative frequency of '1'.

## 7.2 RANDOM VARIABLES AND PROBABILITY DISTRIBUTIONS

In many cases it is more convenient to define the outcome of an experiment as a set of numbers rather than the actual elements of the sample space. So we define a *random variable* as a *function* defined on a sample space. For example, if $\Omega = (H, T)$ for a coin tossing experiment, we may choose to say we get a number '1' when a head occurs and '0' when the tail occurs, i.e. we 'map' from the sample space to a 'range space' or a *new* sample space as shown in Figure 7.8. We may write the function such that $X(H) = 1$ and $X(T) = 0$. More generally, for any element $\omega_i$ in $\Omega$, we define a function $X(\omega_i)$.

Note that the number of elements of $\Omega$ and the number of values taken by $X(\omega_i)$ need not be the same. For an example, toss two coins and record the outcomes and define the random variable $X$ as the number of heads occurring. This is shown in Figure 7.9.

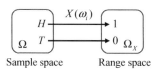

Sample space            Range space

**Figure 7.8**    A random variable $X$ that maps from a sample space $\Omega$ to a range space $\Omega_X$

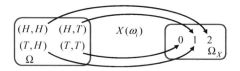

**Figure 7.9**    An example of a random variable $X$

We note that the values taken by a random variable are denoted $x_i$, i.e. $X(\omega_i) = x_i$, and the notation $X(\omega_i)$ is often abbreviated to $X$. In many cases the sample space and the range space 'fuse' together, e.g. when the outcome is already a number (rolling a die, recording a voltage, etc.).

There are two types of random variable. If the sample space $\Omega_X$ consists of discrete elements, i.e. countable, $X$ is said to be a *discrete random variable*, e.g. rolling a die. If $\Omega_X$ consists of 'continuous' values, i.e. uncountable (or non-denumerable), then $X$ is a *continuous random variable*, e.g. the voltage fluctuation on an 'analogue' meter. Some processes may be mixed, e.g. a binary signal in noise.

### *Probability Distributions for Discrete Random Variables*

For a discrete random variable $X$ (which takes on only a discrete set of values $x_1, x_2, \dots$), the probability distribution of $X$ is characterized by specifying the probabilities that the random variable $X$ is equal to $x_i$, for every $x_i$, i.e.

$$P[X = x_i] \quad \text{for } x_i = x_1, x_2, \dots \tag{7.19}$$

where $P[X = x_i]$ describes the *probability distribution* of a discrete random variable $X$ and satisfies $\sum_i P[X = x_i] = 1$, e.g. for rolling a die, the probability distribution is as shown in Figure 7.10.

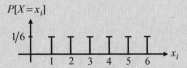

**Figure 7.10** Probability distribution for rolling a die

### *The Cumulative Distribution*

Random variables have a (cumulative) distribution function (cdf). This is the probability of a random variable $X$ taking a value less than or equal to $x$. This is described by $F(x)$ where

$$F(x) = P[X \leq x] = \text{Prob}[X \text{ taking on a value up to and including } x] \tag{7.20}$$

For a discrete random variable there are jumps in the function $F(x)$ as shown in Figure 7.11 (for rolling a die).

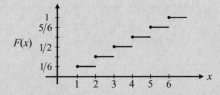

**Figure 7.11** Cumulative distribution function for rolling a die

Since probabilities are non-negative the cumulative distribution function is monotonic non-decreasing.

### Continuous distributions

For a continuous process, the sample space is infinite and non-denumerable. So the probability that $X$ takes the value $x$ is zero, i.e. $P[X = x] = 0$. Whilst technically correct this is not particularly useful, since $X$ will take specific values. So a more useful approach is to think of the probability of $X$ lying within *intervals* on the $x$-axis, i.e. $P[X > a]$, $P[a < X \leq b]$, etc.

We start by considering the distribution function $F(x) = P[X \leq x]$. $F(x)$ must have a general shape such as the graph shown in Figure 7.12.

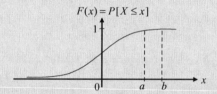

**Figure 7.12**    An example of a distribution function for a continuous process

From Figure 7.12, some properties of $F(x)$ are:

(a) $F(-\infty) = 0, \quad F(\infty) = 1$                                                                (7.21)
(b) $F(x_2) \geq F(x_1)$   for $x_2 \geq x_1$                                             (7.22)
(c) $P[a < X \leq b] = P[X \leq b] - P[X \leq a] = F(b) - F(a)$   for $a < b$     (7.23)

### Probability Density Functions

Using the properties of distribution function $F(x)$, the probability of $X$ lying in an interval $x$ to $x + \delta x$ can be written as

$$P[x < X \leq x + \delta x] = F(x + \delta x) - F(x) \tag{7.24}$$

which shrinks to zero as $\delta x \to 0$. However, consider $P[x < X \leq x + \delta x]/\delta x$. This is the probability of lying in a band (width $\delta x$) divided by that bandwidth. Then, if the quantity $\lim_{\delta x \to 0} P[x < X \leq x + \delta x]/\delta x$ exists it is called the *probability density function* (pdf) which is denoted $p(x)$ and is (from Equation (7.24))

$$p(x) = \lim_{\delta x \to 0} \frac{P[x < X \leq x + \delta x]}{\delta x} = \frac{dF(x)}{dx} \tag{7.25}$$

From Equation (7.25) it follows that

$$F(x) = \int_{-\infty}^{x} p(u)\,du \tag{7.26}$$

Some properties of the probability density function $p(x)$ are:

(a)  $p(x) \geq 0$                                                                                                    (7.27)
    i.e. the probability density function is non-negative;

(b)  $\int_{-\infty}^{\infty} p(x)dx = 1$                                                                             (7.28)
    i.e. the area under the probability density function is unity;

(c)  $P[a < X \leq b] = \int_{a}^{b} p(x)dx$                                                                          (7.29)

    As an example of Equation (7.29), $P[a < X \leq b]$ can be found by evaluating the shaded area shown in Figure 7.13.

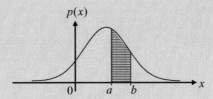

**Figure 7.13**   An example of a probability density function

    Note that we can also define the probability density function for a discrete random variable if the properties of delta functions are used. For example, the probability density function for rolling a die is shown in Figure 7.14.

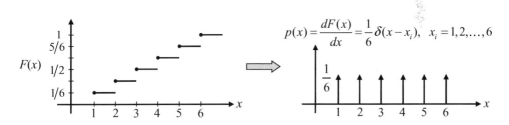

**Figure 7.14**   Probability density function for rolling a die

### Joint Distributions

The above descriptions involve only a single random variable $X$. This is a univariate process. Now, consider a process which involves two random variables (say $X$ and $Y$), i.e. a bivariate process. The probability that $X \leq x$ occurs *jointly* with $Y \leq y$ is

$$P[X \leq x \cap Y \leq y] = F(x, y)$$                                                                              (7.30)

Note that $F(-\infty, y) = F(x, -\infty) = 0$, $F(\infty, \infty) = 1$, $F(x, \infty) = F(x)$ and $F(\infty, y) = F(y)$. Similar to the univariate case the 'joint probability density function' is defined as

$$p(x, y) = \lim_{\substack{\delta x \to 0 \\ \delta y \to 0}} \frac{P[x < X \le x + \delta x \cap y < Y \le y + \delta y]}{\delta x \delta y} = \frac{\partial^2 F(x, y)}{\partial x \partial y} \tag{7.31}$$

and

$$F(x, y) = \int_{-\infty}^{x} \int_{-\infty}^{y} p(u, v) dv du \tag{7.32}$$

Note that

$$\int_{-\infty}^{\infty} \int_{-\infty}^{\infty} p(x, y) dy dx = 1 \quad \text{and} \quad \int_{-\infty}^{x} \int_{-\infty}^{\infty} p(u, v) dv du = F(x)$$

hence

$$p(x) = \int_{-\infty}^{\infty} p(x, y) dy \tag{7.33}$$

This is called a 'marginal' probability density function.

These ideas may be extended to $n$ random variables, $X_1, X_2, \ldots, X_n$, i.e. we may define $p(x_1, x_2, \ldots, x_n)$. We shall only consider univariate and bivariate processes in this book.

## 7.3 EXPECTATIONS OF FUNCTIONS OF A RANDOM VARIABLE

So far, we have used probability distributions to describe the properties of random variables. However, rather than using probability distributions, we often use *averages*. This introduces the concept of the *expectation* of a process.

Consider a discrete random variable $X$ which can assume any values $x_1, x_2, \ldots$ with probabilities $p_1, p_2, \ldots$. If $x_i$ occurs $n_i$ times in $N$ trials of an experiment, then the average value of $X$ is

$$\bar{x} = \frac{1}{N} \sum_i n_i x_i \tag{7.34}$$

where $\bar{x}$ is called the sample mean. Since $n_i/N = f_i$ (the empirical probability of occurrence of $x_i$), Equation (7.34) can be written as $\bar{x} = \sum_i x_i f_i$. As $N \to \infty$, the empirical probability approaches the theoretical probability. So the expression for $\bar{x}$ becomes $\sum_i x_i p_i$ and this defines the theoretical mean value of $X$.

For a continuous process, the probability $p_i$ may be replaced by the probability density multiplied by the bandwidth, i.e. $p_i \to p(x_i)\delta x_i$. So $\sum_i x_i p_i$ becomes $\sum_i x_i p(x_i)\delta x_i$

which as $\delta x_i \to 0$ is written $\int_{-\infty}^{\infty} xp(x)dx$. This defines the theoretical mean value of $X$ which we write as $E[X]$, the *expected value* of $X$, i.e.

$$E[X] = \int_{-\infty}^{\infty} xp(x)dx \qquad (7.35)$$

This is the 'mean value' or the 'first moment' of a random variable $X$. More generally, the expectation operation generalizes to functions of a random variable. For example, if $Y = g(X)$, i.e. as shown in Figure 7.15,

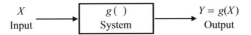

$$\begin{array}{ccc}
X & g(\ ) & Y = g(X) \\
\text{Input} & \text{System} & \text{Output}
\end{array}$$

**Figure 7.15**   System with random input and random output

then the expected (or average) value of $Y$ is

$$E[Y] = E[g(X)] = \int_{-\infty}^{\infty} g(x)p(x)dx \qquad (7.36)$$

For a discrete process, this becomes

$$E[g(X)] = \sum_i g(x_i)p_i \qquad (7.37)$$

This may be extended to functions of several random variables. For example, in a bivariate process with random variables $X$ and $Y$, if $W = g(X, Y)$, then the expected value of $W$ is

$$E[W] = E[g(X, Y)] = \int_{-\infty}^{\infty} \int_{-\infty}^{\infty} g(x, y)p(x, y)dxdy \qquad (7.38)$$

***Moments of a Random Variable***

The probability density function $p(x)$ contains the complete information about the probability characteristics of $X$, but it is sometimes useful to summarize this information in a few numerical parameters – the so-called *moments of a random variable*. The first and second moments are given below:

(a) First moment (mean value):

$$\mu_x = E[X] = \int_{-\infty}^{\infty} xp(x)dx \qquad (7.39)$$

(b) Second moment (mean square value):

$$E[X^2] = \int_{-\infty}^{\infty} x^2 p(x)dx \qquad (7.40)$$

Note that, instead of using Equation (7.40), the 'central moments' (moments about the mean) are usually used. The second moment about the mean is the called the *variance*, which is written as

$$\text{Var}(X) = \sigma_x^2 = E[(X - \mu_x)^2] = \int_{-\infty}^{\infty} (x - \mu_x)^2 p(x)dx \qquad (7.41)$$

where $\sigma_x = \sqrt{\text{Var}(X)}$ is called the *standard deviation*, and is the *root mean square* (rms) of a 'zero' mean variable.

In many cases, the above two moments $\mu_x$ and $\sigma_x^2$ are the most important measures of a random variable $X$. However, the third and fourth moments are useful in considerations of processes that are non-Gaussian (discussed later in this chapter).

The first moment $\mu_x$ is a measure of 'location' of $p(x)$ on the $x$-axis; the variance $\sigma_x^2$ is a measure of *dispersion* or spread of $p(x)$ relative to $\mu_x$. The following few examples illustrate this.

### Some 'Well-known' Distributions

**A Uniform Distribution** (Figure 7.16)
This is often used to model the errors involved in measurement (see quantization noise discussed in Chapter 5).

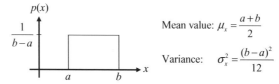

**Figure 7.16**   Probability density function of a uniform distribution

**Rayleigh Distribution** (Figure 7.17)
This is used in fatigue analysis, e.g. to model cyclic stresses.

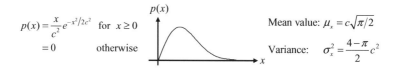

**Figure 7.17**   Probability density function of a Rayleigh distribution

**Gaussian Distribution (Normal Distribution)**
This is probably the most important distribution, since many practical processes can be approximated as Gaussian (see a statement of the central limit theorem below). If a random variable $X$ is normally distributed, then its probability distribution is completely described by two parameters, its mean value $\mu_x$ and variance $\sigma_x^2$ (or standard deviation $\sigma_x$), and the probability density function of a Gaussian distribution is given by

$$p(x) = \frac{1}{\sigma_x\sqrt{2\pi}}e^{-(x-\mu_x)^2/2\sigma_x^2} \tag{7.42}$$

If $\mu_x = 0$ and $\sigma_x^2 = 1$, then it is called the 'standard normal distribution'. For $\mu_x = 0$, some examples of the Gaussian distribution are shown in Figure 7.18.

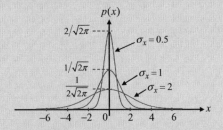

**Figure 7.18**  Probability density functions of Gaussian distribution

The importance of the Gaussian distribution is illustrated by a particular property: let $X_1, X_2, \ldots$ be independent random variables that have their own probability distributions; then the sum of random variables, $S_n = \sum_{k=1}^{n} X_k$, tends to have a Gaussian distribution as $n$ gets large, regardless of their individual distribution of $X_k$. This is a version of the so-called *central limit theorem*. Moreover, it is interesting to observe the speed with which this occurs as $n$ increases.[M7.2]

For a Gaussian bivariate process (random variables $X$ and $Y$), the joint probability density function is written as

$$p(x, y) = \frac{1}{2\pi} \cdot \frac{1}{|S|^{1/2}} \exp\left[-\frac{1}{2}(\mathbf{v} - \boldsymbol{\mu_v})^{\mathrm{T}}\mathbf{S}^{-1}(\mathbf{v} - \boldsymbol{\mu_v})\right] \tag{7.43}$$

where

$$\mathbf{S} = \begin{bmatrix} \sigma_x^2 & \sigma_{xy} \\ \sigma_{xy} & \sigma_y^2 \end{bmatrix}, \quad \mathbf{v} = \begin{bmatrix} x \\ y \end{bmatrix} \quad \text{and} \quad \boldsymbol{\mu_v} = \begin{bmatrix} \mu_x \\ \mu_y \end{bmatrix}$$

Also $\mu_x = E[X]$, $\mu_y = E[Y]$, $\sigma_x^2 = E[(X - \mu_x)^2]$, $\sigma_y^2 = E[(Y - \mu_y)^2]$ and $\sigma_{xy} = E[(X - \mu_x)(Y - \mu_y)]$ (this is discussed shortly).

**Bivariate Processes**[M7.3]

The concept of moments generalizes to bivariate processes, essentially based on Equation (7.38). For example, the expected value of the product of two variables $X$ and $Y$ is

$$E[X\,Y] = \int_{-\infty}^{\infty} \int_{-\infty}^{\infty} xy p(x, y) dx dy \qquad (7.44)$$

This is a generalization of the second moment (see Equation (7.40)). If we centralize the process (i.e. subtract the mean from each) then

$$\text{Cov}(X, Y) = \sigma_{xy} = E[(X - \mu_x)(Y - \mu_y)] = \int_{-\infty}^{\infty} \int_{-\infty}^{\infty} (x - \mu_x)(y - \mu_y) p(x, y) dx dy$$

$$(7.45)$$

$E[X\,Y]$ is called the *correlation* between $X$ and $Y$, and $\text{Cov}(X, Y)$ is called the *covariance* between $X$ and $Y$. They are related by

$$\text{Cov}(X, Y) = E[X\,Y] - \mu_x \mu_y = E[X\,Y] - E[X]E[Y] \qquad (7.46)$$

Note that the covariance and correlation are the same if $\mu_x = \mu_y = 0$. Some definitions for jointly distributed random variables are given below.

$X$ and $Y$ are:

(a) *uncorrelated* if $E[X\,Y] = E[X]\,E[Y]$ (or $\text{Cov}(X, Y) = 0$)

   (note that, for *zero-mean* variables, if $X$ and $Y$ are uncorrelated, then $E[X\,Y] = 0$);

(b) *orthogonal* if $E[X\,Y] = 0$;

(c) *independent* (*statistically*) if $p(x, y) = p(x)p(y)$.

Note that, if $X$ and $Y$ are *independent* they are *uncorrelated*. However, uncorrelated random variables are *not* necessarily independent. For example, Let $X$ be a random variable uniformly distributed over the range $-1$ to $1$. Note that the mean value $E[X] = 0$. Let another random variable $Y = X^2$. Then obviously $p(x, y) \neq p(x)p(y)$, i.e. $X$ and $Y$ are dependent (if $X$ is known, $Y$ is also known). But $\text{Cov}(X, Y) = E[X\,Y] - E[X]E[Y] = E[X^3] = 0$ shows that they are uncorrelated (and also orthogonal). Note that they are related *nonlinearly*.

An important measure called the *correlation coefficient* is defined as

$$\rho_{xy} = \frac{\text{Cov}(X, Y)}{\sigma_x \sigma_y} = \frac{E[(X - \mu_x)(Y - \mu_y)]}{\sigma_x \sigma_y} \qquad (7.47)$$

This is a measure (or degree) of a *linear* relationship between two random variables, and the correlation coefficient has values in the range $-1 \leq \rho_{xy} \leq 1$. If $|\rho_{xy}| = 1$, then two random variables $X$ and $Y$ are 'fully' related in a linear manner, e.g. $Y = aX + b$, where $a$ and $b$ are constants. If $\rho_{xy} = 0$, there is no linear relationship between $X$ and $Y$. Note that the correlation coefficient detects only linear relationships between $X$ and $Y$. Thus, even if $\rho_{xy} = 0$, $X$ and

$Y$ can be related in a nonlinear fashion (see the above example, i.e. $X$ and $Y = X^2$, where $X$ is uniformly distributed on $-1$ to 1).

## Some Important Properties of Moments

(a) $E[aX + b] = aE[X] + b$ ($a$, $b$ are some constants) $\qquad$ (7.48)
(b) $E[aX + bY] = aE[X] + bE[Y]$ $\qquad$ (7.49)
(c) $\text{Var}(X) = E[X^2] - \mu_x^2 = E[X^2] - E^2[X]$ $\qquad$ (7.50)

> *Proof:* $\text{Var}(X) = E[(X - \mu_x)^2] = E[X^2 - 2\mu_x X + \mu_x^2]$
> $\qquad\qquad = E[X^2] - 2\mu_x E[X] + \mu_x^2 = E[X^2] - \mu_x^2$

(d) $\text{Var}(aX + b) = a^2 \text{Var}(X)$ $\qquad$ (7.51)
(e) $\text{Cov}(X, Y) = E[XY] - \mu_x \mu_y = E[XY] - E[X]E[Y]$ $\qquad$ (7.52)
(f) $\text{Var}(X + Y) = \text{Var}(X) + \text{Var}(Y) + 2\text{Cov}(X, Y)$ $\qquad$ (7.53)

> *Proof:* $\text{Var}(X + Y) = E[(X + Y)^2] - E^2[(X + Y)]$
> $\qquad\qquad = E[X^2 + 2XY + Y^2] - E^2[X] - 2E[X]E[Y] - E^2[Y]$
> $\qquad\qquad = (E[X^2] - E^2[X]) + (E[Y^2] - E^2[Y]) + 2(E[XY] - E[X]E[Y])$
> $\qquad\qquad = \text{Var}(X) + \text{Var}(Y) + 2\text{Cov}(X, Y)$

Note that, if $X$ and $Y$ are independent or uncorrelated, $\text{Var}(X + Y) = \text{Var}(X) + \text{Var}(Y)$.

## Higher Moments

We have seen that the first and second moments are sufficient to describe the probability distribution of a Gaussian process. For a non-Gaussian process, some useful information about the probability density function of the process can be obtained by considering higher moments of the random variable.

The generalized $k$th moment is defined as

$$M_k' = E[X^k] = \int_{-\infty}^{\infty} x^k p(x) dx \qquad (7.54)$$

The $k$th moment about the mean (central moment) is defined as

$$M_k = E[(X - \mu_x)^k] = \int_{-\infty}^{\infty} (x - \mu_x)^k p(x) dx \qquad (7.55)$$

In engineering, the third and fourth moments are widely used. For example, the third moment about the mean, $E[(X - \mu_x)^3]$, is a measure of asymmetry of a probability distribution, so it is called the *skewness*. In practice, the coefficient of skewness is more often used, and is defined as

$$\gamma_1 = \frac{E[(X - \mu_x)^3]}{\sigma_x^3} \tag{7.56}$$

Note that, in many texts, Equation (7.56) is simply referred to as skewness. Also note that $\gamma_1 = 0$ for a Gaussian distribution since it has a symmetric probability density function. Typical skewed probability density functions are shown in Figure 7.19. Such asymmetry could arise from signal 'clipping'.

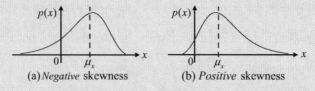

(a) *Negative* skewness       (b) *Positive* skewness

**Figure 7.19**   Skewed probability density functions

The fourth moment about the mean, $E[(X - \mu_x)^4]$, measures the degree of *flattening* of a probability density function *near its mean*. Similar to the skewness, the coefficient of *kurtosis* (or simply the kurtosis) is defined as

$$\gamma_2 = \frac{E[(X - \mu_x)^4]}{\sigma_x^4} - 3 \tag{7.57}$$

where '−3' is introduced to make $\gamma_2 = 0$ for a Gaussian distribution (i.e. $E[(X - \mu_x)^4]/\sigma_x^4 = 3$ for a Gaussian distribution, thus $E[(X - \mu_x)^4]/\sigma_x^4$ is often used and examined with respect to the value 3).

A distribution with positive kurtosis $\gamma_2 > 0$ is called *leptokurtic* (more peaky than Gaussian), and a distribution with negative kurtosis $\gamma_2 < 0$ is called *platykurtic* (more flattened than Gaussian). This is illustrated in Figure 7.20.

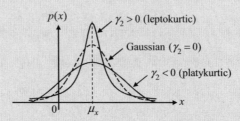

**Figure 7.20**   Probability density functions with different values of kurtosis

Since $\gamma_1 = 0$ and $\gamma_2 = 0$ for a Gaussian process, the third and fourth moments (or $\gamma_1$ and $\gamma_2$) can be used for detecting non-Gaussianity. These higher moments may also be used to detect (or characterize) nonlinearity since nonlinear systems exhibit non-Gaussian responses.

The kurtosis (fourth moment) is widely used as a measure in machinery condition monitoring – for example, early damage in rolling elements of machinery often results in vibration signals whose kurtosis value is significantly increased owing to the impacts occurring because of the faults in such rotating systems.

As a further example, consider a large machine (in good condition) that has many components generating different types of (periodic and random) vibration. In this case, the vibration signal measured on the surface of the machine may have a probability distribution similar to a Gaussian distribution (by the central limit theorem). Later in the machine's operational life, one of the components may produce a repetitive transient signal (possibly due to a bearing fault). This impact produces wide excursions and more oscillatory behaviour and changes the probability distribution from Gaussian to one that is leptokurtic (see Figure 7.20). The detection of the non-Gaussianity can be achieved by monitoring the kurtosis (see MATLAB Example 7.4). Note that, if there is severe damage, i.e. many components are faulty, then the measured signal may become Gaussian again.

### *Computational Considerations of Moments (Digital Form)*

We now indicate some ways in which the moments described above might be estimated from measured data. No attempt is made at this stage to give measures of the accuracy of these estimates. This will be discussed later in Chapter 10.

Suppose we have a set of data $(x_1, x_2, \ldots, x_N)$ collected from $N$ measurements of a random variable $X$. Then the *sample mean* $\bar{x}$ (which estimates the arithmetic mean, $\mu_x$) is computed as

$$\bar{x} = \frac{1}{N} \sum_{n=1}^{N} x_n \tag{7.58}$$

For the estimation of the variance $\sigma_x^2$, one may use the formula

$$s_x^2 = \frac{1}{N} \sum_{n=1}^{N} (x_n - \bar{x})^2 \tag{7.59}$$

However, this estimator usually underestimates the variance, so it is a *biased* estimator. Note that $\bar{x}$ is present in the formula, thus the divisor $N - 1$ is more frequently used. This gives an *unbiased* sample variance, i.e.

$$s_x^2 = \frac{1}{N - 1} \sum_{n=1}^{N} (x_n - \bar{x})^2 \tag{7.60}$$

where $s_x$ is the sample standard deviation. Since

$$\sum_{n=1}^{N} (x_n - \bar{x})^2 = \sum_{n=1}^{N} x_n^2 - 2N\bar{x}^2 + N\bar{x}^2$$

the following computationally more efficient form is often used:

$$s_x^2 = \frac{1}{N - 1} \left[ \left( \sum_{n=1}^{N} x_n^2 \right) - N\bar{x}^2 \right] \tag{7.61}$$

The above estimation can be generalized, i.e. the $k$th sample (raw) moment is defined as

$$m'_k = \frac{1}{N} \sum_{n=1}^{N} x_n^k \tag{7.62}$$

Note that $m'_1 = \bar{x}$ and $m'_2$ is the mean square value of the sample. Similarly the $k$th sample central moment is defined as

$$m_k = \frac{1}{N} \sum_{n=1}^{N} (x_n - \bar{x})^k \tag{7.63}$$

Note that $m_1 = 0$ and $m_2$ is the (biased) sample variance. As in the above equation, the divisor $N$ is usually used for the sample moments. For the estimation of skewness and kurtosis coefficients, the following biased estimators are often used:

$$\text{Skew} = \frac{1}{N} \sum_{n=1}^{N} (x_n - \bar{x})^3 \Big/ s_x^3 \tag{7.64}$$

$$\text{Kurt} = \left( \frac{1}{N} \sum_{n=1}^{N} (x_n - \bar{x})^4 \Big/ s_x^4 \right) - 3 \tag{7.65}$$

where the sample standard deviation

$$s_x = \sqrt{\frac{1}{N} \sum_{n=1}^{N} (x_n - \bar{x})^2}$$

is used.

Finally, for bivariate processes, the sample covariance is computed by either

$$s_{xy} = \frac{1}{N} \sum_{n=1}^{N} (x_n - \bar{x})(y_n - \bar{y}) = \frac{1}{N} \left[ \left( \sum_{n=1}^{N} x_n y_n \right) - N\overline{xy} \right] \text{ (biased estimator)} \tag{7.66}$$

or

$$s_{xy} = \frac{1}{N-1} \sum_{n=1}^{N} (x_n - \bar{x})(y_n - \bar{y}) = \frac{1}{N-1} \left[ \left( \sum_{n=1}^{N} x_n y_n \right) - N\overline{xy} \right]$$

$$\text{(unbiased estimator)} \tag{7.67}$$

Note that, although we have distinguished the biased and unbiased estimators (the divisor is $N$ or $N - 1$), their differences are usually insignificant if $N$ is 'large enough'.

## 7.4 BRIEF SUMMARY

1. The relative frequency (or empirical probability) of event $E$ is

$$f_E = \frac{n_E}{N}$$

2. A random variable is a function defined on a sample space, i.e. a random variable $X$ maps from the sample space $\Omega$ to a range space $\Omega_X$ such that $X(\omega_i) = x_i$. There are two types of random variable: a discrete random variable ($\Omega_X$ consists of discrete elements) and a continuous random variable ($\Omega_X$ consists of continuous values).

3. The *central limit theorem* (roughly speaking) states that the sum of independent random variables (that have *arbitrary* probability distributions) $S_n = \sum_{k=1}^{n} X_k$ becomes normally distributed (Gaussian) as $n$ gets large.

4. The moments of a random variable are summarized in Table 7.1.

**Table 7.1**  Summary of moments

| Moment (central) | Estimator | Measures |
|---|---|---|
| 1st moment: $\mu_x = E[X]$ | $\bar{x} = \dfrac{1}{N} \sum_{n=1}^{N} x_n$ | Mean (location) |
| 2nd moment: $\sigma_x^2 = E[(X - \mu_x)^2]$ | $s_x^2 = \dfrac{1}{N-1} \sum_{n=1}^{N} (x_n - \bar{x})^2$ | Variance (spread or dispersion) |
| 3rd moment: $M_3 = E[(X - \mu_x)^3]$ | $m_3 = \dfrac{1}{N} \sum_{n=1}^{N} (x_n - \bar{x})^3$ | Degree of asymmetry (skewness) $\gamma_1 = \dfrac{E[(X - \mu_x)^3]}{\sigma_x^3}$ |
| 4th moment: $M_4 = E[(X - \mu_x)^4]$ | $m_4 = \dfrac{1}{N} \sum_{n=1}^{N} (x_n - \bar{x})^4$ | Degree of flattening (kurtosis) $\gamma_2 = \dfrac{E[(X - \mu_x)^4]}{\sigma_x^4} - 3$ |

5. The *correlation* of $X$ and $Y$ is defined as $E[X Y]$, and the *covariance* of $X$ and $Y$ is defined as

$$\mathrm{Cov}(X, Y) = \sigma_{xy} = E[(X - \mu_x)(Y - \mu_y)]$$

These are related by

$$\mathrm{Cov}(X, Y) = E[X Y] - \mu_x \mu_y = E[X Y] - E[X]E[Y]$$

6. Two random variables $X$ and $Y$ are *uncorrelated* if

$$E[X Y] = E[X] E[Y] \, (\text{or } \mathrm{Cov}(X, Y) = 0)$$

7. The *correlation coefficient* is defined as

$$\rho_{xy} = \frac{\text{Cov}(X, Y)}{\sigma_x \sigma_y} = \frac{E[(X - \mu_x)(Y - \mu_y)]}{\sigma_x \sigma_y}$$

This is a measure of a *linear* relationship between two random variables. If $|\rho_{xy}| = 1$, then two random variables $X$ and $Y$ are 'fully' related linearly. If $\rho_{xy} = 0$, they are not linearly related at all.

## 7.5 MATLAB EXAMPLES

***Example 7.1:* Relative frequency $f_E = n_E/N$ as an estimate of $P(E)$**

In this MATLAB example, we consider an experiment of tossing a coin, and observe how the relative frequency changes as the number of trials ($N$) increases.

| Line | MATLAB code | Comments |
|---|---|---|
| 1<br>2 | clear all<br>rand('state',0); | Initialize the random number generator. The MATLAB function 'rand' generates *uniformly* distributed random numbers, while 'randn' is used to generate *normally* distributed random numbers. |
| 3<br>4 | X=round(rand(1,1000)); % 1: head, 0: tail<br>id_head=find(X==1); id_tail=find(X==0); | Define the random variable X whose elements are either 1 or 0, and 1000 trials are performed. We regard 1 as the head and 0 as the tail. Find indices of head and tail. |
| 5<br>6<br>7 | N=ones(size(X));<br>head=N; head(id_tail)=0;<br>tail=N; tail(id_head)=0; | The vector 'head' has ones that correspond to the elements of vector X with 1, and the vector 'tail' has ones that correspond to the elements of vector X with 0. |
| 8<br>9 | fr_head=cumsum(head)./cumsum(N);<br>fr_tail=cumsum(tail)./cumsum(N); | Calculate the relative frequencies of head and tail. The MATLAB function 'cumsum(N)' generates a vector whose elements are the cumulative sum of the elements of N. |
| 10<br>11<br>12<br>13<br>14 | figure(1)<br>plot(fr_head)<br>xlabel('\itN \rm(Number of trials)')<br>ylabel('Relative frequency (head)')<br>axis([0 length(N) 0 1]) | Plot the relative frequency of head. |
| 15<br>16<br>17<br>18<br>19 | figure(2)<br>plot(fr_tail)<br>xlabel('\itN \rm(Number of trials)')<br>ylabel('Relative frequency (tail)')<br>axis([0 length(N) 0 1]) | Plot the relative frequency of tail. |

**Results**

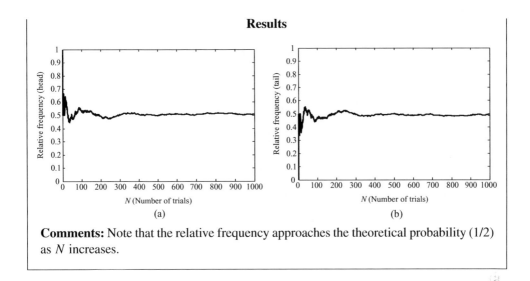

(a)                                                    (b)

**Comments:** Note that the relative frequency approaches the theoretical probability (1/2) as $N$ increases.

---

*Example 7.2:* **Demonstration of the central limit theorem**

The sum of independent random variables, $S_n = \sum_{k=1}^{n} X_k$, becomes normally distributed as $n$ gets large, regardless of individual distribution of $X_k$.

| Line | MATLAB code | Comments |
|------|-------------|----------|
| 1 | clear all | Initialize the random number generator, and generate |
| 2 | rand('state',1); | a matrix X whose elements are drawn from a |
| 3 | X=rand(10,5000); | uniform distribution on the unit interval. The matrix is 10×5000; we regard this as 10 independent random variables with a sample length 5000. |
| 4 | S1=X(1,:); | Generate the sum of random variables, e.g. S5 is the |
| 5 | S2=sum(X(1:2,:)); | sum of five random variables. |
| 6 | S5=sum(X(1:5,:)); | In this example, we consider four cases: S1, S2, S5 |
| 7 | S10=sum(X); | and S10. |
| 8 | nbin=20; N=length(X); |  |
| 9 | [n1 s1]=hist(S1, nbin); | Define the number of bins for the histogram. Then, |
| 10 | [n2 s2]=hist(S2, nbin); | calculate the frequency counts and bin locations for |
| 11 | [n5 s5]=hist(S5, nbin); | S1, S2, S5 and S10. |
| 12 | [n10 s10]=hist(S10, nbin); |  |
| 13 | figure(1) | Plot the histograms of S1, S2, S5 and S10. A |
| 14 | bar(s1, n1/N) | histogram is a graph that shows the distribution of |
| 15 | xlabel('\itS\rm_1') | data. In the histogram, the number of counts is |
| 16 | ylabel('Relative frequency'); axis([0 1 0 0.14]) | normalized by N. |

```
17      figure(2)
18      bar(s2, n2/N)
19      xlabel('\itS\rm_2')
20      ylabel('Relative frequency');
        axis([0 2 0 0.14])
21      figure(3)
22      bar(s5, n5/N)
23      xlabel('\itS\rm_5')
24      ylabel('Relative frequency');
        axis([0.4 4.7 0 0.14])
25      figure(4)
26      bar(s10, n10/N)
27      xlabel('\itS\rm_1_0')
28      ylabel('Relative frequency');
        axis([1.8 8 0 0.14])
```

Examine how the distribution changes as the number of sum $n$ increases.

## Results

(a) Number of sums: 1

(b) Number of sums: 2

(c) Number of sums: 5

(d) Number of sums: 10

**Comments:** Note that it quickly approaches a Gaussian distribution.

***Example 7.3:*** **Correlation coefficient as a measure of the linear relationship between two random variables** *X* **and** *Y*

Consider the correlation coefficient (i.e. Equation (7.47))

$$\rho_{xy} = \frac{\text{Cov}(X, Y)}{\sigma_x \sigma_y} = \frac{E[(X - \mu_x)(Y - \mu_y)]}{\sigma_x \sigma_y}$$

We shall compare three cases: (a) linearly related, $|\rho_{xy}| = 1$; (b) not linearly related, $|\rho_{xy}| = 0$; (c) partially linearly related, $0 < |\rho_{xy}| < 1$.

| Line | MATLAB code | Comments |
|------|-------------|----------|
| 1 | clear all | Initialize the random number generator, |
| 2 | randn('state' ,0); | and define a random variable X. |
| 3 | X=randn(1,1000); | Then, define a random variable Y1 that is |
| 4 | a=2; b=3; Y1=a*X+b; % fully related | linearly related to X, i.e. Y1 = aX+b. |
| 5 | Y2=randn(1,1000); % unrelated | Define another random variable Y2 |
| 6 | Y3=X+Y2; % partially related | which is not linearly related to X. Also, define a random variable Y3 which is partially linearly related to X. |
| 7 | N=length(X); | |
| 8 | s_xy1=sum((X-mean(X)).* (Y1-mean(Y1)))/(N-1); | Calculate the covariance of two random variables, Cov(X, Y1), Cov(X, Y2) and |
| 9 | s_xy2=sum((X-mean(X)).* (Y2-mean(Y2)))/(N-1); | Cov(X, Y3). See Equation (7.67) for a computational |
| 10 | s_xy3=sum((X-mean(X)).* (Y3-mean(Y3)))/(N-1); | formula. |
| 11 | r_xy1=s_xy1/(std(X)*std(Y1)) | Calculate the correlation coefficient for |
| 12 | r_xy2=s_xy2/(std(X)*std(Y2)) | each case. The results are: |
| 13 | r_xy3=s_xy3/(std(X)*std(Y3)) | r_xy1 = 1 (fully linearly related), r_xy2 = −0.0543 ($\approx$ 0, not linearly related), r_xy3 = 0.6529 (partially linearly related). |
| 14 | figure(1) | The degree of linear relationship between |
| 15 | plot(X,Y1, '.') | two random variables is visually |
| 16 | xlabel('\itX'); ylabel('\itY\rm1') | demonstrated. First, plot Y1 versus X; this gives a straight line. |
| 17 | figure(2) | Plot Y2 versus X; the result shows that |
| 18 | plot(X,Y2, '.') | two random variables are not related. |
| 19 | xlabel('\itX'); ylabel('\itY\rm2') | |
| 20 | figure(3) | Plot Y3 versus X; the result shows that |
| 21 | plot(X,Y3, '.') | there is some degree of linear |
| 22 | xlabel('\itX'); ylabel('\itY\rm3') | relationship, but not fully related. |

**Results**

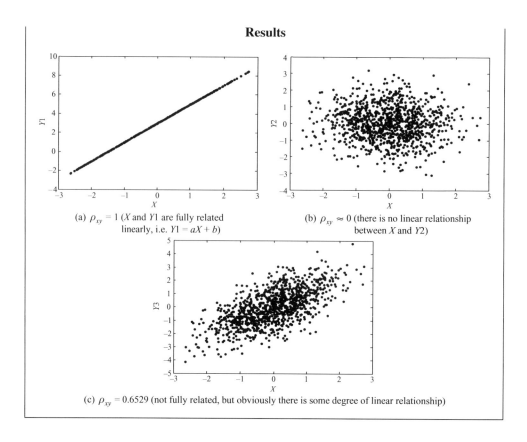

(a) $\rho_{xy} = 1$ ($X$ and $Y1$ are fully related
linearly, i.e. $Y1 = aX + b$)

(b) $\rho_{xy} \approx 0$ (there is no linear relationship
between $X$ and $Y2$)

(c) $\rho_{xy} = 0.6529$ (not fully related, but obviously there is some degree of linear relationship)

---

*Example 7.4:* **Application of the kurtosis coefficient to the machinery condition monitoring**

$$\text{Kurtosis coefficient} : \gamma_2 = \frac{E[(X - \mu_x)^4]}{\sigma_x^4} - 3$$

In this example, we use a 'real' measured signal. Two acceleration signals are stored in the file 'bearing_fault.mat':[1] one is measured on a rotating machine in good working order, and the other is measured on the same machine but with a faulty bearing that results in a series of spiky transients. Both are measured at a sampling rate of 10 kHz and are recorded for 2 seconds. The signals are then high-pass filtered with a cut-on frequency of 1 kHz to remove the rotating frequency component and its harmonics.

Since the machine has many other sources of (random) vibration, in 'normal' condition, the high-pass-filtered signal can be approximated as Gaussian, thus the kurtosis coefficient has a value close to zero, i.e. $\gamma_2 \approx 0$.

---

[1] The data files can be downloaded from the Companion Website (www.wiley.com/go/shin_hammond).

However, if the bearing is faulty, then the signal becomes non-Gaussian due to the transient components in the signal, and its distribution will be more peaky (*near its mean*) than Gaussian, i.e. $\gamma_2 > 0$ (leptokurtic).

| Line | MATLAB code | Comments |
|---|---|---|
| 1 | clear all | Load the measured signal, and let x be |
| 2 | load bearing_fault | the signal in good condition, y the |
| 3 | x=br_good; y=br_fault; | signal with a bearing fault. |
| 4 | N=length(x); | |
| 5 | kur_x=(sum((x-mean(x)).^4)/N)/(std(x,1)^4)-3 | Calculate the kurtosis coefficients of |
| 6 | kur_y=(sum((y-mean(y)).^4)/N)/(std(y,1)^4)-3 | both signals (see Equation (7.65)). The results are: kur_x = 0.0145 (i.e. $\gamma_2 \approx 0$) and kur_y = 1.9196 (i.e. leptokurtic). |
| 7 | [nx x1]=hist(x,31); | Calculate the frequency counts and bin |
| 8 | [ny y1]=hist(y,31); | locations for signals x and y. |
| 9 | figure(1); subplot(2,1,1) | |
| 10 | plot(t,x) | |
| 11 | xlabel('Time (s)'); ylabel('\itx\rm(\itt\rm)') | Plot the signal x, and compare with the |
| 12 | subplot(2,1,2) | corresponding histogram. |
| 13 | bar(x1, nx/N) | |
| 14 | xlabel('\itx'); ylabel('Relative frequency') | |
| 15 | axis([-1 1 0 0.2]) | |
| 16 | figure(2); subplot(2,1,1) | |
| 17 | plot(t,y) | |
| 18 | xlabel('Time (s)'); ylabel('\ity\rm(\itt\rm)') | Plot the signal y, and compare with the |
| 19 | subplot(2,1,2) | corresponding histogram. Also |
| 20 | bar(y1, ny/N) | compare with the signal x. |
| 21 | xlabel('\ity'); ylabel('Relative frequency') | |
| 22 | axis([-1 1 0 0.2]) | |

### Results

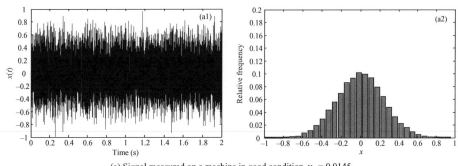

(a) Signal measured on a machine in good condition, $\gamma_2 = 0.0145$

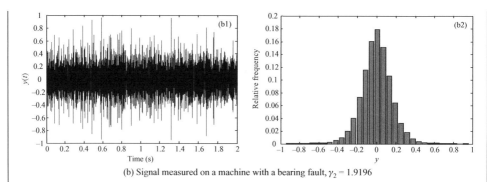

(b) Signal measured on a machine with a bearing fault, $\gamma_2 = 1.9196$

**Comments:** In this example, we have treated the measured time signal as a random variable. *Time-dependent* random variables (stochastic processes) are discussed in Chapter 8.

# 8

# Stochastic Processes; Correlation Functions and Spectra

## Introduction

In the previous chapter, we did not include 'time' in describing random processes. We shall now deal with measured signals which are time dependent, e.g. acoustic pressure fluctuations at a point in a room, a record of a vibration signal measured on a vehicle chassis, etc. In order to describe such (random) signals, we now extend our considerations of the previous chapter to a time-dependent random variable.

We introduce this by a simple example. Let us create a time history by tossing a coin every second, and for each 'head' we record a unit value and for each 'tail' we record a zero. We hold these ones and zeros for a second until the next coin toss. A sample record might look Figure 8.1.

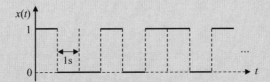

**Figure 8.1**   A sample time history created from tossing a coin

The sample space is $(H, T)$, the range space for $X$ is $(1, 0)$ and we have introduced time by parameterizing $X(\omega)$ as $X_t(\omega)$, i.e. for each $t$, $X$ is a random variable defined on a sample space. Now, we drop $\omega$ and write $X(t)$, and refer to this as a random function of time (shorthand for a random variable defined on a sample space indexed by time).

*Fundamentals of Signal Processing for Sound and Vibration Engineers*
K. Shin and J. K. Hammond.    © 2008 John Wiley & Sons, Ltd

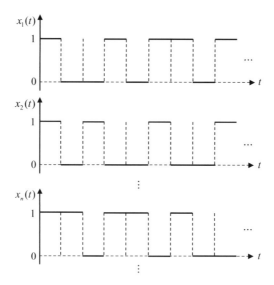

**Figure 8.2**   An example of an ensemble

We shall carry over the ideas introduced in the last chapter to these time series which display uncertainty referred to as stochastic processes. The temporal aspects require us to bring in some additional definitions and concepts.

Figure 8.1 depicts a single 'realization' of the stochastic process $X(t)$ (obtained by the coin tossing experiment). It could be finite in length or infinite, i.e. $-\infty < t < \infty$. Its random character introduces us to the concepts (or necessity) of replicating the experiments, i.e. producing additional realizations of it, which we could imagine as identical experiments run in parallel as shown in Figure 8.2.

The set of such realizations is called an *ensemble* (whether finite or infinite). This is sometimes written as $\{X(t)\}$ where $-\infty < t < \infty$.

## 8.1  PROBABILITY DISTRIBUTION ASSOCIATED WITH A STOCHASTIC PROCESS

We now consider a probability density function for a stochastic process. Let $x$ be a particular value of $X(t)$; then the distribution function at time $t$ is defined as

$$F(x, t) = P[X(t) \leq x] \tag{8.1}$$

and

$$P[x < X(t) \leq x + \delta x] = F(x + \delta x, t) - F(x, t) \tag{8.2}$$

Since

$$\lim_{\delta x \to 0} \frac{P[x < X(t) \leq x + \delta x]}{\delta x} = \lim_{\delta x \to 0} \frac{F(x + \delta x, t) - F(x, t)}{\delta x} = \frac{d F(x, t)}{dx}$$

the probability density function can be written as

$$p(x, t) = \frac{dF(x, t)}{dx} \qquad (8.3)$$

Note that the probability density function $p(x, t)$ for a stochastic process is time dependent, i.e. it evolves with time as shown in Figure 8.3.

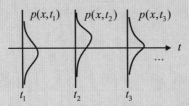

**Figure 8.3**   Evolution of the probability density function of a stochastic process

Alternatively, we may visualize this as below. We project the entire ensemble onto a single diagram and set up a gate as shown in Figure 8.4.

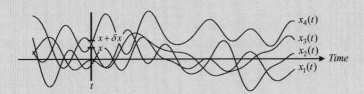

**Figure 8.4**   A collection of time histories

Now, we count the number of signals falling within the gate (say, $k$). Also we count the total number of signals (say, $N$). Then the relative frequency of occurrence of $X(t)$ in the gate at time $t$ is $k/N$. So, as $N$ gets large, we might say that $P[x < X(t) \le x + \delta x]$ is estimated by $k/N$ (for large $N$), so that

$$p(x, t) = \lim_{\delta x \to 0} \frac{P[x < X(t) \le x + \delta x]}{\delta x} = \lim_{\substack{N \to \infty \\ \delta x \to 0}} \frac{k}{N \delta x} \qquad (8.4)$$

It is at this point that the temporal evolution of the process introduces concepts additional to those in Chapter 7. We could conceive of describing how a process might change as time evolves, or how a process relates to itself at different times. We could do this by defining joint probability density functions by setting up additional gates.

For example, for two gates at times $t_1$ and $t_2$ this can be described pictorially as in Figure 8.5. Let $k_2$ be the number of signals falling within *both* gates in the figure. Then, the relative frequency $k_2/N$ estimates the joint probability for large $N$, i.e.

$$P[x_1 < X(t_1) \le x_1 + \delta x_1 \cap x_2 < X(t_2) \le x_2 + \delta x_2] \approx \frac{k_2}{N} \qquad (8.5)$$

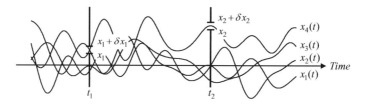

**Figure 8.5**   Pictorial description of the joint probability density function

Thus, the joint probability density function is written as

$$p(x_1, t_1; x_2, t_2) = \lim_{\delta x_1, \delta x_2 \to 0} \frac{P[x_1 < X(t_1) \le x_1 + \delta x_1 \cap x_2 < X(t_2) \le x_2 + \delta x_2]}{\delta x_1 \delta x_2}$$

$$= \lim_{\substack{N \to \infty \\ \delta x_1, \delta x_2 \to 0}} \frac{k_2}{N \delta x_1 \delta x_2} \tag{8.6}$$

Also, the joint distribution function is $F(x_1, t_1; x_2, t_2) = P[X(t_1) \le x_1 \cap X(t_2) \le x_2]$, so Equation (8.6) can be rewritten as

$$p(x_1, t_1; x_2, t_2) = \frac{\partial^2 F(x_1, t_1; x_2, t_2)}{\partial x_1 \partial x_2} \tag{8.7}$$

For a 'univariate' stochastic process, Equation (8.7) can be generalized to the $k$th-order joint probability density function as

$$p(x_1, t_1; x_2, t_2; \ldots; x_k, t_k) \tag{8.8}$$

However, we shall only consider the first and second order, i.e. $p(x, t)$ and $p(x_1, t_1; x_2, t_2)$.

## 8.2 MOMENTS OF A STOCHASTIC PROCESS

As we have defined the moments for random variables in Chapter 7, we now define moments for stochastic processes. The only difference is that 'time' is involved now, i.e. the moments of a stochastic process are time dependent. The first and second moments are as follows:

(a) First moment (mean):

$$\mu_x(t) = E[X(t)] = \int_{-\infty}^{\infty} x p(x, t) dx \tag{8.9}$$

(b) Second moment (mean square):

$$E[X^2(t)] = \int_{-\infty}^{\infty} x^2 p(x, t) dx \tag{8.10}$$

(c) Second central moment (variance):

$$\text{Var}(X(t)) = \sigma_x^2(t) = E[(X(t) - \mu_x(t))^2] = \int_{-\infty}^{\infty} (x - \mu_x(t))^2 p(x, t)dx \quad (8.11)$$

Note that $E[(X(t) - \mu_x(t))^2] = E[X^2(t)] - \mu_x^2(t)$, i.e.

$$\sigma_x^2(t) = E[X^2(t)] - \mu_x^2(t) \quad (8.12)$$

### Ensemble Averages

We noted the concept of the ensemble earlier, i.e. replications of the realizations of the process. We now relate the expected value operator $E$ to an ensemble average. Consider the ensemble shown in Figure 8.6.

Then, from the ensemble, we may estimate the mean by using the formula

$$\bar{x}(t) = \frac{1}{N} \sum_{n=1}^{N} X_n(t) \quad (8.13)$$

We now link Equation (8.13) to the theoretical average as follows. First, for a particular time $t$, group signals according to level (e.g. the gate defined by $x$ and $x + \delta x$). Suppose all $X_i(t)$ in the range $x_1$ and $x_1 + \delta x_1$ are grouped and the number of signals in the group is counted (say $k_1$). Then, repeating this for other groups, the mean value can be estimated from

$$\bar{x}(t) \approx x_1 \frac{k_1}{N} + x_2 \frac{k_2}{N} + \cdots = \sum_i x_i \frac{k_i}{N} \quad (8.14)$$

where $k_i/N$ is the relative frequency associated with the $i$th gate ($x_i$ to $x_i + \delta x_i$). Now, as $N \to \infty$, $k_i/N \to p(x_i, t)\delta x_i$, so

$$\lim_{N \to \infty} \frac{1}{N} \sum_{n=1}^{N} X_n(t) \to \int_{-\infty}^{\infty} xp(x, t)dx \quad (8.15)$$

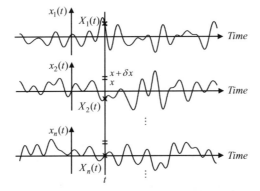

**Figure 8.6** An example of ensemble average

Thus, an average across the ensemble (the *infinite* set) is identified with the theoretical average, $\mu_x(t)$, i.e.

$$\mu_x(t) = E[X(t)] = \lim_{N \to \infty} \frac{1}{N} \sum_{n=1}^{N} X_n(t) \tag{8.16}$$

So, the operator $E[\ ]$ may be interpreted as the *E*xpectation or *E*nsemble average.

## 8.3 STATIONARITY

As we have seen in previous sections, the probability properties of a stochastic process are dependent upon time, i.e. they vary with time. However, to simplify the situation, we often assume that those statistical properties are in a 'steady state', i.e. they do not change *under a shift in time*. For example:

(a) $p(x, t) = p(x)$. This means that $\mu_x(t) = \mu_x$ and $\sigma_x^2(t) = \sigma_x^2$, i.e. the mean and variance are constant.

(b) $p(x_1, t_1; x_2, t_2) = p(x_1, t_1 + T; x_2, t_2 + T)$, i.e. $p(x_1, t_1; x_2, t_2)$ is a function of time difference $(t_2 - t_1)$ only, and does not explicitly depend on individual times $t_1$ and $t_2$.

(c) $p(x_1, t_1; x_2, t_2; \ldots; x_k, t_k) = p(x_1, t_1 + T; x_2, t_2 + T; \ldots; x_k, t_k + T)$ for all $k$.

If a process satisfies only two conditions (a) and (b), then we say it is *weakly stationary* or simply *stationary*. If the process satisfies the third condition also, i.e. all the *k*th-order joint probability density functions are invariant under a shift in time, then we say it is *completely* stationary. In this book, we assume that processes satisfy at least two conditions (a) and (b), i.e. we shall only consider stationary processes. Typical records of non-stationary and stationary data may be as shown in Figure 8.7.

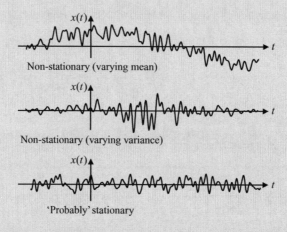

**Figure 8.7**   Typical 'sample' of non-stationary and stationary processes

In general all practical processes are non-stationary, thus the assumption of stationarity is only an approximation. However, in many practical situations, this assumption gives a sufficiently close approximation. For example, if we consider a vibration signal measured on a car body when the car is driven at varying speeds on rough roads, then the signal is obviously non-stationary since the statistical properties vary depending on the types of road and speed. However, if we locate a road whose surface is much the same over a 'long' stretch and drive the car over it at constant speed, then we might expect the vibration signal to have similar characteristics over much of its duration, i.e. 'approximately' stationary.

As we shall see later, the assumption of stationarity is very important, especially when we do not have an ensemble of data. In many situations, we have to deal with only a single record of data rather than a set of records. In such a case, we cannot perform the average across the *ensemble*, but we may average along *time*, i.e. we perform a *time average* instead of ensemble average. By implication, *stationarity* is a necessary condition for the time average to be meaningful. (Note that, for stationary processes, the statistical properties are independent of time.) The problem of deciding whether a process is stationary or not is often difficult and generally relies on prior information, though observations and statistical tests on time histories can be helpful (Priestley, 1981; Bendat and Piersol, 2000).

## 8.4 THE SECOND MOMENTS OF A STOCHASTIC PROCESS; COVARIANCE (CORRELATION) FUNCTIONS

### *The Autocovariance (Autocorrelation) Function*

As defined in Equation (8.11), the variance of a random variable for a stochastic process is written $\sigma_x^2(t) = E\left[(X(t) - \mu_x(t))^2\right]$. However, a simple generalization of the right hand side of this equation introduces an interesting concept, when written as $E[(X(t_1) - \mu_x(t_1))(X(t_2) - \mu_x(t_2))]$. This is the *autocovariance function* defined as

$$C_{xx}(t_1, t_2) = E[(X(t_1) - \mu_x(t_1))(X(t_2) - \mu_x(t_2))] \tag{8.17}$$

Similar to the covariance of two random variables defined in Chapter 7, the autocovariance function measures the 'degree of association' of the signal at time $t_1$ with *itself* at time $t_2$. If the mean value is not subtracted in Equation (8.17), it is called the *autocorrelation function* as given by

$$R_{xx}(t_1, t_2) = E[X(t_1)X(t_2)] \tag{8.18}$$

Note that, sometimes, the *normalized* autocovariance function, $C_{xx}(t_1, t_2)/[\sigma_x(t_1)\sigma_x(t_2)]$, is called the autocorrelation function, and it is also sometimes called an *autocorrelation coefficient*. Thus, care must be taken with the terminology.

If we limit our interest to stationary processes, since the statistical properties remain the same under a shift of time, Equation (8.17) can be simplified as

$$C_{xx}(t_2 - t_1) = E[(X(t_1) - \mu_x)(X(t_2) - \mu_x)] \tag{8.19}$$

Note that this is now a function of the time difference $(t_2 - t_1)$ only. By letting $t_1 = t$ and

$t_2 = t + \tau$, it can be rewritten as

$$C_{xx}(\tau) = E[(X(t) - \mu_x)(X(t + \tau) - \mu_x)] \tag{8.20}$$

where $\tau$ is called the lag. Note that when $\tau = 0$, $C_{xx}(0) = \text{Var}(X(t)) = \sigma_x^2$. Similarly, the autocorrelation function for a stationary process is

$$R_{xx}(\tau) = E[X(t)X(t + \tau)] \tag{8.21}$$

Note that $R_{xx}(\tau)$ is a continuous function of $\tau$ for a continuous stochastic process, and $C_{xx}(\tau)$ and $R_{xx}(\tau)$ are related such that

$$C_{xx}(\tau) = R_{xx}(\tau) - \mu_x^2 \tag{8.22}$$

### Interpretation of the Autocorrelation Function in Terms of the Ensemble

In Section 8.2, we have already seen that the mean value might be defined as an ensemble average (see Equation (8.16)), i.e.

$$\mu_x(t) = \lim_{N \to \infty} \frac{1}{N} \sum_{n=1}^{N} X_n(t)$$

We now apply the same principle to the autocorrelation function for a stationary process. For simplicity, we assume that the mean value is zero, i.e. we set $\mu_x = 0$.

For the $n$th record, we form $X_n(t)X_n(t + \tau)$ as shown in Figure 8.8, and average this product over all records, i.e. an ensemble average.

Then, we can write the autocorrelation function as

$$R_{xx}(\tau) = E[X(t)X(t + \tau)] = \lim_{N \to \infty} \frac{1}{N} \left[ \sum_{n=1}^{N} X_n(t)X_n(t + \tau) \right] \tag{8.23}$$

Since we assumed that $\mu_x = 0$, the autocorrelation function at zero lag is $R_{xx}(0) = \text{Var}(X(t)) = \sigma_x^2$. Also, as $\tau$ increases, it may be reasonable to say that the average

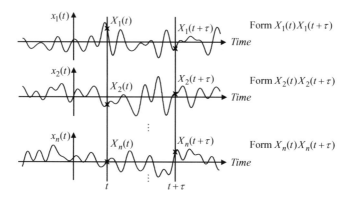

**Figure 8.8**   Ensemble average for the autocorrelation function

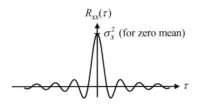

**Figure 8.9**  A typical autocorrelation function

$E[X(t)X(t + \tau)]$ should approach zero, since the values of $X(t)$ and $X(t + \tau)$ for large lags (time separations) are 'less associated (related)' if the process is random. Thus, the general shape of the autocorrelation function $R_{xx}(\tau)$ may be drawn as in Figure 8.9. Note that, as can be seen from the figure, the autocorrelation function is an *even* function of $\tau$ since $E[X(t)X(t + \tau)] = E[X(t - \tau)X(t)]$.

We note that the autocorrelation function does not always decay to zero. An example of when this does not happen is when the signal has a periodic form (see Section 8.6).

### The Cross-covariance (Cross-correlation) Function

If we consider two stochastic processes $\{X(t)\}$ and $\{Y(t)\}$ simultaneously, e.g. an input–output process, then we may generalize the above joint moment. Thus, the *cross-covariance function* is defined as

$$C_{xy}(t_1, t_2) = E[(X(t_1) - \mu_x(t_1))(Y(t_2) - \mu_y(t_2))] \tag{8.24}$$

and, if the mean values are not subtracted, the *cross-correlation function* is defined as

$$R_{xy}(t_1, t_2) = E[X(t_1)Y(t_2)] \tag{8.25}$$

Equation (8.24) or (8.25) is a measure of the association between the signal $X(t)$ at time $t_1$ and the signal $Y(t)$ at time $t_2$, i.e. it is a measure of *cross*-association. If we assume both signals are stationary, then $C_{xy}(t_1, t_2)$ or $R_{xy}(t_1, t_2)$ is a function of time difference $t_2 - t_1$. Then, as before, letting $t_1 = t$ and $t_2 = t + \tau$, the equations can be rewritten as

$$C_{xy}(\tau) = E[(X(t) - \mu_x)(Y(t + \tau) - \mu_y)] \tag{8.26}$$

and

$$R_{xy}(\tau) = E[X(t)Y(t + \tau)] \tag{8.27}$$

where their relationship is

$$C_{xy}(\tau) = R_{xy}(\tau) - \mu_x\mu_y \tag{8.28}$$

Also, the ensemble average interpretation becomes

$$R_{xy}(\tau) = E[X(t)Y(t + \tau)] = \lim_{N \to \infty} \frac{1}{N}\left[\sum_{n=1}^{N} X_n(t)Y_n(t + \tau)\right] \tag{8.29}$$

We shall consider examples of this later, but note here that $R_{xy}(\tau)$ can have a general shape (i.e. neither even, nor odd) and $R_{xy}(\tau) = R_{yx}(-\tau)$.

The cross-correlation (or cross-covariance) function is one of the most important concepts in signal processing, and is applied to various practical problems such as estimating time delays in a system: radar systems are a classical example; leak detection in buried plastic pipe is a more recent application (Gao *et al.*, 2006). Moreover, as we shall see later, together with the autocorrelation function, it can be directly related to the system identification problem.

### Properties of Covariance (Correlation) Functions

We now list some properties of covariance (correlation) functions; the examples to follow in Section 8.6 will serve to clarify these properties:

(a) *The autocovariance (autocorrelation) function:* First, we define the autocorrelation coefficient as

$$\rho_{xx}(\tau) = \frac{C_{xx}(\tau)}{\sigma_x^2} \left( = \frac{R_{xx}(\tau)}{R_{xx}(0)} \text{ for zero mean} \right) \tag{8.30}$$

where $\rho_{xx}(\tau)$ is the normalized (non-dimensional) form of the autocovariance function.
(i) $C_{xx}(\tau) = C_{xx}(-\tau); R_{xx}(\tau) = R_{xx}(-\tau)$ (i.e. the autocorrelation function is 'even')

$$\tag{8.31}$$

(ii) $\rho_{xx}(0) = 1; C_{xx}(0) = \sigma_x^2 (= R_{xx}(0)$ for zero mean) $\tag{8.32}$
(iii) $|C_{xx}(\tau)| \leq \sigma_x^2; |R_{xx}(\tau)| \leq R_{xx}(0)$, thus $-1 \leq \rho_{xx}(\tau) \leq 1$ $\tag{8.33}$

*Proof:* $E[(X(t) \pm X(t+\tau))^2] = E[X^2(t) + X^2(t+\tau) \pm 2X(t)X(t+\tau)] \geq 0$, thus $2R_{xx}(0) \geq 2|R_{xx}(\tau)|$ which gives the above result.

(b) *The cross-covariance (cross-correlation) function:* We define the cross-correlation coefficient as

$$\rho_{xy}(\tau) = \frac{C_{xy}(\tau)}{\sigma_x \sigma_y} \left( = \frac{R_{xy}(\tau)}{\sqrt{R_{xx}(0)R_{yy}(0)}} \text{ for zero mean} \right) \tag{8.34}$$

(i) $C_{xy}(-\tau) = C_{yx}(\tau); R_{xy}(-\tau) = R_{yx}(\tau)$ (neither odd nor even in general) $\tag{8.35}$
(ii) $|C_{xy}(\tau)|^2 \leq \sigma_x^2 \sigma_y^2; |R_{xy}(\tau)|^2 \leq R_{xx}(0)R_{yy}(0)$, thus $-1 \leq \rho_{xy}(\tau) \leq 1$ $\tag{8.36}$

*Proof:* For real values $a$ and $b$,

$$E[(aX(t) + bY(t+\tau))^2] = E[a^2 X^2(t) + b^2 Y^2(t+\tau) + 2abX(t)Y(t+\tau)] \geq 0$$

i.e. $a^2 R_{xx}(0) + 2abR_{xy}(\tau) + b^2 R_{yy}(0) \geq 0$, or if $b \neq 0$

$$(a/b)^2 R_{xx}(0) + 2(a/b)R_{xy}(\tau) + R_{yy}(0) \geq 0$$

The left hand side is a quadratic equation in $a/b$, and this may be rewritten as

$$[R_{xx}(0)(a/b) + R_{xy}(\tau)]^2 \geq R_{xy}^2(\tau) - R_{xx}(0)R_{yy}(0)$$

For *any* values of $a/b$, this inequality must be satisfied. Thus

$$R_{xy}^2(\tau) - R_{xx}(0)R_{yy}(0) \le 0$$

and so the result follows.

(iii) If $X(t)$ and $Y(t)$ are uncorrelated, $C_{xy}(\tau) = 0$; $R_{xy}(\tau) = \mu_x \mu_y$.

Note that the above correlation coefficients are particularly useful when $X(t)$ and $Y(t)$ have different scales. Although we have distinguished the covariance functions and correlation functions, their difference is the presence of mean values only. In most practical situations, the mean values are usually subtracted prior to some processing of data, so the correlation functions and the covariance functions are the same in effect. Consequently, the 'correlation' functions are often preferably used in engineering.

## 8.5 ERGODICITY AND TIME AVERAGES

The moments discussed in previous sections are based on the theoretical probability distributions of the stochastic processes and have been interpreted as ensemble averages, i.e. we need an infinite number of records whose statistical properties are identical. However, in general, ensemble averaging is not feasible as we usually have only a single realization (record) of limited length. Then, the only way to perform the average is along the time axis, i.e. a *time average* may be used in place of an ensemble average. The question is: do time averages along *one* record give the same results as an ensemble average? The answer is 'sometimes', and when they do, such averages are said to be *ergodic*.

Note that we *cannot* simply refer to a process as ergodic. Ergodicity must be related directly to the particular average in question, e.g. mean value, autocorrelation function and cross-correlation function, etc. Anticipating a result from statistical estimation theory, we can state that stationary processes are ergodic with respect to the mean and covariance functions. Thus, for example, the mean value can be written as

$$\mu_x = \lim_{T \to \infty} \frac{1}{T} \int_0^T x(t)\,dt \qquad (8.37)$$

i.e. the time average over *any* single time history will give the same value as the ensemble average $E[X(t)]$.

If we consider a signal with a finite length $T$, then the *estimate* of the mean value can be obtained by

$$\hat{\mu}_x = \bar{x} = \frac{1}{T} \int_0^T x(t)\,dt \qquad (8.38)$$

or if the signal is digitized using samples every $\Delta$ seconds so that $T = N\Delta$, then

$$\bar{x} = \frac{1}{N\Delta} \sum_{n=0}^{N-1} x(n\Delta)\Delta$$

thus

$$\bar{x} = \frac{1}{N} \sum_{n=0}^{N-1} x(n\Delta) \qquad (8.39)$$

Note that the mean value $\bar{x}$ is a single number characterizing the offset (or d.c. level) as being the same over the whole signal.

If the offset changes at some point (i.e. a simple type of non-stationary signal), e.g. at $t = T_1$ as shown in Figure 8.10, then the 'estimate' of the mean value using all $T$ seconds will produce a mean for the whole record – whereas it might have been preferable to split up the averaging into two segment to obtain $\bar{x}_1$ and $\bar{x}_2$.

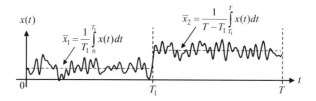

**Figure 8.10**   A varying mean non-stationary signal

This idea may be generalized to estimate a 'drifting' or 'slowly varying' mean value by using *local* averaging. The problem with local averaging (or local smoothing) is that, of necessity, fewer sample values are used in the computation and so the result is subject to more fluctuation (variability). Accordingly, if one wants to 'track' some feature of a non-stationary process then there is a trade-off between the need to have a local (short) average to follow the trends and a long enough segment so that sample fluctuations are not too great. The details of the estimation method and estimator errors will be discussed in Chapter 10.

Similar to the mean value, the estimate of time-averaged mean square value is (we follow the notation of Bendat and Piersol, 2000)

$$\overline{x^2} = \hat{\psi}_x^2 = \frac{1}{T} \int_0^T x^2(t)dt \qquad (8.40)$$

and in digital form

$$\hat{\psi}_x^2 = \frac{1}{N} \sum_{n=0}^{N-1} x^2(n\Delta) \qquad (8.41)$$

The root mean square (rms) is the positive square root of this. Also, the variance of the signal can be estimated as

$$\hat{\sigma}_x^2 = \frac{1}{T} \int_0^T (x(t) - \bar{x})^2 dt \tag{8.42}$$

In digital form, the unbiased estimator is

$$\hat{\sigma}_x^2 = \frac{1}{N-1} \sum_{n=0}^{N-1} (x(n\Delta) - \bar{x})^2 \tag{8.43}$$

where $\hat{\sigma}_x$ is the estimate of the standard deviation.

For the joint moments, the ensemble averages can also be replaced by the time averages if they are ergodic such that, for example, the cross-covariance function is

$$C_{xy}(\tau) = \lim_{T \to \infty} \frac{1}{T} \int_0^T (x(t) - \mu_x)(y(t+\tau) - \mu_y) dt \tag{8.44}$$

i.e. the time average shown above is equal to $E[(X(t) - \mu_x)(Y(t+\tau) - \mu_y)]$ and holds for *any* member of the ensemble. The (unbiased) estimate of the cross-covariance function is

$$\hat{C}_{xy}(\tau) = \frac{1}{T-\tau} \int_0^{T-\tau} (x(t) - \bar{x})(y(t+\tau) - \bar{y}) dt \qquad 0 \le \tau < T$$

$$= \frac{1}{T - |\tau|} \int_{|\tau|}^T (x(t) - \bar{x})(y(t+\tau) - \bar{y}) dt \qquad -T < \tau \le 0 \tag{8.45}$$

In Equation (8.45), if the divisor $T$ is used, it is called the biased estimate. Since $\hat{C}_{xy}(\tau) = \hat{C}_{yx}(-\tau)$ we may need to define the $\hat{C}_{xy}(\tau)$ for positive $\tau$ only. The corresponding digital version can be written as

$$\hat{C}_{xy}(m\Delta) = \frac{1}{N-m} \sum_{n=0}^{N-m-1} (x(n\Delta) - \bar{x})(y((n+m)\Delta) - \bar{y}) \quad 0 \le m \le N-1 \tag{8.46}$$

In this section, we have only defined unbiased estimators. Other estimators and corresponding errors will be discussed in Chapter 10. Based on Equations (8.44)–(8.46), the same form of equations can be used for the autocovariance function $C_{xx}(\tau)$ by simply replacing $y$ with $x$, and for correlation functions $R_{xx}(\tau)$ and $R_{xy}(\tau)$ by omitting the mean values in the expressions.

## 8.6 EXAMPLES

We now demonstrate several examples to illustrate probability density functions and covariance (correlation) functions.

### Probability Distribution of a Sine Wave[M8.1]

A sine wave $x(t) = A \sin(\omega t + \theta)$ may be considered random if the phase angle $\theta$ is random, i.e. $\theta$ is now a random variable, and so each realization has a phase drawn from some probability density function $p(\theta)$ ($A$ and $\omega$ are known constants).

For a fixed value of $t$ we shall compute the probability density function of $x$. To do this, we work from the first principles. We want $p(x, t) = dF(x, t)/dx$ where $F(x, t) = P[X(t) \leq x]$. Let us first calculate $F(x, t)$ and then differentiate with respect to $x$. We shall assume that

$$p(\theta) = \frac{1}{2\pi} \quad \text{for } 0 \leq \theta \leq 2\pi$$

$$= 0 \quad \text{otherwise} \tag{8.47}$$

i.e. $\theta$ is uniformly distributed. Then the distribution function is

$$F(\theta) = \begin{cases} 0 & \theta < 0 \\ \dfrac{\theta}{2\pi} & 0 \leq \theta \leq 2\pi \\ 1 & \theta > 2\pi \end{cases} \tag{8.48}$$

Since it is a stationary process, an arbitrary value of $t$ can be used, i.e. $F(x, t) = F(x)$ and $p(x, t) = p(x)$, so let $t = 0$ for convenience. Then

$$F(x) = P[X \leq x] = P[A \sin \theta \leq x] = P\left[\sin \theta \leq \frac{x}{A}\right] \tag{8.49}$$

This condition is equivalent to

$$P\left[\theta \leq \sin^{-1}\left(\frac{x}{A}\right)\right] = F(\theta)|_{\theta=\sin^{-1}(x/A)} \quad \text{and}$$

$$P\left[\pi - \sin^{-1}\left(\frac{x}{A}\right) < \theta \leq 2\pi\right] = F(\theta)|_{\theta=2\pi} - F(\theta)|_{\theta=\pi-\sin^{-1}(x/A)} \tag{8.50}$$

Note that, in the above 'equivalent' probability condition, the values of $\sin^{-1}(x/A)$ are defined in the range $-\pi/2$ to $\pi/2$. Also $F(\theta)|_\theta$ is allowed to have negative values. Then, the distribution function $F(x, t)$ becomes

$$F(x, t) = F(x) = \frac{1}{2} + \frac{1}{\pi} \sin^{-1}\left(\frac{x}{A}\right) \tag{8.51}$$

and this leads to the probability density function as

$$p(x) = \frac{dF(x)}{dx} = \frac{1}{\pi} \frac{1}{\sqrt{A^2 - x^2}} \tag{8.52}$$

which has the U shape of the probability density function of a sine wave as shown in Figure 8.11.

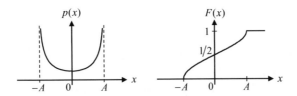

**Figure 8.11**   Probability density function and distribution function of a sine wave

As an alternative method, we demonstrate the use of a time average for the probability density calculation (assuming 'ergodicity' for the probability density function). Consider the sine wave above (and set the phase to zero for convenience, i.e. $\theta = 0$ for a particular realization). Then, we set up a gate $(x \leq x(t) \leq x + dx)$ as shown in Figure 8.12, and evaluate the time spent within the gate. Then

$$p(x)dx \approx \text{probability of lying in the gate} \approx \frac{\sum dt_i}{T} \tag{8.53}$$

where $T =$ the total record length, and $\sum dt_i =$ total time in the gate.

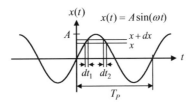

**Figure 8.12**   A sine wave with a gate

For the sine wave, we take one period, i.e. $T = T_P$. Since $dx = \omega A \cos(\omega t)dt$, it follows that

$$dt = \frac{dx}{\omega A \cos(\omega t)} = \frac{dx}{\omega A \sqrt{1 - (x/A)^2}} \tag{8.54}$$

Also, let $dt_1 = dt_2 = dt$, so

$$p(x)dx = \frac{2dt}{T_P} = \frac{2dx}{(2\pi/\omega)\omega A\sqrt{1 - (x/A)^2}} = \frac{dx}{\pi\sqrt{A^2 - x^2}} \tag{8.55}$$

which is the same result as Equation (8.52), i.e.

$$p(x) = \frac{1}{\pi\sqrt{A^2 - x^2}}$$

### The Autocorrelation (Autocovariance) Function

### A Sine Wave[M8.2]

Let $x(t) = A \sin(\omega t + \theta)$ where $\theta$ is a random variable with a uniform probability density function as discussed above. For any fixed value of $t$, let

$$x(t) = A \sin(\omega t + \theta) = x_1(\theta)$$
$$x(t + \tau) = A \sin[\omega(t + \tau) + \theta] = x_2(\theta) \tag{8.56}$$

The mean value is

$$\mu_x = E[x(t)] = E[x_1(\theta)] = \int_{-\infty}^{\infty} A \sin(\omega t + \theta) p(\theta) d\theta = 0 \tag{8.57}$$

Then the autocorrelation function becomes

$$R_{xx}(\tau) = E[x(t)x(t+\tau)] = E[x_1(\theta)x_2(\theta)]$$

$$= \int_{-\infty}^{\infty} A^2 \sin(\omega t + \theta) \sin[\omega(t+\tau) + \theta] p(\theta) d\theta$$

$$= \frac{A^2}{2\pi} \int_{0}^{2\pi} \frac{1}{2} [\cos(\omega\tau) - \cos(2\omega t + \omega\tau + 2\theta)] d\theta$$

$$= \frac{A^2}{2} \cos(\omega\tau) \tag{8.58}$$

which is a cosine function as shown in Figure 8.13. Note that this is an example where the autocorrelation does not decay to zero as $\tau \to \infty$.

Assuming 'ergodicity' for the autocorrelation function, the *time average* for a single trace $x(t) = A \sin(\omega t + \theta)$ gives the same result:

$$R_{xx}(\tau) = \lim_{T \to \infty} \frac{1}{T} \int_{0}^{T} x(t)x(t+\tau)dt = \lim_{T \to \infty} \frac{1}{2T} \int_{-T}^{T} x(t)x(t+\tau)dt$$

$$= \lim_{T \to \infty} \frac{1}{2T} \int_{-T}^{T} A^2 \sin(\omega t + \theta) \sin(\omega t + \omega\tau + \theta)dt$$

$$= \lim_{T \to \infty} \frac{A^2}{2T} \int_{-T}^{T} \frac{1}{2} [\cos(\omega\tau) - \cos(2\omega t + \omega\tau + 2\theta)]dt = \frac{A^2}{2} \cos(\omega\tau) \tag{8.59}$$

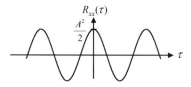

**Figure 8.13**   Autocorrelation function of a sine wave

It is appropriate to emphasize the meaning of the autocorrelation function by considering $R_{xx}(\tau)$ as the 'average' of the product $x(t)x(t + \tau)$. The term $x(t + \tau)$ is a shifted version of $x(t)$, so this product is easily visualized as the original time history $x(t)$ which is 'overlaid' by a shifted version of itself (when $\tau$ is positive the shift is to the left). So the product highlights the 'match' between $x(t)$ and itself shifted by $\tau$ seconds.

For the sine wave, which is periodic, this matching is 'perfect' for $\tau = 0$. When $\tau = 1/4$ period (i.e. $\tau = \pi/2\omega$), the positive and negative matches cancel out when integrated. When $\tau = 1/2$ period this is perfect matching again but with a sign reversal, and so on. This shows that the periodic signal has a periodic autocorrelation function. Note that, as can be seen in Equation (8.59), the autocorrelation (autocovariance) function does not depend on $\theta$, i.e. it is 'phase blind'.

### Asynchronous Random Telegraph Signal
Consider a time function that switches between two values $+a$ and $-a$ as shown in Figure 8.14. The crossing times $t_i$ are random and we assume that it is modelled as a Poisson process with a rate parameter $\lambda$. Then, the probability of $k$ crossings in time $\tau$ is

$$p_k = \frac{e^{-\lambda|\tau|}(\lambda|\tau|)^k}{k!} \tag{8.60}$$

where $\lambda$ is the number of crossings per unit time.

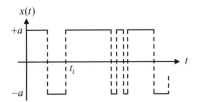

**Figure 8.14**   Asynchronous random telegraph signal

If we assume that the process is in steady state, i.e. $t \to \infty$, then $P[X(t) = a] = P[X(t) = -a] = 1/2$. So the mean value $\mu_x = E[x(t)] = 0$. And the product $x(t)x(t + \tau)$ is either $a^2$ or $-a^2$, i.e. it is $a^2$ if the number of crossings is even in time $\tau$ and $-a^2$ if the number of crossings is odd in time $\tau$. The total probability for $a^2$ (i.e. an even number of crossings occurs) is $\sum_{k=0}^{\infty} p_{2k}$, and the total probability for $-a^2$ is $\sum_{k=0}^{\infty} p_{2k+1}$. Thus, the autocorrelation function becomes

$$R_{xx}(\tau) = E\left[x(t)x(t + \tau)\right] = \sum_{k=0}^{\infty}\left[a^2 p_{2k} - a^2 p_{2k+1}\right]$$

$$= a^2 e^{-\lambda|\tau|}\left[\sum_{k=0}^{\infty}\left(\frac{(\lambda|\tau|)^{2k}}{(2k)!} - \frac{(\lambda|\tau|)^{2k+1}}{(2k+1)!}\right)\right] = a^2 e^{-\lambda|\tau|}\left[\sum_{k=0}^{\infty}\frac{(-\lambda|\tau|)^k}{k!}\right]$$

$$= a^2 e^{-2\lambda|\tau|} \tag{8.61}$$

which is an exponentially decaying function as shown in Figure 8.15, where the decay rate is controlled by the parameter $\lambda$.

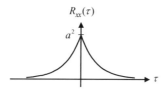

**Figure 8.15**   Autocorrelation function of an asynchronous random telegraph signal

**White Noise**
This is a very useful *theoretical* concept that has many desirable features in signal processing. In the previous example of the Poisson process, the autocorrelation function is defined as $R_{xx}(\tau) = a^2 e^{-2\lambda|\tau|}$. Note that, as the parameter $\lambda$ gets larger (i.e. the number of crossings per unit time increases), $R_{xx}(\tau)$ becomes narrower. We may relate this to the concept of white noise by considering a limiting form. As $\lambda \to \infty$, the process is very erratic and $R_{xx}(\tau)$ becomes 'spike-like'. In order that $R_{xx}(\tau)$ does not 'disappear completely' we can allow the value $a$ to become large in compensation. This gives an idea of a 'completely erratic' random process whose autocorrelation (autocovariance) function is like a delta function, and the process that has this property is called *white noise*, i.e.

$$\text{Autocorrelation function of white noise: } R_{xx}(\tau) = k\delta(\tau) \qquad (8.62)$$

An example of the autocorrelation function of white noise is shown in MATLAB Example 8.10 which demonstrates some important aspects of correlation analysis related to system identification. Note, however, that in continuous time such processes cannot occur in practice, and can only be approximated. Thus, we often refer to 'band-limited white noise' whose spectral density function is constant within a band as we shall see in Section 8.7.

**Synchronous Random Telegraph Signal**
Consider a switching signal where now the signal can only change sign at 'event points' spaced $\Delta$ seconds apart. At each event point the signal may *switch or not* (with equal probability) as shown in Figure 8.16.

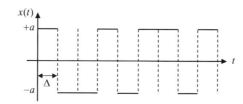

**Figure 8.16**   Synchronous random telegraph signal

Since this has equal probability, the mean value is $\mu_x = 0$. We shall calculate $R_{xx}(\tau)$ using time averages. For $0 \le \tau \le \Delta$, the product $x(t)x(t+\tau)$ is

$$x(t)x(t+\tau) = a^2 \text{ for a fraction } \frac{\Delta - \tau}{\Delta} \text{ of the time}$$

$$= a^2 \text{ for a fraction } \frac{1}{2}\left(\frac{\tau}{\Delta}\right) \text{ of the time}$$

$$= -a^2 \text{ for a fraction } \frac{1}{2}\left(\frac{\tau}{\Delta}\right) \text{ of the time} \tag{8.63}$$

Thus, the autocorrelation function becomes

$$R_{xx}(\tau) = \lim_{T \to \infty} \frac{1}{T} \int_0^T x(t)x(t+\tau)dt = a^2\left(1 - \frac{\tau}{\Delta}\right) \quad 0 \le \tau \le \Delta \tag{8.64}$$

Note that for $|\tau| > \Delta$, the probabilities of $a^2$ and $-a^2$ are the same, so

$$R_{xx}(\tau) = 0 \quad |\tau| > \Delta \tag{8.65}$$

As a result, the autocorrelation function is as shown in Figure 8.17.

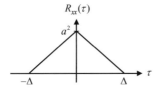

**Figure 8.17**   Autocorrelation function of a synchronous random telegraph signal

**A Simple Practical Problem**[M8.3]
To demonstrate an application of the autocorrelation function, consider the simple acoustic problem shown in Figure 8.18. The signal at the microphone may be written as

$$x(t) = as(t - \Delta_1) + bs(t - \Delta_2) \tag{8.66}$$

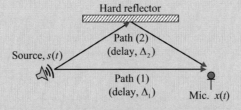

**Figure 8.18**   A simple acoustic example

We assume that the source signal is broadband, i.e. $R_{ss}(\tau)$ is narrow. By letting $\Delta = \Delta_2 - \Delta_1$ the autocorrelation function of the microphone signal $x(t)$ is

$$
\begin{aligned}
R_{xx}(\tau) &= E\left[x(t)x(t+\tau)\right]\\
&= E\left[(as(t-\Delta_1)+bs(t-\Delta_2))(as(t-\Delta_1+\tau)+bs(t-\Delta_2+\tau))\right]\\
&= (a^2+b^2)R_{ss}(\tau)+abR_{ss}(\tau-(\Delta_2-\Delta_1))+abR_{ss}(\tau+(\Delta_2-\Delta_1))\\
&= (a^2+b^2)R_{ss}(\tau)+abR_{ss}(\tau-\Delta)+abR_{ss}(\tau+\Delta) \qquad (8.67)
\end{aligned}
$$

That is, it consists of the autocorrelation function of the source signal and its shifted versions as shown in Figure 8.19. For this particular problem, the relative time delay $\Delta = \Delta_2 - \Delta_1$ can be identified from the autocorrelation function of $x(t)$, and also the relative distance can be found if the speed of sound is multiplied by $\Delta$.

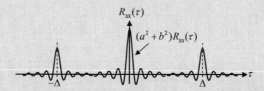

**Figure 8.19**   Autocorrelation function for time delay problem

We shall see later that if we also measure the source signal, then $\Delta_1$ can also be found by using the cross-correlation function. Thus, the complete transmission paths can be identified as long as $R_{ss}(\tau)$ is narrow compared with the relative time delay $\Delta = \Delta_2 - \Delta_1$. This will be demonstrated through a MATLAB example in Chapter 9.

**The Autocorrelation (Autocovariance) Function of Non-stochastic Processes**
It is worth noting that the time average definition may be utilized with non-random (i.e. deterministic) functions and even for transient phenomena. In such cases we may or may *not* use the divisor $T$.

1. *A square wave:* Consider a square periodic signal as shown in Figure 8.20. This function is periodic, so the autocorrelation (autocovariance) function will be *periodic*, and we use the autocorrelation function as

$$
R_{xx}(\tau) = \frac{1}{T_P}\int_0^{T_P} x(t)x(t+\tau)\,dt \qquad (8.68)
$$

To form $R_{xx}(\tau)$, we sketch $x(t+\tau)$ and 'slide it over $x(t)$' to form the integrand. Then, it can be easily verified that $R_{xx}(\tau)$ is a triangular wave as shown in Figure 8.21.

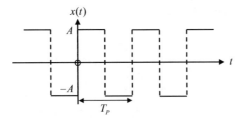

**Figure 8.20**   A square periodic signal

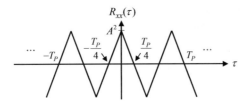

**Figure 8.21**   Autocorrelation function of a square wave

2. *A transient signal:* In such a case there is no point in dividing by $T$, so the autocorrelation function for a transient signal is defined as

$$R_{xx}(\tau) = \int_{-\infty}^{\infty} x(t)x(t+\tau)dt \qquad (8.69)$$

We note an important link with the frequency domain, i.e. if

$$x(t) = \int_{-\infty}^{\infty} X(f)e^{j2\pi ft}df \quad \text{and} \quad X(f) = \int_{-\infty}^{\infty} x(t)e^{-j2\pi ft}dt$$

then the Fourier transform of $R_{xx}(\tau)$ is

$$F\{R_{xx}(\tau)\} = \int_{-\infty}^{\infty} R_{xx}(\tau)e^{-j2\pi f\tau}d\tau$$

$$= \int_{-\infty}^{\infty}\int_{-\infty}^{\infty} x(t)x(t+\tau)e^{-j2\pi f\tau}d\tau dt \quad (\text{let } t_1 = t+\tau)$$

$$= \int_{-\infty}^{\infty} x(t_1)e^{-j2\pi ft_1}dt_1 \int_{-\infty}^{\infty} x(t)e^{j2\pi ft}dt$$

$$= X(f)X^*(f) = |X(f)|^2 \qquad (8.70)$$

Thus, the following relationship holds:

$$|X(f)|^2 = \int_{-\infty}^{\infty} R_{xx}(\tau) e^{-j2\pi f\tau} d\tau \tag{8.71}$$

$$R_{xx}(\tau) = \int_{-\infty}^{\infty} |X(f)|^2 e^{j2\pi f\tau} df \tag{8.72}$$

i.e. the 'energy spectral density' and the autocorrelation function are Fourier pairs. This will be discussed further in Section 8.7.

### The Cross-correlation (Cross-covariance) Function

**Two Harmonic Signals**[M8.4]

Consider the two functions

$$
\begin{aligned}
x(t) &= A\sin(\omega t + \theta_x) + B \\
y(t) &= C\sin(\omega t + \theta_y) + D\sin(n\omega t + \phi)
\end{aligned}
\tag{8.73}
$$

We form the cross-correlation function using the time average, i.e.

$$
\begin{aligned}
R_{xy}(\tau) &= \lim_{T\to\infty} \frac{1}{T} \int_0^T x(t) y(t+\tau) dt \\
&= \frac{1}{2} AC \cos\left[\omega\tau - (\theta_x - \theta_y)\right]
\end{aligned}
\tag{8.74}
$$

and compare this with the autocorrelation functions which are given as

$$
\begin{aligned}
R_{xx}(\tau) &= \frac{A^2}{2} \cos(\omega\tau) + B^2 \\
R_{yy}(\tau) &= \frac{C^2}{2} \cos(\omega\tau) + \frac{D^2}{2} \cos(n\omega\tau)
\end{aligned}
\tag{8.75}
$$

Note that the cross-correlation function finds the components in $y(t)$ that match or fit $x(t)$. More importantly, the cross-correlation *preserves the relative phase* $(\theta_x - \theta_y)$, i.e. it detects the delay that is associated with the 'correlated (in a linear manner)' components of $x(t)$ and $y(t)$.

Once again an intuitive idea of what a cross-correlation reveals arises from the visualization of the product $x(t)y(t+\tau)$ as looking at the match between $x(t)$ and the shifted version of $y(t)$. In the above example the oscillation with frequency $\omega$ in $y(t)$ matches that in $x(t)$, but the harmonic $n\omega$ does not. So the cross-correlation reveals this match and also the phase shift (delay) between these components.

## A Signal Buried in Noise[M8.5]

Consider a signal buried in noise, i.e. $y(t) = s(t) + n(t)$, as shown in Figure 8.22.

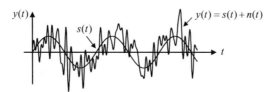

**Figure 8.22** A sinusoidal signal buried in noise

We assume that the noise and signal are uncorrelated: for example, $s(t)$ is a sine wave and $n(t)$ is wideband noise. Then, the cross-correlation function of the signal $s(t)$ and noise $n(t)$ is $R_{sn}(\tau) = E[s(t)n(t + \tau)] = \mu_s \mu_n$, i.e. $C_{sn}(\tau) = E[(s(t) - \mu_s)(n(t + \tau) - \mu_n)] = 0$. Note that the cross-covariance function of two uncorrelated signals is zero for all $\tau$. Thus, the autocorrelation function of $y(t)$ becomes

$$R_{yy}(\tau) = E\left[(s(t) + n(t))(s(t + \tau) + n(t + \tau))\right]$$
$$= E\left[s(t)s(t + \tau)\right] + E\left[n(t)n(t + \tau)\right] + 2\mu_s \mu_n \qquad (8.76)$$

Assuming that the mean values are zero, this is

$$R_{yy}(\tau) = R_{ss}(\tau) + R_{nn}(\tau) \qquad (8.77)$$

Since the autocorrelation function of the noise $R_{nn}(\tau)$ decays very rapidly (see Equation (8.62)), the autocorrelation function of the signal $R_{ss}(\tau)$ will dominate for larger values of $\tau$, as shown in Figure 8.23. This demonstrates a method of identifying sinusoidal components embedded in noise.

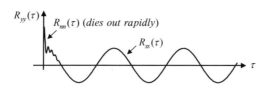

**Figure 8.23** Autocorrelation function of a sinusoidal signal buried in noise

## Time Delay Problem[M8.6]

Consider a wheeled vehicle moving over rough terrain as shown in Figure 8.24. Let the time function (profile) experienced by the leading wheel be $x(t)$ and that by the trailing wheel be $y(t)$. Also let the autocorrelation function of $x(t)$ be $R_{xx}(\tau)$. We now investigate the properties of the cross-correlation function $R_{xy}(\tau)$.

Assume that the vehicle moves at a constant speed $V$. Then, $y(t) = x(t - \Delta)$ where $\Delta = L/V$. So the cross-correlation function is

$$R_{xy}(\tau) = E[x(t)y(t + \tau)] = E[x(t)x(t + \tau - \Delta)]$$
$$= R_{xx}(\tau - \Delta) \qquad (8.78)$$

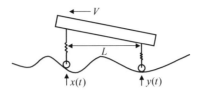

**Figure 8.24**   A wheeled vehicle moving over rough terrain

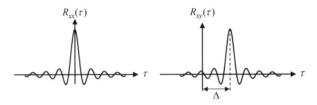

**Figure 8.25**   Autocorrelation and cross-correlation functions for time delay problem

That is, the cross-correlation function $R_{xy}(\tau)$ becomes a delayed version of $R_{xx}(\tau)$ as shown in Figure 8.25. The cross-correlation function detects the time delay between the two signals.

The detection of time delay using the cross-correlation function has been applied to many problems, e.g. radar systems, acoustic source localization, mechanical fault detection, pipe leakage detection, earthquake location, etc. The basic concept of using the cross-correlation function for a simplified radar system is demonstrated in MATLAB Example 8.7.

## 8.7  SPECTRA

So far we have discussed stochastic processes in the time domain. We now consider frequency domain descriptors. In Part I, Fourier methods were applied to deterministic phenomena, e.g. periodic and transient signals. We shall now consider Fourier methods for stationary random processes.

Consider a truncated sample function $x_T(t)$ of a random process $x(t)$ as shown in Figure 8.26, i.e.

$$x_T(t) = x(t) \quad |t| < T/2$$
$$= 0 \qquad \text{otherwise}$$

(8.79)

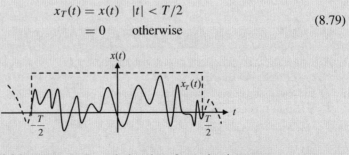

**Figure 8.26**   A truncated sample function of a stochastic process

We shall consider the decomposition of the power of this sample function in the frequency domain. As seen from the figure, the truncated signal $x_T(t)$ is pulse-like, i.e. it can be regarded as a transient signal. Thus, it can be represented by the Fourier integral

$$x_T(t) = \int_{-\infty}^{\infty} X_T(f)e^{j2\pi ft}df \tag{8.80}$$

Since the total energy of the signal $\int_{-\infty}^{\infty} x_T^2(t)dt$ tends to infinity as $T$ gets large, we shall consider the average power of the signal, i.e.

$$\frac{1}{T} \int_{-\infty}^{\infty} x_T^2(t)dt$$

Then, by Parseval's theorem it can be shown that

$$\frac{1}{T} \int_{-\infty}^{\infty} x_T^2(t)dt = \frac{1}{T} \int_{-T/2}^{T/2} x_T^2(t)dt = \frac{1}{T} \int_{-\infty}^{\infty} |X_T(f)|^2 df = \int_{-\infty}^{\infty} \frac{1}{T}|X_T(f)|^2 df \tag{8.81}$$

where the quantity $|X_T(f)|^2/T$ is called the *raw (or sample) power spectral density*, which is denoted as

$$\hat{S}_{xx}(f) = \frac{1}{T}|X_T(f)|^2 \tag{8.82}$$

Note that the power of the signal in a data segment of length $T$ is

$$\frac{1}{T} \int_{-\infty}^{\infty} x_T^2(t)dt = \int_{-\infty}^{\infty} \hat{S}_{xx}(f)df \tag{8.83}$$

Now, as $T \to \infty$ Equation (8.81) can be written as

$$\lim_{T\to\infty} \frac{1}{T} \int_{-T/2}^{T/2} x_T^2(t)dt = \int_{-\infty}^{\infty} \lim_{T\to\infty} \frac{|X_T(f)|^2}{T}df \tag{8.84}$$

Note that the left hand side of the equation is the average power of the sample function, thus it may be tempting to define $\lim_{T\to\infty}|X_T(f)|^2/T$ as the power spectral density. However we shall see (later) that $\hat{S}_{xx}(f)$ does not converge (in a statistical sense) as $T \to \infty$, which is the reason that the term 'raw' is used. In Chapter 10, we shall see that $\hat{S}_{xx}(f)$ evaluated from a larger data length is just as erratic as for the shorter data length, i.e. the estimate $\hat{S}_{xx}(f)$ cannot be improved simply by using more data (even for $T \to \infty$). We shall also see later (in Chapter 10) that the *standard deviation of the estimate is as great as the quantity being estimated!* That is, it is independent of data length and equal to the *true* spectral density as follows:

$$\text{Var}\left(\hat{S}_{xx}(f)\right) = S_{xx}^2(f) \quad \left(\text{or} \quad \frac{\text{Var}\left(\hat{S}_{xx}(f)\right)}{S_{xx}^2(f)} = 1\right) \tag{8.85}$$

In fact, we have now come across an estimate for which ergodicity does not hold, i.e. $\hat{S}_{xx}(f)$ is *not* ergodic. So some method of reducing the variability is required.

We do this by *averaging* the raw spectral density to remove the erratic behaviour. Consider the following average

$$E\left[\lim_{T\to\infty}\frac{1}{T}\int_{-T/2}^{T/2}x_T^2(t)dt\right] = E\left[\int_{-\infty}^{\infty}\lim_{T\to\infty}\frac{|X_T(f)|^2}{T}df\right] \tag{8.86}$$

Assuming *zero mean values*, the left hand side of Equation (8.86) is the variance of the process, thus it can be written as

$$\text{Var}(x(t)) = \sigma_x^2 = \int_{-\infty}^{\infty}S_{xx}(f)df \tag{8.87}$$

where

$$S_{xx}(f) = \lim_{T\to\infty}\frac{E\left[|X_T(f)|^2\right]}{T} \tag{8.88}$$

This function is called the *power spectral density function* of the process, and it states that the average power of the process (the variance) is decomposed in the frequency domain through the function $S_{xx}(f)$, which has a clear physical interpretation. Furthermore there is a direct relationship with the autocorrelation function such that

$$S_{xx}(f) = \int_{-\infty}^{\infty}R_{xx}(\tau)e^{-j2\pi f\tau}d\tau \tag{8.89}$$

$$R_{xx}(\tau) = \int_{-\infty}^{\infty}S_{xx}(f)e^{j2\pi f\tau}df \tag{8.90}$$

These relations are sometimes called the Wiener–Khinchin theorem.

Note that, if $\omega$ is used, the equivalent result is

$$S_{xx}(\omega) = \int_{-\infty}^{\infty}R_{xx}(\tau)e^{-j\omega\tau}d\tau \tag{8.91}$$

$$R_{xx}(\tau) = \frac{1}{2\pi}\int_{-\infty}^{\infty}S_{xx}(\omega)e^{j\omega\tau}d\omega \tag{8.92}$$

Similar to the Fourier transform pair, the location of the factor $2\pi$ may be interchanged or replaced with $1/\sqrt{2\pi}$ for symmetrical form. The proof of the above Fourier pair

(Equations (8.89)–(8.92)) needs some elements discussed in Chapter 10, so this will be justified later.

Note that the function $S_{xx}(f)$ is an *even* function of frequency and is sometimes called the two-sided power spectral density. If $x(t)$ is in volts, $S_{xx}(f)$ has units of volts$^2$/Hz. Often a one-sided power spectral density is defined as

$$
\begin{aligned}
G_{xx}(f) &= 2S_{xx}(f) & f > 0 \\
&= S_{xx}(f) & f = 0 \\
&= 0 & f < 0
\end{aligned}
\tag{8.93}
$$

### *Examples of Power Spectral Density Functions*[1]

(a) If $R_{xx}(\tau) = k\delta(\tau)$, $k > 0$, i.e. white noise, then

$$
S_{xx}(f) = \int_{-\infty}^{\infty} R_{xx}(\tau)e^{-j2\pi f\tau}\,d\tau = \int_{-\infty}^{\infty} k\delta(\tau)e^{-j2\pi f\tau}\,d\tau = ke^{-j2\pi f\cdot 0} = k
\tag{8.94}
$$

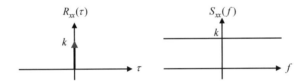

**Figure 8.27**   Power spectral density of white noise

Note that a 'narrow' autocorrelation function results in a broadband spectrum (Figure 8.27).

(b) If $R_{xx}(\tau) = \sigma_x^2 e^{-\lambda|\tau|}$, $\lambda > 0$, then

$$
S_{xx}(f) = \int_{-\infty}^{\infty} R_{xx}(\tau)e^{-j2\pi f\tau}\,d\tau = \int_{-\infty}^{\infty} \sigma_x^2 e^{-\lambda|\tau|}e^{-j2\pi f\tau}\,d\tau
$$

$$
= \sigma_x^2 \left[ \int_{-\infty}^{0} e^{\lambda\tau}e^{-j2\pi f\tau}\,d\tau + \int_{0}^{\infty} e^{-\lambda\tau}e^{-j2\pi f\tau}\,d\tau \right] = \frac{2\lambda\sigma_x^2}{\lambda^2 + (2\pi f)^2}
\tag{8.95}
$$

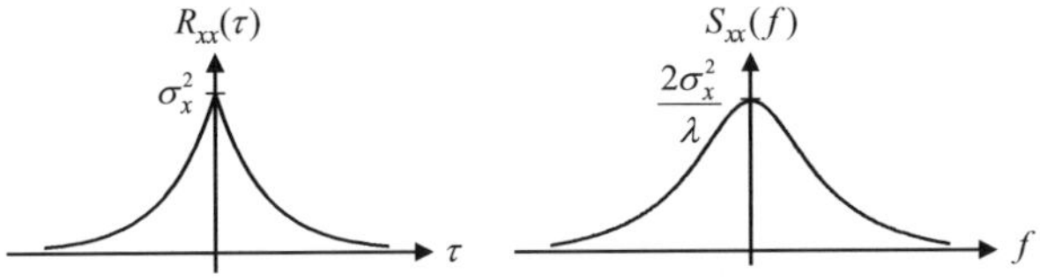

**Figure 8.28**   Exponentially decaying autocorrelation and corresponding power spectral density

The exponentially decaying autocorrelation function results in a mainly low-frequency power spectral density function (Figure 8.28).

[1] See examples in Section 4.3 and compare.

(c) If $R_{xx}(\tau) = (A^2/2)\cos(2\pi f_0 \tau)$, then

$$S_{xx}(f) = \int_{-\infty}^{\infty} R_{xx}(\tau)e^{-j2\pi f\tau}d\tau = \frac{A^2}{2}\int_{-\infty}^{\infty}\frac{1}{2}\left(e^{j2\pi f_0\tau} + e^{-j2\pi f_0\tau}\right)e^{-j2\pi f\tau}d\tau$$

$$= \frac{A^2}{4}\int_{-\infty}^{\infty}\left(e^{-j2\pi(f-f_0)\tau} + e^{-j2\pi(f+f_0)\tau}\right)d\tau = \frac{A^2}{4}\delta(f - f_0) + \frac{A^2}{4}\delta(f + f_0)$$

$$(8.96)$$

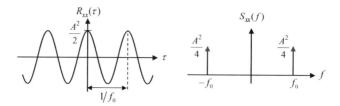

**Figure 8.29**   Sinusoidal autocorrelation and corresponding power spectral density

An oscillatory autocorrelation function corresponds to spikes in the power spectral density function (Figure 8.29).

(d) *Band-limited white noise*: If the power spectral density function is

$$S_{xx}(f) = a \quad -B < f < B$$
$$= 0 \quad \text{otherwise}$$
$$(8.97)$$

then the corresponding autocorrelation function (shown in Figure 8.30) is

$$R_{xx}(\tau) = \int_{-\infty}^{\infty} S_{xx}(f)e^{j2\pi f\tau}df = \int_{-B}^{B} ae^{j2\pi f\tau}df = 2aB\frac{\sin(2\pi B\tau)}{2\pi B\tau} \qquad (8.98)$$

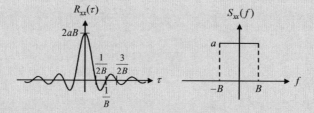

**Figure 8.30**   Autocorrelation and power spectral density of band-limited white noise

### The Cross-spectral Density Function

Generalizing the Wiener–Khinchin theorem, the cross-spectral density function is 'defined' as

$$S_{xy}(f) = \int_{-\infty}^{\infty} R_{xy}(\tau) e^{-j2\pi f\tau} d\tau \qquad (8.99)$$

with inverse

$$R_{xy}(\tau) = \int_{-\infty}^{\infty} S_{xy}(f) e^{j2\pi f\tau} df \qquad (8.100)$$

As with the power spectral density function, if $\omega$ is used in place of $f$, then

$$S_{xy}(\omega) = \int_{-\infty}^{\infty} R_{xy}(\tau) e^{-j\omega\tau} d\tau \qquad (8.101)$$

$$R_{xy}(\tau) = \frac{1}{2\pi} \int_{-\infty}^{\infty} S_{xy}(\omega) e^{j\omega\tau} d\omega \qquad (8.102)$$

Alternatively, $S_{xy}(f)$ is defined as

$$S_{xy}(f) = \lim_{T \to \infty} \frac{E[X_T^*(f) Y_T(f)]}{T} \qquad (8.103)$$

where $X_T(f)$ and $Y_T(f)$ are Fourier transforms of truncated functions $x_T(t)$ and $y_T(t)$ defined for $|t| < T/2$ (see Figure 8.26).

The equivalence of Equations (8.99) and (8.103) may be justified in the same manner as for the power spectral density functions as discussed in Chapter 10.

In general, the cross-spectral density function is *complex* valued, i.e.

$$S_{xy}(f) = |S_{xy}(f)| e^{j \arg S_{xy}(f)} \qquad (8.104)$$

This can be interpreted as the frequency domain equivalent of the cross-correlation function. That is, $|S_{xy}(f)|$ is the *cross-amplitude spectrum* and it shows whether frequency components in one signal are 'associated' with large or small amplitude at the same frequency in the other signal, i.e. it is the measure of association of amplitude in $x$ and $y$ at frequency $f$; $\arg S_{xy}(f)$ is the *phase spectrum* and this shows whether frequency components in one signal 'lag' or 'lead' the components at the same frequency in the other signal, i.e. it shows lags/leads (or phase difference) between $x$ and $y$ at frequency $f$.

### Properties of the Cross-spectral Density Function

(a) An important property is

$$\boxed{S_{xy}(f) = S_{yx}^*(f)} \qquad (8.105)$$

This can be easily proved using the fact that $R_{xy}(\tau) = R_{yx}(-\tau)$.

(b) A one-sided cross-spectral density function $G_{xy}(f)$ is defined as

$$
\begin{aligned}
G_{xy}(f) &= 2S_{xy}(f) & f &> 0 \\
&= S_{xy}(f) & f &= 0 \\
&= 0 & f &< 0
\end{aligned}
\tag{8.106}
$$

(c) The coincident spectral density (co-spectrum) and quadrature spectral density (quad-spectrum) are defined as (Bendat and Piersol, 2000)

$$
G_{xy}(f) = 2 \int_{-\infty}^{\infty} R_{xy}(\tau)e^{-j2\pi f\tau}\,d\tau = C_{xy}(f) - jQ_{xy}(f) \quad f \geq 0
\tag{8.107}
$$

where $C_{xy}(f)$ and $Q_{xy}(f)$ are called the co-spectra and quad-spectra, respectively (the reason for these names is explained later in the example given in page 249). $C_{xy}(f)$ is an even function of $f$ and $Q_{xy}(f)$ is an odd function of $f$. Also, by writing $R_{xy}(\tau)$ as the sum of even and odd parts (see Equation (3.18)), then

$$
C_{xy}(f) = 2 \int_{0}^{\infty} \left[ R_{xy}(\tau) + R_{yx}(\tau) \right] \cos(2\pi f\tau)d\tau = C_{xy}(-f)
\tag{8.108}
$$

$$
Q_{xy}(f) = 2 \int_{0}^{\infty} \left[ R_{xy}(\tau) - R_{yx}(\tau) \right] \sin(2\pi f\tau)d\tau = -Q_{xy}(-f)
\tag{8.109}
$$

Similar to $S_{xy}(f)$, the one-sided cross-spectral density function $G_{xy}(f)$ can be written as

$$
G_{xy}(f) = \left| G_{xy}(f) \right| e^{j \arg G_{xy}(f)}
\tag{8.110}
$$

Then, it can be shown that

$$
\left| G_{xy}(f) \right| = \sqrt{C_{xy}^2(f) + Q_{xy}^2(f)}
\tag{8.111}
$$

and

$$
\arg G_{xy}(f) = -\tan^{-1}\left( \frac{Q_{xy}(f)}{C_{xy}(f)} \right)
\tag{8.112}
$$

(d) For the phase of $S_{xy}(f) = \left| S_{xy}(f) \right| e^{j \arg S_{xy}(f)}$, let $\theta_x(f)$ and $\theta_y(f)$ be the phase components at frequency $f$ corresponding to $x(t)$ and $y(t)$, respectively. Then, $\arg S_{xy}(f)$ gives the phase difference such that

$$
\arg S_{xy}(f) = -\left[ \theta_x(f) - \theta_y(f) \right]
\tag{8.113}
$$

which is $\theta_y(f) - \theta_x(f)$. Note that it is *not* $\theta_x(f) - \theta_y(f)$ (see Equation (8.103) where $X_T^*(f)$ is multiplied by $Y_T(f)$, and also compare it with Equations (8.73) and (8.74)). Thus, in some texts, $S_{xy}(f)$ is defined as

$$S_{xy}(f) = \left|S_{xy}(f)\right| e^{-j\theta_{xy}(f)} \tag{8.114}$$

where $\theta_{xy}(f) = \theta_x(f) - \theta_y(f)$. So, care must be taken with the definitions.

(e) A useful inequality satisfied by $S_{xy}(f)$ is

$$\left|S_{xy}(f)\right|^2 \le S_{xx}(f)S_{yy}(f) \tag{8.115}$$

or

$$\left|G_{xy}(f)\right|^2 \le G_{xx}(f)G_{yy}(f) \tag{8.116}$$

The proof of this result is given in Appendix B. We shall see later that this is a particularly useful result – we shall define, in Chapter 9, the 'coherence function' as $\left|S_{xy}(f)\right|^2/(S_{xx}(f)S_{yy}(f))$) which is a *normalized* cross-spectral density function.

### Examples of Cross-spectral Density Functions

Two examples are as follows:

(a) Consider two functions (see also Equation (8.73) in Section 8.6)[M8.8]

$$x(t) = A \sin(2\pi pt + \theta_x)$$
$$y(t) = C \sin(2\pi pt + \theta_y) + D \sin(n2\pi pt + \phi) \tag{8.117}$$

In the previous section, it was shown that the cross-correlation function is

$$R_{xy}(\tau) = \frac{1}{2}AC \cos\left[2\pi p\tau - (\theta_x - \theta_y)\right] \tag{8.118}$$

Let $\theta_{xy} = \theta_x - \theta_y$; then

$$R_{xy}(\tau) = \frac{1}{2}AC \cos(2\pi p\tau - \theta_{xy}) \tag{8.119}$$

The cross-spectral density function is

$$S_{xy}(f) = \int\limits_{-\infty}^{\infty} R_{xy}(\tau)e^{-j2\pi f\tau}d\tau = \frac{AC}{4}\int\limits_{-\infty}^{\infty}\left(e^{j(2\pi p\tau - \theta_{xy})} + e^{-j(2\pi p\tau - \theta_{xy})}\right)e^{-j2\pi f\tau}d\tau$$

$$= \frac{AC}{4}\int\limits_{-\infty}^{\infty}\left(e^{-j2\pi(f-p)\tau}e^{-j\theta_{xy}} + e^{-j2\pi(f+p)\tau}e^{j\theta_{xy}}\right)d\tau$$

$$= \frac{AC}{4}\left[\delta(f-p)e^{-j\theta_{xy}} + \delta(f+p)e^{j\theta_{xy}}\right] \tag{8.120}$$

and the one-sided cross-spectral density function is

$$G_{xy}(f) = \frac{AC}{2}\delta(f - p)e^{-j\theta_{xy}} \tag{8.121}$$

Thus, it can be seen that

$$|G_{xy}(f)| = \underbrace{\frac{AC}{2}\delta(f - p)}_{\text{Amplitude association}} \tag{8.122}$$

and

$$\arg G_{xy}(f) = -\theta_{xy} = \underbrace{-(\theta_x - \theta_y)}_{\text{Phase difference}} \tag{8.123}$$

From $G_{xy}(f)$ in Equation (8.121), we see that the co-spectra and quad-spectra are

$$C_{xy}(f) = \frac{AC}{2}\delta(f - p)\cos\theta_{xy} \tag{8.124}$$

$$Q_{xy}(f) = \frac{AC}{2}\delta(f - p)\sin\theta_{xy} \tag{8.125}$$

Since $x(t) = A\sin(2\pi pt + \theta_x) = A\sin(2\pi pt + \theta_y + \theta_{xy})$, Equation (8.117) can be written as

$$
\begin{aligned}
x(t) &= A\sin(2\pi pt + \theta_y)\cos\theta_{xy} + A\cos(2\pi pt + \theta_y)\sin\theta_{xy} \\
y(t) &= C\sin(2\pi pt + \theta_y) + D\sin(n2\pi pt + \phi)
\end{aligned}
\tag{8.126}
$$

Comparing Equation (8.124) and (8.126), we see that $C_{xy}(f)$ measures the correlation of the *in-phase* components, i.e. between $A\sin(2\pi pt + \theta_y)$ and $C\sin(2\pi pt + \theta_y)$, thus it is called the *coincident* spectrum. Similarly, $Q_{xy}(f)$ measures the correlation between sine and cosine components ($A\cos(2\pi pt + \theta_y)$ and $C\sin(2\pi pt + \theta_y)$), i.e. quadrature components, thus it is called the *quadrature* spectrum.

(b) Consider the wheeled vehicle example shown previously, in Figure 8.24 shown again here[M8.9]

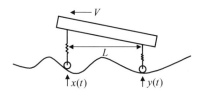

**Figure 8.24**   A wheeled vehicle moving over rough terrain

We have seen that $R_{xy}(\tau) = R_{xx}(\tau - \Delta)$, where $\Delta = L/V$. So the cross-spectral density function is

$$S_{xy}(f) = \int_{-\infty}^{\infty} R_{xx}(\tau - \Delta)e^{-j2\pi f \tau} d\tau = \int_{-\infty}^{\infty} R_{xx}(u)e^{-j2\pi f(u+\Delta)} du$$

$$= e^{-j2\pi f \Delta} \int_{-\infty}^{\infty} R_{xx}(u)e^{-j2\pi fu} du$$

$$= e^{-j2\pi f \Delta} S_{xx}(f) \tag{8.127}$$

This shows that the frequency component $f$ in the signal $y(t)$ lags that component in $x(t)$ by phase angle $2\pi f \Delta$. This is obvious from simple considerations: for example, if $x(t) = A \cos(\omega t)$ then $y(t) = A \cos[\omega(t - \Delta)] = A \cos(\omega t - \omega \Delta)$, i.e. the lag angle is $\omega \Delta = 2\pi f \Delta$.

### Comments on the Time Delay Problem

At this point, it may be worth relating the time delay problem to the pure delay discussed in Chapter 4, where we defined the group delay as $t_g = -d\phi(\omega)/d\omega$. We saw that a pure delay (say, delay time is $\Delta$ for all frequencies) produces a constant group delay, i.e. $t_g = \Delta$.

The above time delay problem can be considered as identifying a pure delay system. To see this more clearly, rewrite Equation (8.127) as

$$S_{xy}(f) = H(f)S_{xx}(f) \tag{8.128}$$

where $H(f)$ is the frequency response function which can be written as

$$H(f) = \frac{S_{xy}(f)}{S_{xx}(f)} = e^{-j2\pi f \Delta} \tag{8.129}$$

i.e. $H(f)$ is a pure delay system. Note that we are identifying the system by performing the ratio $S_{xy}(f)/S_{xx}(f)$. We shall compare the results of using arg $S_{xy}(f)$ and arg $H(f)$ in MATLAB Example 8.9. We shall also demonstrate a simple system identification problem by performing $H(f) = S_{xy}(f)/S_{xx}(f)$ in MATLAB Example 8.10. More details of system identification will be discussed in the next chapter.

## 8.8 BRIEF SUMMARY

1. A stochastic process $X(t)$ is stationary if it satisfies two conditions: (i) $p(x, t) = p(x)$ and (ii) $p(x_1, t_1; x_2, t_2) = p(x_1, t_1 + T; x_2, t_2 + T)$.
2. If certain averages of the process are ergodic, then ensemble averages can be replaced by time averages.

3. Autocovariance and autocorrelation functions are defined by

$$C_{xx}(\tau) = E[(X(t) - \mu_x)(X(t + \tau) - \mu_x)]$$
$$R_{xx}(\tau) = E[X(t)X(t + \tau)]$$

where $C_{xx}(\tau) = R_{xx}(\tau) - \mu_x^2$. The corresponding time averages are

$$C_{xx}(\tau) = \lim_{T \to \infty} \frac{1}{T} \int_0^T (x(t) - \mu_x)(x(t + \tau) - \mu_x)dt$$

$$R_{xx}(\tau) = \lim_{T \to \infty} \frac{1}{T} \int_0^T x(t)x(t + \tau)dt$$

4. Cross-covariance and cross-correlation functions are defined by

$$C_{xy}(\tau) = E[(X(t) - \mu_x)(Y(t + \tau) - \mu_y)]$$
$$R_{xy}(\tau) = E[X(t)Y(t + \tau)]$$

where $C_{xy}(\tau) = R_{xy}(\tau) - \mu_x\mu_y$. The corresponding time averages are

$$C_{xy}(\tau) = \lim_{T \to \infty} \frac{1}{T} \int_0^T (x(t) - \mu_x)(y(t + \tau) - \mu_y)dt$$

$$R_{xy}(\tau) = \lim_{T \to \infty} \frac{1}{T} \int_0^T x(t)y(t + \tau)dt$$

5. An unbiased estimate of the cross-covariance function is

$$\hat{C}_{xy}(\tau) = \frac{1}{T - \tau} \int_0^{T-\tau} (x(t) - \bar{x})(y(t + \tau) - \bar{y})dt \quad 0 \le \tau < T$$

where $\hat{C}_{xy}(-\tau) = \hat{C}_{yx}(\tau)$. The corresponding digital form is

$$\hat{C}_{xy}(m\Delta) = \frac{1}{N - m} \sum_{n=0}^{N-m-1} (x(n\Delta) - \bar{x})(y((n + m)\Delta) - \bar{y}) \quad 0 \le m \le N - 1$$

6. The autocorrelation functions of a periodic signal and a transient signal are, respectively,

$$R_{xx}(\tau) = \frac{1}{T_P} \int_0^{T_P} x(t)x(t + \tau)dt \quad \text{and} \quad R_{xx}(\tau) = \int_{-\infty}^{\infty} x(t)x(t + \tau)dt$$

7. The autocorrelation function of white noise is

$$R_{xx}(\tau) = k\delta(\tau)$$

8. The cross-correlation function of two uncorrelated signals $s(t)$ and $n(t)$ is

$$R_{sn}(\tau) = E\left[s(t)n(t+\tau)\right] = 0 \text{ (assuming zero mean values)}$$

9. The power spectral density and cross-spectral density functions are

$$S_{xx}(f) = \lim_{T\to\infty} \frac{E\left[|X_T(f)|^2\right]}{T} \quad\text{and}\quad S_{xy}(f) = \lim_{T\to\infty} \frac{E\left[X_T^*(f)Y_T(f)\right]}{T}$$

and the corresponding raw (or sample) spectral density functions are

$$\hat{S}_{xx}(f) = \frac{1}{T}|X_T(f)|^2 \quad\text{and}\quad \hat{S}_{xy}(f) = \frac{1}{T}X_T^*(f)Y_T(f)$$

10. The Wiener–Khinchin theorem is

$$S_{xx}(f) = \int_{-\infty}^{\infty} R_{xx}(\tau)e^{-j2\pi f\tau}d\tau \quad\text{and}\quad R_{xx}(\tau) = \int_{-\infty}^{\infty} S_{xx}(f)e^{j2\pi f\tau}df$$

Also,

$$S_{xy}(f) = \int_{-\infty}^{\infty} R_{xy}(\tau)e^{-j2\pi f\tau}d\tau \quad\text{and}\quad R_{xy}(\tau) = \int_{-\infty}^{\infty} S_{xy}(f)e^{j2\pi f\tau}df$$

11. The cross-spectral density function is complex valued, i.e.

$$S_{xy}(f) = \left|S_{xy}(f)\right|e^{j \arg S_{xy}(f)} \quad\text{and}\quad S_{xy}(f) = S_{yx}^*(f)$$

where $|S_{xy}(f)|$ is the measure of association of amplitude in $x$ and $y$ at frequency $f$, and arg $S_{xy}(f)$ shows lags/leads (or phase difference) between $x$ and $y$ at frequency $f$.

## 8.9 MATLAB EXAMPLES

*Example 8.1:* **Probability density function of a sine wave**

The theoretical probability density function of a sine wave $x(t) = A\sin(\omega t + \theta)$ is

$$p(x) = \frac{1}{\pi\sqrt{A^2 - x^2}}$$

In this MATLAB example, we compare the histograms resulting from the ensemble average and the time average. For the ensemble average $\theta$ is a random variable and $t$ is fixed, and for the time average $\theta$ is fixed and $t$ is a time variable.

| Line | MATLAB code | Comments |
|------|-------------|----------|
| 1 | clear all | Define the amplitude and frequency of a sine wave. |
| 2 | A=2; w=1; t=0; | For the ensemble average, let time $t = 0$. |
| 3 | rand('state',1); | Initialize the random number generator, and then |
| 4 | theta=rand(1,20000)*2*pi; | generate $\theta$ which is uniformly distributed on the |
| 5 | x1=A*sin(w*t+theta); | range 0 to $2\pi$. The number of elements of $\theta$ is |
|   |   | 20 000. |
|   |   | Also generate a sequence x1 which can be |
|   |   | considered as an ensemble (only for the specified |
|   |   | time, $t = 0$). |
| 6 | nbin=20; N1=length(x1); | Define the number of bins for the histogram. Then |
| 7 | [n1 s1]=hist(x1,nbin); | calculate the frequency counts and bin locations. |
| 8 | figure(1) % Ensemble average | Plot the histogram of x1. Note that it has a U shape |
| 9 | bar(s1, n1/N1) | as expected. One may change the number of |
| 10 | xlabel('\itx\rm_1') | elements of $\theta$, and compare the results. |
| 11 | ylabel('Relative frequency') |   |
| 12 | t=0:0.01:(2*pi)/w-0.01; | For the time average, $\theta$ is set to zero and a sine |
| 13 | x2=A*sin(w*t); | wave (x2) is generated for one period. |
| 14 | [n2 s2]=hist(x2, nbin); | Calculate the frequency counts and bin locations for |
| 15 | N2=length(x2); | x2. |
| 16 | figure(2) % Time average | Plot the histogram of x2. Compare the result with |
| 17 | bar(s2, n2/N2) | the ensemble average. |
| 18 | xlabel('\itx\rm_2') |   |
| 19 | ylabel('Relative frequency') |   |

## Results

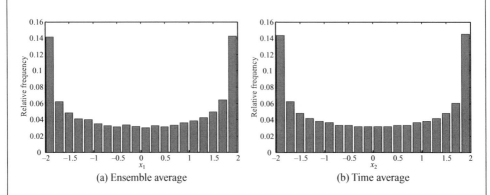

(a) Ensemble average          (b) Time average

**Comments:** The two results are very similar and confirm the theoretical probability density function. This illustrates that the process is ergodic with respect to the estimation of the probability density function.

---

***Example 8.2:*** **Autocorrelation function of a sine wave**

We compare the autocorrelation functions of a sinusoidal signal $x(t) = A \sin(\omega t + \theta)$, resulting from the ensemble average and the time average. The theoretical autocorrelation function is

$$R_{xx}(\tau) = \frac{A^2}{2} \cos(\omega\tau)$$

For the ensemble average $\theta$ is a random variable and $t$ is fixed, and for the time average $\theta$ is fixed and $t$ is a time variable.

| Line | MATLAB code | Comments |
|---|---|---|
| 1 | clear all | Define the amplitude and frequency of a sine |
| 2 | A=2; w=2*pi*1; t=0; fs=100; | wave. For the ensemble average, let time $t = 0$. |
| 3 | rand('state',1); | Also define the sampling rate. |
| 4 | theta=rand(1,5000)*2*pi; | Initialize the random number generator, and then |
| 5 | x1=A*sin(w*t+theta); | generate $\theta$ which is uniformly distributed on the range 0 to $2\pi$. The number of elements of $\theta$ is 5000. Then generate a sequence x1 which can be considered as an ensemble (only for the specified time, $t = 0$). |
| 6 | Rxx1=[]; maxlags=5; | Define an empty matrix (Rxx1) which is used in |
| 7 | for tau=-maxlags:1/fs:maxlags; | the 'for' loop, and define the maximum lag |
| 8 |   tmp=A*sin(w*(t+tau)+theta); | (5 seconds) for the calculation of the |
| 9 |   tmp=mean(x1.*tmp); | autocorrelation function. |
| 10 |   Rxx1=[Rxx1 tmp]; | The 'for' loop calculates the autocorrelation |
| 11 | end | function Rxx1 based on the ensemble average. |
| 12 | tau=-maxlags:1/fs:maxlags; | The variable 'tau' is the lag in seconds (Line 12). |
| 13 | Rxx=A^2/2*cos(w*tau); | Calculate the theoretical autocorrelation function Rxx. This is used for comparison. |
| 14 | figure(1) % Ensemble average | Plot the autocorrelation function Rxx1 obtained |
| 15 | plot(tau,Rxx1,tau,Rxx, 'r:') | by ensemble average (solid line), and compare |
| 16 | xlabel('Lag (\it\tau)') | this with the theoretical autocorrelation function |
| 17 | ylabel('Autocorrelation') | Rxx (dashed line). |
| 18 | t=0:1/fs:20-1/fs; | For the time average, $\theta$ is set to zero and a sine |
| 19 | x2=A*sin(w*t); | wave (x2) is generated for 20 seconds. The |
| 20 | [Rxx2, tau2]=xcorr(x2,x2,maxlags*fs, 'unbiased'); | MATLAB function 'xcorr(y,x)' estimates the cross-correlation function between x and y, i.e. |
| 21 | tau2=tau2/fs; | $R_{xy}(\tau)$ (note that it is not $R_{yx}(\tau)$). In this MATLAB code, the number of maximum lag (maxlags*fs) is also specified, and the unbiased estimator is used.<br>The variable 'tau2' is the lag in seconds (Line 21). |
| 22 | figure(2) % Time average | Plot the autocorrelation function Rxx2 obtained |
| 23 | plot(tau2,Rxx2,tau,Rxx, 'r:') | by time average (solid line), and compare this |
| 24 | xlabel('Lag (\it\tau)') | with the theoretical autocorrelation function Rxx |
| 25 | ylabel('Autocorrelation') | (dashed line). |

**Results**

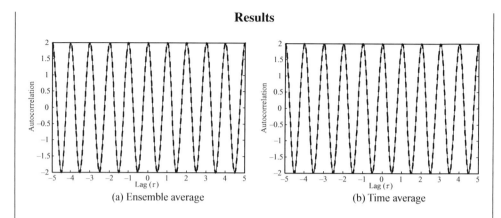

(a) Ensemble average    (b) Time average

**Comments:** The two results are almost identical and very close to the theoretical autocorrelation function. This demonstrates that the process is ergodic with respect to the estimation of the autocorrelation function.

---

*Example 8.3:* **Autocorrelation function of an echoed signal**

Consider the following echoed signal (see Equation (8.66) and Figure 8.18):

$$x(t) = s_1(t) + s_2(t) = as(t - \Delta_1) + bs(t - \Delta_2)$$

In this example, we use a sine burst signal as the source signal, and demonstrate that the autocorrelation function $R_{xx}(\tau)$ detects the relative time delay $\Delta = \Delta_2 - \Delta_1$.

We shall also consider the case that some additive noise is present in the signal.

| Line | MATLAB code | Comments |
|------|-------------|----------|
| 1 | clear all | Define the parameters of the above equation. The |
| 2 | a=2; b=1; fs=200; | sampling rate is chosen as 200 Hz. Note that the |
|   | delta1=1; delta2=2.5; | relative time delay is 1.5 seconds. |
| 3 | t=0:1/fs:0.5-1/fs; | Define the time variable up to 0.5 seconds, and |
| 4 | s=sin(2*pi*10*t); | generate the 10 Hz sine burst signal. |
| 5 | N=4*fs; | Generate signals $s_1(t) = as(t - \Delta_1)$ and |
| 6 | s1=[zeros(1,delta1*fs) a*s]; | $s_2(t) = bs(t - \Delta_2)$ up to 4 seconds. Then combine |
|   | s1=[s1 zeros(1,N-length(s1))]; | these to make the signal $x(t)$. |
| 7 | s2=[zeros(1,delta2*fs) b*s]; | |
|   | s2=[s2 zeros(1,N-length(s2))]; | |
| 8 | x = s1+s2; | |
| 9 | % randn('state',0); | This is for later use. Uncomment these lines then. |
| 10 | % noise = 1*std(s)*randn(size(x)); | Initialize the random number generator, then |
| 11 | % x=x+noise; | generate the Gaussian white noise whose variance is the same as the source signal $s(t)$, i.e. the signal-to-noise ratio is 0 dB. |

| | | |
|---|---|---|
| 12 | L=length(x); t=[0:L-1]/fs; | Define the time variable again according to the |
| 13 | maxlags=2.5*fs; | length of the signal $x(t)$. |
| 14 | [Rxx, tau]=xcorr(x,x,maxlags); | The maximum lag of the autocorrelation function |
| 15 | tau=tau/fs; | is set to 2.5 seconds. The variable 'tau' is the lag in seconds. |
| | | Note that the autocorrelation function is not normalized in this case because the signal is transient. |
| 16 | figure(1) | Plot the signal $x(t)$. |
| 17 | plot(t,x) | Later, compare this with the noisy signal. |
| 18 | xlabel('Time (s)'); ylabel('\itx\rm(\itt\rm)') | |
| 19 | axis([0 4 -4 4]) | |
| 20 | figure(2) | Plot the autocorrelation function $R_{xx}(\tau)$. |
| 21 | plot(tau,Rxx) | Note its symmetric structure, and the peak values |
| 22 | xlabel('Lag (\it\tau)'); ylabel('Autocorrelation') | occur at $R_{xx}(0)$, $R_{xx}(\Delta)$ and $R_{xx}(-\Delta)$. Run this MATLAB program again for the noisy |
| 23 | axis([-2.5 2.5 -300 300]) | signal (uncomment Lines 9–11, and compare $R_{xx}(\tau)$ with the corresponding time signal). |

## Results

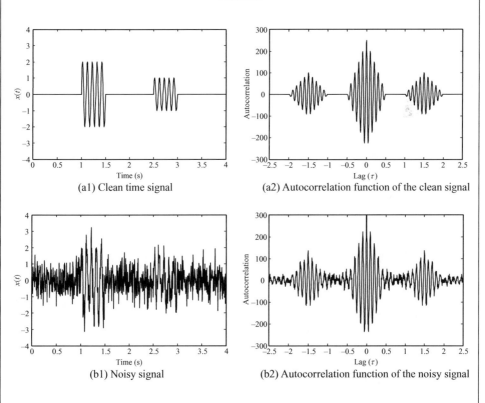

(a1) Clean time signal

(a2) Autocorrelation function of the clean signal

(b1) Noisy signal

(b2) Autocorrelation function of the noisy signal

**Comments:** Comparing Figures (b1) and (b2), it can be seen that the autocorrelation function is much cleaner than the corresponding time signal, and detects the relative time delay even if a significant amount of noise is present. This is because the source signal and the noise are not correlated, and the white noise contributes to $R_{xx}(\tau)$ at zero lag only (in theory). This noise reduction will be demonstrated again in MATLAB Example 8.5.

---

*Example 8.4:* **Cross-correlation function**

Consider two signals (see Equation (8.73))

$$x(t) = A \sin(\omega t + \theta_x) + B$$
$$y(t) = C \sin(\omega t + \theta_y) + D \sin(n\omega t + \phi)$$

The cross-correlation function is $R_{xy}(\tau) = \frac{1}{2}AC \cos[\omega \tau - (\theta_x - \theta_y)]$ (see Equation (8.74)).

| Line | MATLAB code | Comments |
|---|---|---|
| 1 | clear all | Define the parameters and time variable of the |
| 2 | A=1; B=1; C=2; D=2; | above equation. The sampling rate is chosen as |
|   | thetax=0; thetay=-pi/4; | 200 Hz. |
|   | phi=pi/2; n=2; | Calculate the relative time delay for reference. Note |
| 3 | w=2*pi*1; fs=200; T=100; | that the relative phase is $\theta_x - \theta_y = \pi/4$ that |
|   | t=0:1/fs:T-1/fs; | corresponds to the time delay of 0.125 seconds. |
| 4 | rel_time_delay=(thetax-thetay)/w | Generate signals $x(t)$ and $y(t)$ accordingly. |
| 5 | x=A*sin(w*t+thetax)+B; | |
| 6 | y=C*sin(w*t+thetay) | |
|   | +D*sin(n*w*t+phi); | |
| 7 | maxlag=4*fs; | The maximum lag of the cross-correlation function |
| 8 | [Rxy, tau]=xcorr(y,x,maxlag, | is set to 4 seconds. The unbiased estimator is used |
|   | 'unbiased'); | for the calculation of the cross-correlation function. |
| 9 | tau=tau/fs; | |
| 10 | figure(1) | Plot the signals $x(t)$ and $y(t)$. |
| 11 | plot(t(1:maxlag),x(1:maxlag), | |
|   | t(1:maxlag),y(1:maxlag), 'r') | |
| 12 | xlabel('Time (s)'); | |
|   | ylabel('\itx\rm(\itt\rm) | |
|   | and \ity\rm(\itt\rm)') | |
| 13 | figure(2) | Plot the cross-correlation function $R_{xy}(\tau)$. Note that |
| 14 | plot(tau(maxlag+1:end), | it shows the values for positive lags only in the |
|   | Rxy(maxlag+1:end)); hold on | figure. |
| 15 | plot([rel_time_delay rel_time_delay], | Compare the figure with the theoretical |
|   | [-1.5 1.5], 'r:') | cross-correlation function (i.e. Equation (8.74)). |
| 16 | hold off | |
| 17 | xlabel('Lag (\it\tau)'); | |
|   | ylabel('Cross-correlation') | |

## Results

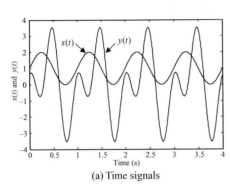

(a) Time signals

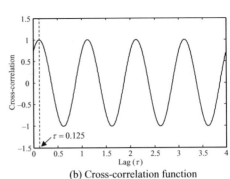

(b) Cross-correlation function

**Comments:** The cross-correlation function finds the components in $y(t)$ that are correlated with the components in $x(t)$ and preserves the relative phase (i.e. time delay).

---

*Example 8.5:* **A signal buried in noise**

Consider the signal buried in noise (see Figure 8.22)

$$y(t) = s(t) + n(t)$$

In this example, we use a sine wave for $s(t)$ and a band-limited white noise for $n(t)$, where $s(t)$ and $n(t)$ are uncorrelated and both have zero mean values. Thus, the cross-correlation between $s(t)$ and $n(t)$ is $E[s(t)n(t+\tau)] = E[n(t)s(t+\tau)] = 0$.

Then, the autocorrelation function is

$$R_{yy}(\tau) = R_{ss}(\tau) + R_{nn}(\tau)$$

It is shown that

$$R_{yy}(\tau) \approx R_{ss}(\tau) \text{ for large } \tau \text{ (see Figure 8.23)}$$

Considering the time-averaged form of correlation functions, we also compare the results for different values of $T$ (total record time).

| Line | MATLAB code | Comments |
|---|---|---|
| 1 | clear all | Define the parameters of a sine wave and the sampling |
| 2 | A=1; w = 2*pi*1; fs=200; | rate. The total record time is specified by 'T'. Initially, |
| 3 | T=100; % T=1000; | we use T = 100 seconds (i.e. 100 periods in total). Later, |
| 4 | t=0:1/fs:T-1/fs; | we shall increase it to 1000 seconds. Also, define the |
| 5 | s=A*sin(w*t); | time variable, and generate the 1 Hz sine wave. |
| 6 | randn('state',0); | Generate the broadband white noise signal. (The |
| 7 | n=randn(size(s)); | frequency band is limited by half the sampling rate, i.e. zero to fs/2 Hz.) |

| | | |
|---|---|---|
| 8 | fc=20; | These lines convert the above broadband white noise |
| 9 | [b,a]=butter(9, fc/(fs/2)); | into a band-limited white noise by filtering with a digital |
| 10 | n = filtfilt(b,a,n); | low-pass filter. |

'fc' is the cut-off frequency of the digital filter. The MATLAB function '[b,a] = butter(9, fc/(fs/2))' designs a ninth-order low-pass digital Butterworth filter (IIR), where 'b' is a vector containing coefficients of a moving average part and 'a' is a vector containing coefficients of an auto-regressive part of the transfer function (see Equation (6.12)).

The MATLAB function 'output = filtfilt(b,a,input)' performs zero-phase digital filtering. (Digital filtering will be briefly mentioned in Appendix H.) The resulting sequence 'n' is the band-limited (zero to 20 Hz) white noise.

| | | |
|---|---|---|
| 11 | n=sqrt(2)*(std(s)/std(n))*n; % SNR=-3dB | Make the noise power twice the signal power, i.e. 'Var(n) = 2×Var(s)'. Note that the signal-to-noise ratio is −3 |
| 12 | y=s+n; | dB. Then, make the noisy signal 'y' by adding n to s. |

| | | |
|---|---|---|
| 13 | maxlags=4*fs; | Calculate the autocorrelation function up to 4 seconds of |
| 14 | [Ryy, tau]=xcorr(y,y,maxlags, 'unbiased'); | lag. |
| 15 | tau=tau/fs; | |

| | | |
|---|---|---|
| 16 | figure(1) | Plot the signals $s(t)$ and $y(t)$ up to 8 seconds on the same |
| 17 | plot(t(1:8*fs),y(1:8*fs), t(1:8*fs),s(1:8*fs), 'r:') | figure. |
| 18 | xlabel('Time (s)') | |
| 19 | ylabel('\its\rm(\itt\rm) and \ity\rm(\itt\rm)') | |

| | | |
|---|---|---|
| 20 | figure(2) | Plot the autocorrelation function $R_{yy}(\tau)$ for the positive |
| 21 | plot(tau(maxlags+1:end), Ryy(maxlags+1:end), 'r') | lags only. Run this MATLAB program again for T = 1000 (change |
| 22 | xlabel('Lag (\it\tau)'); ylabel('Autocorrelation') | the value at Line 3), and compare the results. |
| 23 | axis([0 4 -1.5 1.5]) | |

## Results

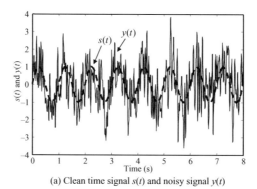

(a) Clean time signal $s(t)$ and noisy signal $y(t)$

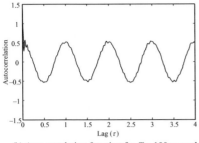
(b) Autocorrelation function for $T = 100$ seconds

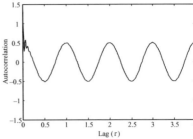
(c) Autocorrelation function for $T = 1000$ seconds

**Comments:** Comparing Figures (b) and (c), it can be seen that as $T$ increases the results improve. Note, however, that it will not be so if the signal is a transient. For example, if the total record time in MATLAB Example 8.3 is increased the result gets worse (Line 5, N = 4*fs; of MATLAB Example 8.3). This is because, after the transient signal dies out, it does not average 'signal × noise' but 'noise × noise'. Thus, we must apply the correlation functions appropriately depending on the nature of the signals.

---

*Example 8.6:* **Application of the cross-correlation function** (time delay problem 1)

Consider the wheeled vehicle example given in Section 8.6 (see Figure 8.24). We assume that the surface profile results in a band-limited time function (or profile) $s(t)$ and the trailing wheel experiences the same profile $\Delta$ seconds later, i.e. $s(t - \Delta)$.
 We measure both these signals, and include uncorrelated broadband noises $n_x(t)$ and $n_y(t)$, i.e.

$$x(t) = s(t) + n_x(t)$$
$$y(t) = s(t - \Delta) + n_y(t)$$

The cross-correlation function $R_{xy}(\tau)$ is (assuming zero mean values)

$$R_{xy}(\tau) = E[(s(t) + n_x(t))\left(s(t - \Delta + \tau) + n_y(t + \tau)\right)]$$
$$= E\left[s(t)s(t + \tau - \Delta)\right] = R_{ss}(\tau - \Delta)$$

| Line | MATLAB code | Comments |
|---|---|---|
| 1 | clear all | The sampling rate is 1000 Hz, and the time variable |
| 2 | fs=1000; T=5; t=0:1/fs:T-1/fs; | is defined up to 5 seconds. |
| 3 | randn('state',0); | Broadband white noise is generated, and then it is |
| 4 | s=randn(size(t)); | low-pass filtered to produce a band-limited white |
| 5 | fc=100; [b,a]=butter(9, fc/(fs/2)); | noise $s(t)$, where the cut-off frequency of the filter is |
| 6 | s=filtfilt(b,a,s); | 100 Hz. |
| 7 | s=s-mean(s); s=s/std(s); | Produce the signal $s(t)$ such that it has zero mean |
|   | % Makes mean(s)=0 & std(s)=1; | value and the standard deviation is one. |

| | | |
|---|---|---|
| 8 | delta=0.2; | Define the delay time $\Delta = 0.2$ seconds, and generate |
| 9 | x=s(delta*fs+1:end); | two sequences that correspond to $s(t)$ and $s(t - \Delta)$. |
| 10 | y=s(1:end-delta*fs); | Generate the broadband white noise $n_x(t)$ and $n_x(y)$. |
| 11 | randn('state',1); | Then add these signals appropriately to make noisy |
| | nx=1*std(s)*randn(size(x)); | measured signals $x(t)$ and $y(t)$. Note that the |
| 12 | randn('state',2); | signal-to-noise ratio is 0 dB for both signals. |
| | ny=1*std(s)*randn(size(y)); | |
| 13 | x=x+nx; y=y+ny; | |
| | | |
| 14 | maxlag1=0.25*fs; | Calculate the autocorrelation function $R_{ss}(\tau)$ and the |
| | maxlag2=0.5*fs; | cross-correlation function $R_{xy}(\tau)$, where the |
| 15 | [Rss, tau1]=xcorr(s,s,maxlag1, | unbiased estimators are used. |
| | 'unbiased'); | |
| 16 | [Rxy, tau2]=xcorr(y,x,maxlag2, | |
| | 'unbiased'); | |
| 17 | tau1=tau1/fs; tau2=tau2/fs; | |
| | | |
| 18 | figure(1) | Plot the autocorrelation function $R_{ss}(\tau)$. |
| 19 | plot(tau1,Rss) | |
| 20 | axis([-0.25 0.25 -0.4 1.2]) | |
| 21 | xlabel('Lag (\it\tau)') | |
| 22 | ylabel('Autocorrelation | |
| | (\itR_s_s\rm(\it\tau\rm))') | |
| | | |
| 23 | figure(2) | Plot the cross-correlation function $R_{xy}(\tau)$, and |
| 24 | plot(tau2(maxlag2+1:end), | compare this with the autocorrelation function |
| | Rxy(maxlag2+1:end)) | $R_{ss}(\tau)$. |
| 25 | axis([0 0.5 -0.4 1.2]) | |
| 26 | xlabel('Lag (\it\tau)') | |
| 27 | ylabel('Cross-correlation | |
| | (\itR_x_y\rm(\it\tau\rm))') | |

### Results

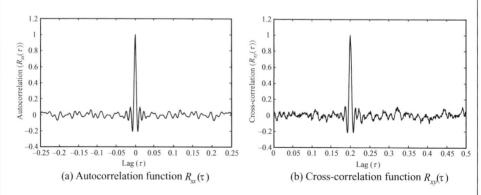

(a) Autocorrelation function $R_{ss}(\tau)$       (b) Cross-correlation function $R_{xy}(\tau)$

**Comments:** Note that, although both signals $x(t)$ and $y(t)$ have a very low SNR (0 dB in this example), the cross-correlation function gives a clear copy of $R_{ss}(\tau)$ at $\tau = 0.2$ seconds.

***Example 8.7:*** **Application of the cross-correlation function** (time delay problem 2)

This example describes the basic concept of using the cross-correlation function in a (radar-like) system. It is similar to the previous Example (MATLAB Example 8.6), except that we shall use a pulse-like signal for this example.

Let $x(t)$ be a pulse transmitted by the radar system, and $y(t)$ be the received signal that contains a reflected pulse from a target such that

$$y(t) = ax(t - \Delta) + n(t)$$

where $n(t)$ is uncorrelated broadband noise. Note that the amplitude of the reflected pulse in $y(t)$ may be very small compared with the original pulse, and so the SNR of the received signal $y(t)$ will also be very low.

To maximize the detectability, a special filter called a 'matched filter' is usually used. The matched filter is known to be an optimal detector while maximizing the SNR of a signal that is buried in noise (Papoulis, 1977; Bencroft, 2002).

If the length of the pulse $x(t)$ is $T$ seconds, the impulse response function of the matched filter is defined by

$$h(t) = x(T - t)$$

i.e. the pulse $x(t)$ is time reversed and shifted. Now, the received signal $y(t)$ is filtered, i.e. $y(t)$ is an input to the matched filter as shown in Figure (a).

(a) Matched filtering

The output signal is the convolution of $y(t)$ and $x(T - t)$. Thus, it follows that

$$out(t) = \int_{-\infty}^{\infty} y(\tau)h(t - \tau)d\tau = \int_{-\infty}^{\infty} y(\tau)x(T - (t - \tau))d\tau = \int_{-\infty}^{\infty} y(\tau)x(T - t + \tau)dt$$

$$= \int_{-\infty}^{\infty} y(\tau)x(\tau + (T - t))d\tau = R_{yx}(T - t) = R_{xy}(t - T)$$

Note that the result is the cross-correlation between the original pulse $x(t)$ and the received signal $y(t)$, which is shifted by the length of the filter $T$.

Assuming zero mean values, the cross-correlation function $R_{xy}(\tau)$ is

$$R_{xy}(\tau) = E\left[x(t)\left(ax(t - \Delta + \tau) + n(t + \tau)\right)\right]$$
$$= aE\left[x(t)x(t + \tau - \Delta)\right] = aR_{xx}(\tau - \Delta)$$

i.e. the cross-correlation function gives the total time delay $\Delta$ between the transmitter and the receiver. Thus, the distance to the target can be estimated by multiplying half of the time delay $\Delta/2$ by the speed of the wave. The filtered output is

$$out(t) = R_{xy}(t-T) = a R_{xx}(t-T-\Delta) = a R_{xx}(t-(T+\Delta))$$

We will compare the two results: the direct cross-correlation function $R_{xy}(\tau)$ and the filtered output $out(t)$. A significant amount of noise will be added to the received signal (SNR is $-6$ dB).

| Line | MATLAB code | Comments |
|------|-------------|----------|
| 1 | clear all | The sampling rate is 200 Hz, and the time |
| 2 | fs=200; | variable is defined up to 1 second. This is the |
| 3 | t=0:1/fs:1; | duration of the pulse. |
| 4 | x=chirp(t,5,1,15); | For the transmitted pulse, a chirp waveform is |
| 5 | h=fliplr(x); % Matched filter | used. The MATLAB function 'chirp(t,5,1,15)' generates a linear swept frequency signal at the time instances defined by 't', where the instantaneous frequency at time 0 is 5 Hz and at time 1 second is 15 Hz. Then, define the matched filter h. The MATLAB function 'fliplr(x)' flips the vector x in the left/right direction. The result is $h(t) = x(T-t)$, where $T$ is 1 second in this case. |
| 6 | figure(1) | Plot the transmitted chirp waveform $x(t)$. |
| 7 | plot(t,x) | |
| 8 | xlabel('Time (s)') | |
| 9 | ylabel('Chirp waveform, \itx\rm(\itt\rm)') | |
| 10 | figure(2) | Plot the impulse response function of the |
| 11 | plot(t,h) | matched filter $h(t)$, and compare with the |
| 12 | xlabel('Time (s)') | waveform $x(t)$. |
| 13 | ylabel('Matched filter, \ith\rm(\itt\rm)') | |
| 14 | delta=2; a=0.1; | Define the total time delay $\Delta = 2$ seconds and |
| 15 | y=[zeros(1,delta*fs) a*x zeros(1,3*fs)]; | the relative amplitude of the reflected waveform |
| 16 | t=[0:length(y)-1]/fs; | $a = 0.1$. |
| 17 | randn('state',0); | Generate the received signal $y(t)$. We assume |
| 18 | noise =2*std(a*x)*randn(size(y)); | that the signal is measured for up to 6 seconds. |
| 19 | y=y+noise; | Define the time variable again according to the signal $y(t)$. Generate the white noise whose standard deviation is twice that of the reflected waveform, then add this to the received signal. The resulting signal has an SNR of $-6$ dB, i.e. the noise power is four times greater than the signal power. |

| | | |
|---|---|---|
| 20 | figure(3) | Plot the noisy received signal $y(t)$. Note |
| 21 | plot(t,y) | that the reflected waveform is completely |
| 22 | xlabel('Time (s)') | buried in noise, and so is not noticeable |
| 23 | ylabel('Received signal, \ity\rm(\itt\rm)') | (see Figure (d) below). |
| 24 | maxlags=5*fs; | Define the maximum lags (up to |
| 25 | [Rxy, tau] =xcorr(y,x,maxlags); | 5 seconds), and calculate the |
| 26 | tau=tau/fs; | cross-correlation function $R_{xy}(\tau)$. Note that $R_{xy}(\tau)$ is not normalized. |
| 27 | figure(4) | Plot the cross-correlation function $R_{xy}(\tau)$. |
| 28 | plot(tau(maxlags+1:end), Rxy(maxlags+1:end)) | Note that the peak occurs at $\tau = 2$. |
| 29 | xlabel('Lag (\it\tau)') | |
| 30 | ylabel('Cross-correlation, \itR_x_y\rm(\it\tau\rm)') | |
| 31 | out=conv(y,h); out=out (1:length(y)); % or out=filter(h,1,y); | Now, calculate $out(t)$ by performing the convolution of $y(t)$ and $h(t)$. The same result can be achieved by 'out=filter(h,1,y)'. Note that 'h' can be considered as an FIR (Finite Impulse Response) digital filter (or an MA system). Then, the elements of 'h' are the coefficients of the MA part of the transfer function (see Equation (6.12)). In this case, there is no coefficient for the auto-regressive part, except '1' in the denominator of the transfer function. |
| 32 | figure(5) | Plot the filtered signal $out(t)$, and compare |
| 33 | plot(t(1:maxlags),out(1:maxlags)) | this with the cross-correlation function |
| 34 | xlabel('Time (s)') | $R_{xy}(\tau)$. Now, the peak occurs at $t = 3$ and |
| 35 | ylabel('Filtered signal, \itout\rm(\itt\rm)') | the shape is exactly same as the cross-correlation function, i.e. $R_{xy}(\tau)$ is delayed by the length of the filter $T$. |

**Results**

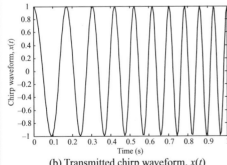

(b) Transmitted chirp waveform, $x(t)$

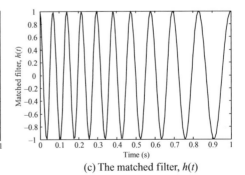

(c) The matched filter, $h(t)$

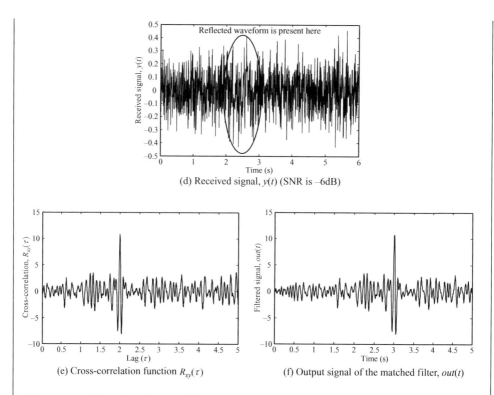

(d) Received signal, $y(t)$ (SNR is −6dB)

(e) Cross-correlation function $R_{xy}(\tau)$

(f) Output signal of the matched filter, $out(t)$

**Comments:** Note that Figure (f) is simply a delayed version of Figure (e). This example demonstrates that the cross-correlation function maximizes the SNR of a signal that is buried in noise.

---

*Example 8.8:* **Cross-spectral density function** (compare with MATLAB Example 8.4)

Consider two signals (see Equation (8.117))

$$x(t) = A \sin(2\pi pt + \theta_x)$$
$$y(t) = C \sin(2\pi pt + \theta_y) + D \sin(n2\pi pt + \phi)$$

These are the same as in MATLAB Example 8.4 except that the constant $B$ is not included here. The cross-correlation function and one-sided cross-spectral density function are (see Equations (8.119) and (8.121))

$$R_{xy}(\tau) = \frac{1}{2} AC \cos(2\pi p\tau - \theta_{xy}) \text{ and } G_{xy}(f) = \frac{AC}{2}\delta(f - p)e^{-j\theta_{xy}}$$

where $\theta_{xy} = \theta_x - \theta_y$.

| Line | MATLAB code | Comments |
|------|-------------|----------|
| 1 | clear all | Same as in MATLAB Example 8.4, except |
| 2 | A=1; C=2; D=2; thetax=0; | that 'T' is increased by 10 times for better |
|   | thetay=-pi/4; phi=pi/2; n=2; | estimation of the cross-correlation function. |
| 3 | p=1; w=2*pi*p; fs=200; | |
|   | T=1000; t=0:1/fs:T-1/fs; | |
| 4 | x=A*sin(w*t+thetax); | |
| 5 | y=C*sin(w*t+thetay)+D*sin(n*w*t+phi); | |
| 6 | maxlag=4*fs; | |
| 7 | [Rxy, tau]=xcorr(y,x,maxlag,'unbiased'); | |
| 8 | tau=tau/fs; | |
| | | |
| 9 | f=fs*(0:maxlag-1)/maxlag; | Define the frequency variable. |
| 10 | Rxy=Rxy(maxlag+1:end-1); | Discard the negative part of $\tau$, i.e. we only |
|   | % makes exactly four periods | take $R_{xy}(\tau)$ for $\tau \geq 0$. This makes it exactly |
| 11 | Sxy=fft(Rxy); | four periods. Then, obtain $S_{xy}(f)$ via the DFT |
|   | | of the cross-correlation function. |
| | | |
| 12 | format long | The MATLAB command 'format long' |
| 13 | thetaxy=thetax-thetay | displays longer digits. Display the value of |
| 14 | ind=find(f==p); | $\theta_{xy} = \theta_x - \theta_y$ which is $\pi/4$, and find the index |
| 15 | arg_Sxy_at_p_Hz=angle(Sxy(ind)) | of frequency $p$ Hz in the vector 'f'. |
|   | | Display the value of arg $S_{xy}(f)$ at $p$ Hz, and |
|   | | compare with the value of $\theta_{xy}$. |

### Results

thetaxy=0.785 398 163 397 45

arg_Sxy_at_p_Hz=−0.785 403 708 042 95

**Comments:** This demonstrates that arg $S_{xy}(f) = -(\theta_x - \theta_y)$. We can see that the longer the data length (T), the better the estimate of $R_{xy}(\tau)$ that results in a better estimate of $S_{xy}(f)$. Note, however, that we estimate $S_{xy}(f)$ by Fourier transforming the product of the estimate of $R_{xy}(\tau)$ and the rectangular window (i.e. the maximum lag is defined when $R_{xy}(\tau)$ is calculated (see Lines 7 and 10 of the MATLAB code)). The role of window functions is discussed in Chapter 10.

---

*Example 8.9:* **Application of the cross-spectral density function** (compare with MATLAB Example 8.6)

Consider the same example as in MATLAB Example 8.6 (the wheeled vehicle), where the measured signal is

$$x(t) = s(t) + n_x(t)$$
$$y(t) = s(t - \Delta) + n_y(t)$$

and the cross-correlation function and the cross-spectral density function are

$$R_{xy}(\tau) = R_{ss}(\tau - \Delta) \text{ and } S_{xy}(f) = H(f)S_{ss}(f) = e^{-j2\pi f \Delta} S_{ss}(f)$$

First, we shall estimate $\Delta$ using $S_{xy}(f)$. Then, by forming the ratio $S_{xy}(f)/S_{xx}(f)$ we shall estimate the frequency response function $H(f)$ from which the time delay $\Delta$ can also be estimated. Note that we are not using $S_{xy}(f)/S_{ss}(f)$, but the following:

$$H_1(f) = \frac{S_{xy}(f)}{S_{xx}(f)} = \frac{S_{xy}(f)}{S_{ss}(f) + S_{n_x n_x}(f)} \tag{1}$$

Since $R_{n_x n_x}(\tau)$ is an even function, $S_{n_x n_x}(f)$ is real valued, i.e. arg $S_{n_x n_x}(f) = 0$. Thus, it can be shown that

$$\arg H(f) = \arg H_1(f) = -2\pi f \Delta$$

Note that $H_1(f)$ may underestimate the magnitude of $H(f)$ depending on the variance of the noise. However, the phase of $H_1(f)$ is *not* affected by uncorrelated noise, i.e. we can see that the phase of $H_1(f)$ is less sensitive to noise than the magnitude of $H_1(f)$. More details of the estimator $H_1(f)$ defined by Equation (1) will be discussed in Chapter 9.

| Line | MATLAB code | Comments |
|---|---|---|
| 1 | clear all | Same as in MATLAB Example 8.6, |
| 2 | fs=500; T=100; t=0:1/fs:T-1/fs; | except that the sampling rate is reduced |
| 3 | randn('state',0); | and the total record time 'T' is increased. |
| 4 | s=randn(size(t)); | Note that the delay time $\Delta = 0.2$ seconds |
| 5 | fc=100; [b,a]=butter(9,fc/(fs/2)); | as before. |
| 6 | s=filtfilt(b,a,s); | Note also that the same number of lags is |
| 7 | s=s-mean(s); s=s/std(s); | used for both autocorrelation and |
|  | % Makes mean(s)=0 & std(s)=1; | cross-correlation functions. |
| 8 | delta=0.2; | |
| 9 | x=s(delta*fs+1:end); | |
| 10 | y=s(1:end-delta*fs); | |
| 11 | randn('state',1); | |
|  | nx=1*std(s)*randn(size(x)); | |
| 12 | randn('state',2); | |
|  | ny=1*std(s)*randn(size(y)); | |
| 13 | x=x+nx; y=y+ny; | |
| 14 | maxlag=fs; | |
| 15 | [Rxx, tau]=xcorr(x,x,maxlag, 'unbiased'); | |
| 16 | [Rxy, tau]=xcorr(y,x,maxlag, 'unbiased'); | |
| 17 | tau=tau/fs; | |
| 18 | f=fs*(0:maxlag-1)/maxlag; | Define the frequency variable. |
| 19 | Rxy_1=Rxy(maxlag+1:end-1); | Discard the negative part of $\tau$, i.e. we |
| 20 | Sxy=fft(Rxy_1); | only take $R_{xy}(\tau)$ for $\tau \geq 0$. If we include the negative part of $\tau$ when the DFT is performed, then the result is a pure delay. If this is the case, we must compensate for this delay. |

Then, obtain the cross-spectral density function via the DFT of the cross-correlation function.

```
21    figure(1)
22    plot(f(1:maxlag/2+1),
      unwrap(angle(Sxy(1:maxlag/2+1))))
23    hold on
24    xlabel('Frequency (Hz)')
25    ylabel('arg\itS_x_y\rm(\itf\rm) (rad)');
      axis([0 fs/2 -160 0])
```

Plot the unwrapped arg $S_{xy}(f)$ up to half the sampling rate. Then hold the figure. We can see the linear phase characteristic up to about 100 Hz. Note that the signal is band-limited (0 to 100 Hz), thus the values above 100 Hz are meaningless.

```
26    ind=find(f==fc);
27    P1=polyfit(f(2:ind),
      unwrap(angle(Sxy(2:ind))),1);
28    format long
29    t_delay1=-P1(1)/(2*pi)
30    plot(f(2:ind), P1(1)*f(2:ind)+P1(2), 'r:');
      hold off
```

Find the index of the cut-off frequency (100 Hz) in the vector 'f'. Then, perform first-order polynomial curve fitting to find the slope of the phase curve. Display the estimated time delay. Plot the results of curve fitting on the same figure, then release the figure.

```
31    N=2*maxlag;
32    f=fs*(0:N-1)/N;
33    Sxx=fft(Rxx(1:N)); Sxy=fft(Rxy(1:N));
34    % Sxx=fft(Rxx(1:N)).
      *exp(i*2*pi.*f*(maxlag/fs));
35    % Sxy=fft(Rxy(1:N)).
      *exp(i*2*pi.*f*(maxlag/fs));
36    H1=Sxy./Sxx;
```

Calculate $S_{xx}(f)$ and $S_{xy}(f)$ using the DFT of $R_{xx}(\tau)$ and $R_{xy}(\tau)$, respectively. Since $R_{xx}(\tau)$ is an even function, we must include the negative part of $\tau$ in order to preserve the symmetric property. Note that the last value of the vector Rxx is not included to pinpoint frequency values in the vector f. Then, estimate the frequency response function $H_1(f) = S_{xy}(f)/S_{xx}(f)$ (Line 36). As mentioned earlier, we must compensate the delay due to the inclusion of the negative part of $\tau$. However, this is not necessary for the estimation of the frequency response function, i.e. the ratio $S_{xy}(f)/S_{xx}(f)$ cancels the delay if $R_{xx}(\tau)$ and $R_{xy}(\tau)$ are delayed by same amount. Lines 34 and 35 compensate for the delay, and can be used in place of Line 33.

```
37    figure(2)
38    plot(f(1:maxlag+1),
      unwrap(angle(H1(1:maxlag+1))))
39    hold on
40    xlabel('Frequency (Hz)')
41    ylabel('arg\itH\rm_1(\itf\rm) (rad)');
      axis([0 fs/2 -160 0])
```

Plot the unwrapped arg $H_1(f)$ up to half the sampling rate. Then hold the figure. Compare this with the previous result.

```
42    ind=find(f==fc);
43    P2=polyfit(f(2:ind),
      unwrap(angle(H1(2:ind))), 1);
44    t_delay2=-P2(1)/(2*pi)
45    plot(f(2:ind), P2(1)*f(2:ind)+P2(2), 'r:');
      hold off
```

Perform first-order polynomial curve fitting as before. Display the estimated time delay, and plot the results of curve fitting on the same figure, then release the figure.

**Results**

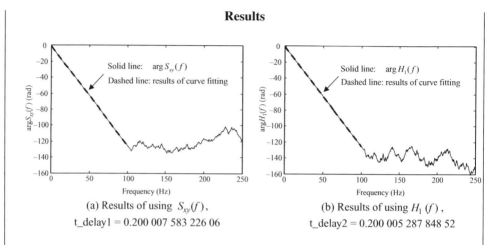

(a) Results of using $S_{xy}(f)$,

t_delay1 = 0.200 007 583 226 06

(b) Results of using $H_1(f)$,

t_delay2 = 0.200 005 287 848 52

**Comments:** Note that the two methods give almost identical results and estimate delay time $\Delta = 0.2$ very accurately.

---

## *Example 8.10:* System identification using spectral density functions

In the previous MATLAB example, we saw that $H_1(f) = S_{xy}(f)/S_{xx}(f)$ estimates the system frequency response function. Although we shall discuss this matter in depth in Chapter 9, a simple example at this stage may be helpful for understanding the role of correlation and spectral density functions.

Consider the input–output relationship of a single-degree-of-freedom system in Figure (a).

$$x(t) \longrightarrow \boxed{h(t) = \frac{A}{\omega_d} e^{-\zeta\omega_n t} \sin \omega_d t} \longrightarrow y(t)$$

(a) A single-degree-of-freedom system

In this example, we use white noise as an input $x(t)$, i.e. $R_{xx}(\tau) = k\delta(\tau)$; then the output $y(t)$ is obtained by $y(t) = h(t) * x(t)$.

$R_{xx}(\tau)$, $R_{yy}(\tau)$, $R_{xy}(\tau)$ and $H_1(f) = S_{xy}(f)/S_{xx}(f)$ are examined for two different values of measurement time. We shall see that the estimation results get better as the total record time $T$ increases.

| Line | MATLAB code | Comments |
|---|---|---|
| 1 | clear all | Define parameters for the impulse |
| 2 | fs=100; t=[0:1/fs:2.5-1/fs]; | response function $h(t)$, and generate a |
| 3 | A=100; zeta=0.03; f=10; wn=2*pi*f;<br>wd=sqrt(1-zeta^2)*wn; | sequence accordingly. Note that the<br>impulse response is truncated at 2.5 |
| 4 | h=(A/wd)*exp(-zeta*wn*t).*sin(wd*t); | seconds. |

| | | |
|---|---|---|
| 5 | randn('state',0); | Generate a white noise signal for input |
| 6 | T=100; % 100 and 2000 | $x(t)$. Note that the variance of $x(t)$ is |
| 7 | x=2*randn(1,T*fs); | four (in theory). Then obtain the output |
| 8 | y=conv(h,x); y=y(1:end-length(h)+1); | signal by convolution of $h(t)$ and $x(t)$. |
| 9 | % y=filter(h,1,x); | First, run this MATLAB program using |
| | | the total record time T = 100 seconds. |
| | | Later run this program again using T |
| | | = 2000, and compare the results. |
| | | Note that the sequence 'h' is an FIR |
| | | filter, and Line 9 can be used instead of |
| | | Line 8. |
| | | |
| 10 | maxlag=2.5*fs; | Calculate the correlation functions. |
| 11 | [Rxx, tau]=xcorr(x,x,maxlag, 'unbiased'); | Note that we define the maximum lag |
| 12 | [Ryy, tau]=xcorr(y,y,maxlag, 'unbiased'); | equal to the length of the filter h. |
| 13 | [Rxy, tau]=xcorr(y,x,maxlag, 'unbiased'); | |
| 14 | tau=tau/fs; | |
| | | |
| 15 | N=2*maxlag; | Calculate the spectral density |
| 16 | f=fs*(0:N-1)/N; | functions. |
| 17 | Sxx=fft(Rxx(1:N)); | Note that different scaling factors are |
| | Syy=fft(Ryy(1:N))/(fs^2); | used for Sxx, Syy and Sxy in order to |
| 18 | Sxy=fft(Rxy(1:N))/fs; | relate to their continuous functions (in |
| 19 | H1=Sxy./Sxx; | relative scale). This is due to the |
| 20 | H=fft(h,N)/fs; | convolution operation in Line 8, i.e. |
| | | the sequence 'y' must be divided by |
| | | 'fs' for the equivalent time domain |
| | | signal $y(t)$. |
| | | Calculate $H_1(f) = S_{xy}(f)/S_{xx}(f)$, |
| | | and also calculate $H(f)$ by the DFT of |
| | | the impulse response sequence. Then |
| | | compare these two results. |
| | | |
| 21 | figure(1) | Plot the autocorrelation function |
| 22 | plot(tau,Rxx) | $R_{xx}(\tau)$. It is close to the delta function |
| 23 | xlabel('Lag (\it\tau)'); | (but note that it is not a 'true' delta |
| | ylabel('\itR_x_x\rm(\it\tau\rm)') | function), and $R_{xx}(0) \approx 4$ which is the |
| | | variance of $x(t)$. |
| | | |
| 24 | figure(2) | Plot the autocorrelation function |
| 25 | plot(tau,Ryy) | $R_{yy}(\tau)$. Note that its shape is reflected |
| 26 | xlabel('Lag (\it\tau)'); | by the impulse response function. |
| | ylabel('\itR_y_y\rm(\it\tau\rm)') | |
| | | |
| 27 | figure(3) | Plot the cross-correlation function |
| 28 | plot(tau,Rxy) | $R_{xy}(\tau)$. Note that its shape resembles |
| 29 | xlabel('Lag (\it\tau)'); | the impulse response function. |
| | ylabel('\itR_x_y\rm(\it\tau\rm)') | |
| | | |
| 30 | figure(4) | Plot the magnitude spectrum of both |
| 31 | plot(f(1:N/2+1), | $H_1(f)$ and $H(f)$ (in dB scale), and |
| | 20*log10(abs(H1(1:N/2+1)))); hold on | compare them. |
| 32 | xlabel('Frequency (Hz)'); | |
| | ylabel('|\itH\rm_1(\itf\rm)| (dB)') | |
| 33 | plot(f(1:N/2+1), 20*log10(abs(H(1:N/2+1))), | |
| | 'r:'); hold off | |

| 34 | figure(5) |
| 35 | plot(f(1:N/2+1), unwrap(angle(H1(1:N/2+1)))); hold on |
| 36 | xlabel('Frequency (Hz)'); ylabel('arg\itH\rm_1(\itf\rm) (rad)') |
| 37 | plot(f(1:N/2+1), unwrap(angle(H(1:N/2+1))), 'r:'); hold off |

Plot the phase spectrum of both $H_1(f)$ and $H(f)$, and compare them. Run this MATLAB program again for $T = 2000$, and compare the results.

## Results

### $T = 100$    $T = 2000$

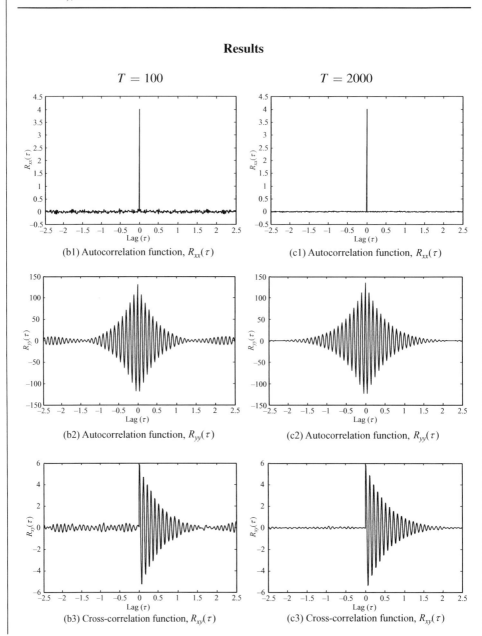

(b1) Autocorrelation function, $R_{xx}(\tau)$

(c1) Autocorrelation function, $R_{xx}(\tau)$

(b2) Autocorrelation function, $R_{yy}(\tau)$

(c2) Autocorrelation function, $R_{yy}(\tau)$

(b3) Cross-correlation function, $R_{xy}(\tau)$

(c3) Cross-correlation function, $R_{xy}(\tau)$

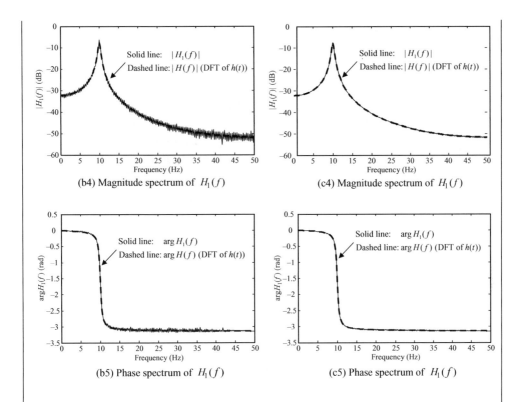

(b4) Magnitude spectrum of $H_1(f)$

(c4) Magnitude spectrum of $H_1(f)$

(b5) Phase spectrum of $H_1(f)$

(c5) Phase spectrum of $H_1(f)$

**Comments:**

1. By comparing the results of using T = 100 and T = 2000, it can be seen that as the length of data (T) increases, i.e. as the number of averages increases, we obtain better estimates of correlation functions and frequency response functions.

   Note particularly that the cross-correlation function $R_{xy}(\tau)$ has a shape similar to the impulse response function $h(t)$. In fact, in the next chapter, we shall see that $R_{xy}(\tau) = kh(\tau)$ where $k$ is the variance of the input white noise. To see this, type the following script in the MATLAB command window (use the result of T = 2000):

   ```
   plot(t,4*h); hold on
   plot(tau(maxlag+1:end), Rxy(maxlag+1:end), 'r:'); hold off
   xlabel('Time (s) and lag (\it\tau\rm)'); ylabel('Amplitude')
   ```

   The results are as shown in Figure (d). Note that $h(t)$ is multiplied by 4 which is the variance of the input white noise. (Note that it is not true white noise, but is band-limited up to 'fs/2', i.e. fs/2 corresponds to $B$ in Equation (8.97) and Figure 8.30.)

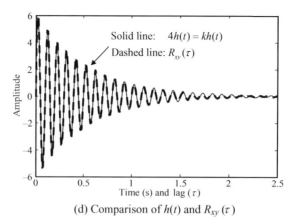

(d) Comparison of $h(t)$ and $R_{xy}(\tau)$

Also note that the autocorrelation function of the output is the scaled version of the autocorrelation function of the impulse response function, i.e. $R_{yy}(\tau) = k R_{hh}(\tau)$. Type the following script in the MATLAB command window to verify this:

```
Rhh=xcorr(h,h,maxlag);
plot(tau,4*Rhh); hold on
plot(tau, Ryy, 'r:'); hold off
xlabel('Lag (\it\tau\rm)'); ylabel('Amplitude')
```

The results are shown in Figure (e). Note that $R_{hh}(\tau)$ is not normalized since $h(t)$ is transient.

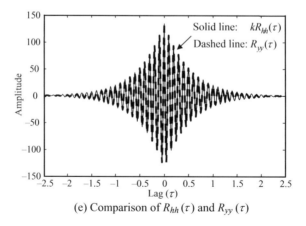

(e) Comparison of $R_{hh}(\tau)$ and $R_{yy}(\tau)$

Note that, in this MATLAB example, the system frequency response function is scaled appropriately to match its continuous function. However, the correlation and spectral density functions are not exactly matched to their continuous functions, they are scaled relatively.

2. In this example, we have estimated the spectral density functions by taking the Fourier transform of correlation functions. However, there are better estimation methods such

as the segment averaging method (also known as Welch's method (Welch, 1967)). Although this will be discussed in Chapter 10, the use of Welch's method is briefly demonstrated. Type the following script in the MATLAB command window (use the result of T =100):

```
Sxx_w=cpsd(x,x, hanning(N),N/2, N, fs, 'twosided');
Sxy_w=cpsd(x,y/fs, hanning(N),N/2, N, fs, 'twosided');
H1_w=Sxy_w./Sxx_w;
figure(1)
plot(f(1:N/2+1), 20*log10(abs(H1_w(1:N/2+1)))); hold on
plot(f(1:N/2+1), 20*log10(abs(H1(1:N/2+1))), 'r:'); hold off
xlabel('Frequency (Hz)'); ylabel('|\itH\rm_1(\itf\rm)| (dB)')
figure(2)
plot(f(1:N/2+1), unwrap(angle(H1_w(1:N/2+1)))); hold on
plot(f(1:N/2+1), unwrap(angle(H1(1:N/2+1))), 'r:'); hold off
xlabel('Frequency (Hz)'); ylabel('arg\itH\rm_1(\itf\rm) (rad)')
```

The MATLAB function 'cpsd' estimates the cross-spectral density function using Welch's method. In this MATLAB script, the spectral density functions are estimated using a Hann window and 50 % overlap. Then, the frequency response function estimated using Welch's method is compared with the previous estimate (shown in Figures (b4) and (b5)). Note that the output sequence is divided by the sampling rate, i.e. 'y/fs' is used in the calculation of cross-spectral density 'Sxy_w' to match to its corresponding continuous function.

The results are shown in Figures (f1) and (f2). Note that the result of using Welch's method is the smoothed estimate. This smoothing reduces the variability, but the penalty for this is the degradation of accuracy due to bias error. In general, the smoothed estimator underestimates peaks and overestimates troughs. The details of bias and random errors are discussed in Chapter 10.

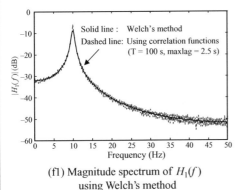

(f1) Magnitude spectrum of $H_1(f)$
using Welch's method

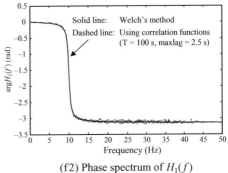

(f2) Phase spectrum of $H_1(f)$
using Welch's method

# 9

# Linear System Response to Random Inputs: System Identification

## Introduction

Having described linear systems and random signals, we are now able to model the response of linear systems to random excitations. We concentrate on single-input, single-output systems, but can also consider how additional inputs in the form of noise on measured signals might affect these characteristics. We shall restrict the systems to be linear and time invariant, and all the signals involved to be stationary random processes. Starting with basic input–output relationships, we introduce the concept and interpretation of the ordinary coherence function. This leads on to the main aim of this book, namely the identification of linear systems based on measurements of input and output.

## 9.1 SINGLE-INPUT, SINGLE-OUTPUT SYSTEMS

Consider the input–output relationship depicted as in Figure 9.1, which describes a linear time-invariant system characterized by an impulse response function $h(t)$, with input $x(t)$ and output $y(t)$.

If the input starts at $t_0$, then the response of the system is

$$y(t) = x(t){*}h(t) = \int_{t_0}^{t} h(t - t_1)x(t_1)dt_1 \qquad (9.1)$$

If we assume that the system is stable and the response to the stationary random input $x(t)$ has reached a steady state, i.e. $y(t)$ is also a stationary process for $t_0 \rightarrow -\infty$, then Equation

*Fundamentals of Signal Processing for Sound and Vibration Engineers*
K. Shin and J. K. Hammond.    © 2008 John Wiley & Sons, Ltd

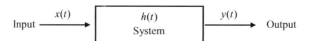

**Figure 9.1**   A single-input, single-output system

(9.1) can be written as

$$y(t) = \int_{-\infty}^{t} h(t-t_1)x(t_1)dt_1 = \int_{0}^{\infty} h(\tau)x(t-\tau)d\tau \qquad (9.2)$$

Whilst Equation (9.2) describes fully how input $x(t)$ is related to the corresponding response $y(t)$, it is more helpful to develop relationships relating the first and second moments of the input and response. We shall do this in both the time and frequency domains. We shall include mean values, (auto and cross-) correlation functions and (power and cross-) spectral density functions.

## Mean Values

If the mean value of input $x(t)$ is $\mu_x$, then the mean value of the output $y(t)$, $\mu_y$, may be obtained by taking expectations of Equation (9.2), i.e.

$$\mu_y = E[y(t)] = E\left[\int_{0}^{\infty} h(\tau)x(t-\tau)d\tau\right] \qquad (9.3)$$

The expectation operation is linear and so the right hand side of Equation (9.3) can be written as

$$\int_{0}^{\infty} h(\tau)E[x(t-\tau)]d\tau = \int_{0}^{\infty} h(\tau)\mu_x d\tau = \mu_x \int_{0}^{\infty} h(\tau)d\tau$$

So it follows that

$$\mu_y = \mu_x \int_{0}^{\infty} h(\tau)d\tau \qquad (9.4)$$

From this it is clear that if the input has a zero mean value then so does the output regardless of the form of $h(\tau)$.

It will be convenient to assume that the signals have zero mean values in what follows. This keeps the equations from becoming unwieldy.

### Autocorrelation Functions

If $x(t)$ is characterized by its autocorrelation function $R_{xx}(\tau)$ it is logical to ask how the output autocorrelation $R_{yy}(\tau)$ is related to the $R_{xx}(\tau)$. So we need to evaluate $R_{yy}(\tau) = E[y(t)y(t + \tau)]$, where $y(t)$ is given by Equation (9.2). This follows, where again we exploit the linear nature of the expected value operator which allows it under integral operations:

$$R_{yy}(\tau) = E[y(t)y(t + \tau)] = E\left[\int_0^\infty \int_0^\infty h(\tau_1)x(t - \tau_1)h(\tau_2)x(t + \tau - \tau_2)d\tau_1 d\tau_2\right]$$

$$= \int_0^\infty \int_0^\infty h(\tau_1)h(\tau_2)E\left[x(t - \tau_1)x(t + \tau - \tau_2)\right]d\tau_1 d\tau_2 \qquad (9.5)$$

Thus,

$$R_{yy}(\tau) = \int_0^\infty \int_0^\infty h(\tau_1)h(\tau_2)R_{xx}(\tau + \tau_1 - \tau_2)d\tau_1 d\tau_2 \qquad (9.6)$$

This is rather complicated and is a difficult equation to evaluate, and we find that the frequency domain equivalent is more useful.

Taking the Fourier transform of Equation (9.6) gives

$$S_{yy}(f) = \int_{-\infty}^\infty R_{yy}(\tau)e^{-j2\pi f\tau}d\tau$$

$$= \int_0^\infty h(\tau_1)e^{j2\pi f\tau_1}d\tau_1 \int_0^\infty h(\tau_2)e^{-j2\pi f\tau_2}d\tau_2 \int_{-\infty}^\infty R_{xx}(\tau + \tau_1 - \tau_2)e^{-j2\pi f(\tau + \tau_1 - \tau_2)}d\tau$$

$$(9.7)$$

Let $\tau + \tau_1 - \tau_2 = u$ in the last integral to yield

$$S_{yy}(f) = |H(f)|^2 S_{xx}(f) \qquad (9.8)$$

where $H(f) = \int_0^\infty h(\tau)e^{-j2\pi f\tau}d\tau$ is the system frequency response function. (Recall that the Fourier transform of the convolution of two functions is the product of their transforms, i.e. $Y(f) = F\{y(t)\} = F\{h(t) * x(t)\} = H(f)X(f)$ which gives $|Y(f)|^2 = |H(f)|^2 |X(f)|^2$, and compare this with above equation.)

We see that the frequency domain expression is much simpler than the corresponding time domain expression. Equation (9.8) describes how the power spectral density of the input is 'shaped' by the frequency response characteristic of the system. The output variance is the area under the $S_{yy}(f)$ curve, i.e. the output variance receives contributions from across the frequency range and the influence of the frequency response function is apparent.

### Cross-relationships

The expression in Equation (9.8) is real valued (there is no phase component), and shows only the magnitude relationship between input and output at frequency $f$. The following expression may be more useful since it includes the phase characteristic of the system.

Let us start with the input–output cross-correlation function $R_{xy}(\tau) = E[x(t)y(t + \tau)]$. Then

$$R_{xy}(\tau) = E[x(t)y(t + \tau)] = E\left[\int_0^\infty x(t)h(\tau_1)x(t + \tau - \tau_1)d\tau_1\right]$$

$$= \int_0^\infty h(\tau_1)E[x(t)x(t + \tau - \tau_1)]d\tau_1 \qquad (9.9)$$

i.e.,

$$R_{xy}(\tau) = \int_0^\infty h(\tau_1)R_{xx}(\tau - \tau_1)d\tau_1 \qquad (9.10)$$

Whilst Equation (9.10) is certainly simpler than Equation (9.6), the frequency domain equivalent is even simpler.

The Fourier transform of Equation (9.10) gives the frequency domain equivalent as

$$S_{xy}(f) = \int_{-\infty}^\infty R_{xy}(\tau)e^{-j2\pi f\tau}d\tau = \int_0^\infty h(\tau_1)e^{-j2\pi f\tau_1}d\tau_1 \int_{-\infty}^\infty R_{xx}(\tau - \tau_1)e^{-j2\pi f(\tau - \tau_1)}d\tau$$

$$\qquad (9.11)$$

thus

$$S_{xy}(f) = H(f)S_{xx}(f) \qquad (9.12)$$

Equation (9.12) contains the phase information of the frequency response function such that $\arg S_{xy}(f) = \arg H(f)$. Thus, this expression is often used as the basis of system identification schemes, i.e. by forming the ratio $S_{xy}(f)/S_{xx}(f)$ to give $H(f)$. Note also that if we restrict ourselves to $f \geq 0$, then we may write the alternative expressions to Equations (9.8) and (9.12) as

$$G_{yy}(f) = |H(f)|^2 G_{xx}(f) \qquad (9.13)$$

and

$$G_{xy}(f) = H(f)G_{xx}(f) \tag{9.14}$$

## Examples

### A First-order Continuous System with White Noise Input[M9.1]

Consider the first-order system shown in Figure 9.2 where the system equation is

$$T\dot{y}(t) + y(t) = x(t) \quad T > 0 \tag{9.15}$$

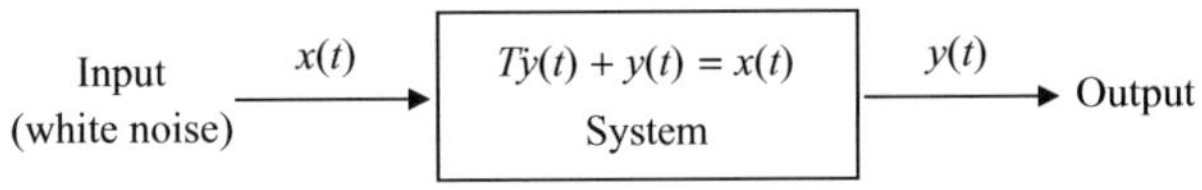

Input $x(t)$ → [ $T\dot{y}(t) + y(t) = x(t)$  System ] → $y(t)$ Output
(white noise)

**Figure 9.2**   A first-order continuous system driven by white noise

We shall assume that $x(t)$ has zero mean value and is 'white', i.e. has a delta function autocorrelation which we write $R_{xx}(\tau) = \sigma_x^2 \delta(\tau)$. The impulse response function of the system is

$$h(t) = \frac{1}{T}e^{-t/T} \quad t \geq 0$$

the transfer function is

$$H(s) = \frac{1}{1 + Ts}$$

and the frequency response function is

$$H(f) = \frac{1}{1 + j2\pi f T} \tag{9.16}$$

Using Equation (9.6), the autocorrelation function of the output is

$$R_{yy}(\tau) = \int_0^\infty \int_0^\infty h(\tau_1)h(\tau_2)\sigma_x^2\delta(\tau + \tau_1 - \tau_2)d\tau_1 d\tau_2$$

$$= \sigma_x^2 \int_0^\infty h(\tau_1) \int_0^\infty h(\tau_2)\delta(\tau + \tau_1 - \tau_2)d\tau_2 d\tau_1$$

$$= \sigma_x^2 \int_0^\infty h(\tau_1)h(\tau + \tau_1)d\tau_1 = \sigma_x^2 R_{hh}(\tau) \tag{9.17}$$

This shows that for a white noise input, the output autocorrelation function is a scaled version of the autocorrelation function formed from the system impulse response function.

From Equation (9.8), the power spectral density function of the output is

$$S_{yy}(f) = |H(f)|^2 S_{xx}(f) = \frac{1}{1 + (2\pi fT)^2}\sigma_x^2 \tag{9.18}$$

Note that

$$R_{yy}(0) = \sigma_y^2 = \sigma_x^2 \int_0^\infty h^2(\tau_1)d\tau_1 = \sigma_x^2 \int_{-\infty}^\infty |H(f)|^2 df = \int_{-\infty}^\infty S_{yy}(f)df$$

i.e. the output variance is shaped by the frequency response characteristic of the system and is spread across the frequencies as shown in Figure 9.3. A filter operating on white noise in this way is often called a 'shaping filter'.

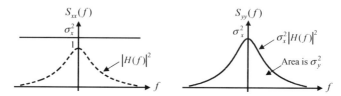

**Figure 9.3**    Power spectral density functions of input and output for the system in Figure 9.2.

## Cross-relationships
Consider the cross-spectral density and the cross-correlation functions which can be written as

$$S_{xy}(f) = H(f)S_{xx}(f) = \frac{1}{1 + j2\pi fT}\sigma_x^2 \tag{9.19}$$

$$R_{xy}(\tau) = \int_0^\infty h(\tau_1)\sigma_x^2\delta(\tau - \tau_1)d\tau_1 = \sigma_x^2 h(\tau) = \frac{\sigma_x^2}{T}e^{-\tau/T} \quad \tau \geq 0$$
$$= 0 \qquad\qquad\qquad\qquad\qquad\qquad \tau < 0 \tag{9.20}$$

From these two equations, it is seen that, if the input is white noise, the cross-spectral density function is just a scaled version of the system frequency response function, and the cross-correlation function is the impulse response function scaled by the variance of the input white noise (see also the comments in MATLAB Example 8.10 in Chapter 8). This result applies generally (directly from Equations (9.10) and (9.12)). Accordingly white noise seems the ideal random excitation for system identification. These results are theoretical and in practice band-limiting limits the accuracy of any identification.

### A First-order Continuous System with a Sinusoidal Input[M9.2]

Consider the same system as in the previous example. If the input is a pure sine function, e.g.
$x(t) = A \sin(2\pi f_0 t + \theta)$ with the autocorrelation function $R_{xx}(\tau) = (A^2/2)\cos(2\pi f_0 \tau)$ (see
Section 8.6), the power spectral density function $S_{xx}(f) = (A^2/4)[\delta(f - f_0) + \delta(f + f_0)]$
and the variance $\sigma_x^2 = A^2/2$, then the power spectral density function of the output is

$$S_{yy}(f) = |H(f)|^2 S_{xx}(f) = \frac{1}{1 + (2\pi f T)^2} \frac{A^2}{4} [\delta(f - f_0) + \delta(f + f_0)] \qquad (9.21)$$

The variance and autocorrelation function of the output are

$$\sigma_y^2 = \int_{-\infty}^{\infty} S_{yy}(f) df = \frac{A^2}{2} \frac{1}{1 + (2\pi f_0 T)^2} \qquad (9.22)$$

$$R_{yy}(\tau) = F^{-1}\{S_{yy}(f)\} = \frac{A^2}{2} \frac{1}{1 + (2\pi f_0 T)^2} \cos(2\pi f_0 \tau) \qquad (9.23)$$

The cross-spectral density and the cross-correlation functions are

$$S_{xy}(f) = H(f)S_{xx}(f) = \frac{1}{1 + j2\pi f T} \frac{A^2}{4} [\delta(f - f_0) + \delta(f + f_0)] \qquad (9.24)$$

$$R_{xy}(\tau) = F^{-1}\{S_{xy}(f)\} = \frac{A^2}{2\sqrt{1 + (2\pi f_0 T)^2}} \sin(2\pi f_0 \tau + \phi) \qquad (9.25)$$

where

$$\phi = \tan^{-1}\left(\frac{1}{2\pi f_0 T}\right)$$

Thus, it can be seen that the response to a sinusoidal input is sinusoidal with the same frequency,
the variance differs (Equation (9.23) with $\tau = 0$) and the cross-correlation (9.25) shows the
phase shift.

### A Second-order Vibrating System

Consider the single-degree-of-freedom system shown in Figure 9.4 where the equation
of motion is

$$m\ddot{y}(t) + c\dot{y}(t) + ky(t) = x(t) \qquad (9.26)$$

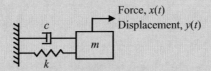

**Figure 9.4**   A single-degree-of-freedom vibration system

The impulse response function of the system is

$$h(t) = \frac{1}{m\omega_d} e^{-\zeta\omega_n t} \sin \omega_d t \quad t \geq 0$$

and the frequency response function is

$$H(f) = \frac{1}{k - m(2\pi f)^2 + jc(2\pi f)}$$

where $\omega_n = \sqrt{k/m}$, $\zeta = c/2m\omega_n$ and $\omega_d = \omega_n\sqrt{1 - \zeta^2}$.

If the input is white noise with the autocorrelation function $R_{xx}(\tau) = \sigma_x^2\delta(\tau)$, then the power spectral density and the cross-spectral density functions of the output are

$$S_{yy}(f) = |H(f)|^2 S_{xx}(f) = \frac{1}{\left[k - m(2\pi f)^2\right]^2 + [c(2\pi f)]^2}\sigma_x^2 \qquad (9.27)$$

and

$$S_{xy}(f) = H(f)S_{xx}(f) = \frac{1}{k - m(2\pi f)^2 + jc(2\pi f)}\sigma_x^2 \qquad (9.28)$$

The cross-correlation function of the output is

$$R_{xy}(\tau) = \sigma_x^2 h(\tau) = \frac{\sigma_x^2}{m\omega_d} e^{-\zeta\omega_n \tau} \sin \omega_d \tau \quad \tau \geq 0$$

$$= 0 \qquad\qquad\qquad\qquad\qquad \tau < 0 \qquad (9.29)$$

which is a scaled version of the impulse response function (see the comments in MATLAB Example 8.10).

This and the other examples given in this section indicate how the correlation functions and the spectral density functions may be used for system identification of single-input, single-output systems (see also some other considerations given in Appendix C). Details of system identification methods are discussed in Section 9.3.

## 9.2 THE ORDINARY COHERENCE FUNCTION

As a measure of the degree of *linear association* between two signals (e.g. input and output signals), the ordinary coherence function is widely used. The ordinary coherence function (or simply the coherence function) between two signals $x(t)$ and $y(t)$ is defined as

$$\gamma_{xy}^2(f) = \frac{|G_{xy}(f)|^2}{G_{xx}(f)G_{yy}(f)} = \frac{|S_{xy}(f)|^2}{S_{xx}(f)S_{yy}(f)} \qquad (9.30)$$

From the inequality property of the spectral density functions given in Chapter 8, i.e. $|S_{xy}(f)|^2 \leq S_{xx}(f)S_{yy}(f)$, it follows from Equation (9.30) that

$$0 \leq \gamma_{xy}^2(f) \leq 1 \qquad (9.31)$$

If $x(t)$ and $y(t)$ are input and output signals, then $S_{yy}(f) = |H(f)|^2 S_{xx}(f)$ and $S_{xy}(f) = H(f)S_{xx}(f)$. So the coherence function for the single-input, single-output system given in Figure 9.1 is shown to be

$$\gamma_{xy}^2(f) = \frac{|H(f)|^2 S_{xx}^2}{S_{xx}(f)|H(f)|^2 S_{xx}(f)} = 1 \qquad (9.32)$$

Thus, it is shown that the coherence function is unity if $x(t)$ and $y(t)$ are *linearly* related. Conversely, if $S_{xy}(f)$ is zero, i.e. the two signals are uncorrelated, then the coherence function is zero. If the coherence function is greater than zero but less than one, then $x(t)$ and $y(t)$ are partially linearly related. Possible departures from linear relationship between $x(t)$ and $y(t)$ include:

1. *Noise* may be present in the measurements of either or both $x(t)$ and $y(t)$.
2. $x(t)$ and $y(t)$ are *not only linearly related* (e.g. they may also be related nonlinearly).
3. $y(t)$ is an output due not only to input $x(t)$ but also to *other inputs*.

Since $\gamma_{xy}^2(f)$ is a function of frequency its appearance across the frequency range can be very revealing. In some ranges it may be close to unity and in others not, e.g. see Figure 9.5, indicating frequency ranges where 'linearity' may be more or less evident.

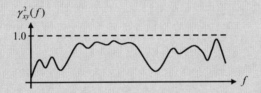

**Figure 9.5**   A typical example of the coherence function

## *Effect of Measurement Noise*

### Case (a) Output Noise
Consider the effect of measurement noise on the output as shown in Figure 9.6, where $y_m(t)$ is a measured signal such that $y_m(t) = y(t) + n_y(t)$. We assume that input $x(t)$ and measurement noise $n_y(t)$ are uncorrelated. Since $y(t)$ is linearly related to $x(t)$, then $y(t)$ and $n_y(t)$ are also uncorrelated. Then, the coherence function between $x(t)$ and $y_m(t)$ is

$$\gamma_{xy_m}^2(f) = \frac{|S_{xy_m}(f)|^2}{S_{xx}(f)S_{y_m y_m}(f)} \qquad (9.33)$$

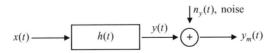

**Figure 9.6**   The effect of measurement noise on the output

where $S_{y_m y_m}(f) = S_{yy}(f) + S_{n_y n_y}(f) = |H(f)|^2 S_{xx}(f) + S_{n_y n_y}(f)$. Using the standard input–output relationship, i.e.

$$S_{xy_m}(f) = S_{xy}(f) + S_{xn_y}(f) = S_{xy}(f) = H(f)S_{xx}(f) \tag{9.34}$$

the coherence function becomes

$$\gamma_{xy_m}^2(f) = \frac{|H(f)|^2 S_{xx}^2(f)}{S_{xx}(f)\left[|H(f)|^2 S_{xx}(f) + S_{n_y n_y}(f)\right]} = \frac{1}{1 + \dfrac{S_{n_y n_y}(f)}{|H(f)|^2 S_{xx}(f)}} \tag{9.35}$$

So

$$\gamma_{xy_m}^2(f) = \frac{1}{1 + \dfrac{S_{n_y n_y}(f)}{S_{yy}(f)}} = \frac{S_{yy}(f)}{S_{yy}(f) + S_{n_y n_y}(f)} = \frac{S_{yy}(f)}{S_{y_m y_m}(f)} \tag{9.36}$$

From Equation (9.36), it can be seen that the coherence function $\gamma_{xy_m}^2(f)$ describes how much of the output power of the measured signal $y_m(t)$ is contributed (linearly) by input $x(t)$. Also, since the noise portion is

$$\frac{S_{n_y n_y}(f)}{S_{y_m y_m}(f)} = \frac{S_{y_m y_m}(f) - S_{yy}(f)}{S_{y_m y_m}(f)} = 1 - \gamma_{xy_m}^2(f)$$

the quantity $1 - \gamma_{xy_m}^2(f)$ is the fractional portion of the output power that is *not* due to input $x(t)$.

Thus, a useful concept, called the *coherent output power (spectral density function)*, is defined as

$$S_{yy}(f) = \gamma_{xy_m}^2(f)S_{y_m y_m}(f) \tag{9.37}$$

which describes the part of the output power fully coherent with the input. In words, the power spectrum of the output that is due to the source is the product of the coherence function between the source and the measured output, and the power spectrum of the measured output. Similarly, the *noise power* (or uncoherent output power) is the part of the output power not coherent with the input, and is

$$S_{n_y n_y}(f) = \left[1 - \gamma_{xy_m}^2(f)\right] S_{y_m y_m}(f) \tag{9.38}$$

The ratio $S_{yy}(f)/S_{n_y n_y}(f)$ is the signal-to-noise ratio at the output at frequency $f$. If this is large then $\gamma_{xy_m}^2(f) \to 1$, and if it is small then $\gamma_{xy_m}^2(f) \to 0$, i.e.

$$\gamma_{xy_m}^2(f) \to 1 \text{ as } \frac{S_{yy}(f)}{S_{n_y n_y}(f)} \to \infty \quad \text{and} \quad \gamma_{xy_m}^2(f) \to 0 \text{ as } \frac{S_{yy}(f)}{S_{n_y n_y}(f)} \to 0 \tag{9.39}$$

**Case (b) Input Noise**
Now consider the effect of measurement noise (assumed uncorrelated with $x(t)$) on the input as shown in Figure 9.7, where $x_m(t)$ is a measured signal such that $x_m(t) = x(t) + n_x(t)$ and $S_{x_m x_m}(f) = S_{xx}(f) + S_{n_x n_x}(f)$. Then, similar to the previous case, the coherence function

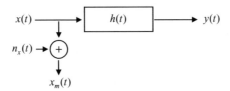

**Figure 9.7**   The effect of measurement noise on the input

between $x_m(t)$ and $y(t)$ is

$$\gamma_{x_m y}^2(f) = \frac{1}{1 + \frac{S_{n_x n_x}(f)}{S_{xx}(f)}} = \frac{S_{xx}(f)}{S_{xx}(f) + S_{n_x n_x}(f)} = \frac{S_{xx}(f)}{S_{x_m x_m}(f)} \tag{9.40}$$

Thus, the input power and noise power can be decomposed as

$$S_{xx}(f) = \gamma_{x_m y}^2(f) S_{x_m x_m}(f) \tag{9.41}$$

and

$$S_{n_x n_x}(f) = \left[1 - \gamma_{x_m y}^2(f)\right] S_{x_m x_m}(f) \tag{9.42}$$

### Case (c) Input and Output Noise
Consider the uncorrelated noise at *both* input and output as shown in Figure 9.8, where $x_m(t)$ and $y_m(t)$ are the measured input and output.

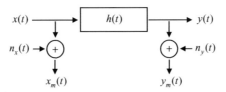

**Figure 9.8**   The effect of measurement noise on both input and output

The noises are assumed mutually uncorrelated and uncorrelated with $x(t)$. Then the coherence function between $x_m(t)$ and $y_m(t)$ becomes

$$\gamma_{x_m y_m}^2(f) = \frac{|S_{x_m y_m}(f)|^2}{S_{x_m x_m}(f) S_{y_m y_m}(f)} = \frac{1}{1 + \frac{S_{n_x n_x}(f)}{S_{xx}(f)} + \frac{S_{n_y n_y}(f)}{S_{yy}(f)} + \frac{S_{n_x n_x}(f) S_{n_y n_y}(f)}{S_{xx}(f) S_{yy}(f)}} \tag{9.43}$$

Note that, in this case, it is not possible to obtain the signal powers $S_{xx}(f)$ and $S_{yy}(f)$ using the measured signals $x_m(t)$ and $y_m(t)$ without knowledge or measurement of the noise. Some comments on the ordinary coherence function are given in Appendix D.

## 9.3  SYSTEM IDENTIFICATION[M9.4, 9.5]

The objective of this section is to show how we can estimate the frequency response function of a linear time-invariant system when the input is a stationary random process.

It is assumed that both input and output are measurable, but may be noise contaminated. In fact, as we have seen earlier, the frequency response function $H(f)$ can be obtained by forming the ratio $S_{xy}(f)/S_{xx}(f)$ (see Equation (9.12)). However, if noise is present on the measured signals $x_m(t)$ and $y_m(t)$ as shown in Figure 9.8, then we may only 'estimate' the frequency response function, e.g. by forming $S_{x_m y_m}(f)/S_{x_m x_m}(f)$. In fact we shall now see that this is only one particular approach to estimating the frequency response from among others.

Figure 9.8 depicts the problem we wish to address. On the basis of making measurements $x_m(t)$ and $y_m(t)$, which are noisy versions of input $x(t)$ and response $y(t)$, we wish to identify the linear system linking $x$ and $y$.

To address this problem we begin by resorting to something very much simpler. Forget for the moment the time histories involved and consider the problem of trying to link two random variables $X$ and $Y$ when measures of this bivariate process are available as pairs $(x_i, y_i)$, $i = 1, 2, \ldots, N$. Suppose we wish to find a *linear* relationship between $x$ and $y$ of the form $y = ax$. We may plot the data as a scatter diagram as shown in Figure 9.9. The parameter $a$ might be found by adjusting the line to 'best-fit' the scatter of points. In this context the points $(x_i, y_i)$ could come from any two variables, but to maintain contact with Figure 9.8 it is convenient to think of $x$ as an input and $y$ as the output. With reference to Figure 9.9, the slope $a$ is the 'gain' relating $x$ to $y$.

To find the slope $a$ that is best means deciding on some objective measure of closeness of fit and selecting the value of $a$ that achieves the 'optimal' closeness. So we need some measure of the 'error' between the line and the data points.

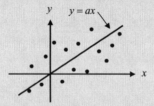

**Figure 9.9**   Scatter diagram relating variable $x$ (input) and $y$ (output)

We choose to depict three errors that characterize the 'distance' of a data point from the line. These are shown in Figure 9.10:

- Case 1: The distance (error) is measured in the $y$ direction and denoted $e_y$. This assumes that offsets in the $x$ direction are not important. Since $x$ is identified with input and $y$ with output, the implication of this is that it is errors on the output that are more important than errors on the input. This is analogous to the system described in Case (a), Section 9.2 (see Figure 9.6).
- Case 2: In this case we reverse the situation and accentuate the importance of offsets (errors) in the $x$ direction, $e_x$, i.e. errors on input are more important than on output. This is analogous to the system described in Case (b), Section 9.2 (see Figure 9.7).
- Case 3: Now we recognize that errors in both $x$ and $y$ directions matter and choose an offset (error) measure normal to the line $e_T$. The subscript $T$ denotes 'total'. This is analogous to the system described in Case (c), Section 9.2 (see Figure 9.8).

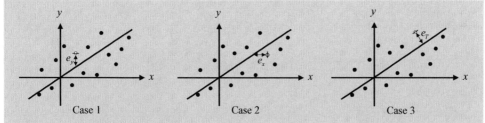

**Figure 9.10**   Scatter diagram; three types of error, $e_y$, $e_x$ and $e_T$

For each of the cases we need to create an objective function or cost function from these measures and find the slope $a$ that minimizes the cost function. There are an unlimited set of such functions and we shall choose the simplest, namely the sum of squared errors. This results in three least squares optimisation problems as follows.

**Case 1:**  Errors in the Output, $e_y$ (i.e. $x_i$ are known exactly but $y_i$ are noisy)

In this case, we will find the parameter $a_1$ that fits the data such that $y = a_1 x$ and minimizes the sum of squares of errors $\sum_{i=1}^{N} (e_y^i)^2$, where the error is defined as $e_y^i = y_i - a_1 x_i$. We form an objective function (or a cost function) as

$$J_1 = \frac{1}{N} \sum_{i=1}^{N} (e_y^i)^2 = \frac{1}{N} \sum_{i=1}^{N} (y_i - a_1 x_i)^2 \qquad (9.44)$$

and minimize $J_1$ with respect to $a_1$. $J_1$ is a quadratic function of $a_1$, and has a single minimum located at the solution of $dJ_1/da_1 = 0$, i.e.

$$\frac{dJ_1}{da_1} = \frac{2}{N} \sum_{i=1}^{N} (y_i - a_1 x_i)(-x_i) = 0 \qquad (9.45)$$

Thus, the parameter $a_1$ is found by

$$a_1 = \frac{\sum_{i=1}^{N} x_i y_i}{\sum_{i=1}^{N} x_i^2} \qquad (9.46)$$

Note that, if the divisor $N$ is used in the numerator and denominator of Equation (9.46), the numerator is the cross-correlation of two variables $x$ and $y$, and the denominator is the variance of $x$ (assuming zero mean value). If $N$ is large then it is logical to write a limiting form for $a_1$ as

$$a_1 = \frac{E[xy]}{E[x^2]} = \frac{\sigma_{xy}}{\sigma_x^2} \text{ (for zero mean)} \qquad (9.47)$$

emphasizing the ratio of the cross-correlation to the input power for this estimator.

**Case 2:**  Errors in the Input, $e_x$ (i.e. $y_i$ are known exactly but $x_i$ are noisy)

Now we find the parameter $a_2$ for $y = a_2 x$. The error is defined as $e_x^i = x_i - y_i/a_2$, and we form an objective function as

$$J_2 = \frac{1}{N} \sum_{i=1}^{N} \left(e_x^i\right)^2 = \frac{1}{N} \sum_{i=1}^{N} \left(x_i - \frac{y_i}{a_2}\right)^2 \tag{9.48}$$

Then, minimizing $J_2$ with respect to $a_2$, i.e.

$$\frac{dJ_2}{da_2} = \frac{2}{N} \sum_{i=1}^{N} \left(x_i - \frac{y_i}{a_2}\right)\left(\frac{y_i}{a_2^2}\right) = 0 \tag{9.49}$$

gives the value of parameter $a_2$ as

$$a_2 = \frac{\sum_{i=1}^{N} y_i^2}{\sum_{i=1}^{N} x_i y_i} \tag{9.50}$$

Note that, in contrast to Equation (9.46), the numerator of Equation (9.50) represents the power of the output and the denominator represents the cross-correlation between $x$ and $y$. Again, taking a limiting case we express this

$$a_2 = \frac{E\left[y^2\right]}{E\left[xy\right]} = \frac{\sigma_y^2}{\sigma_{xy}} \text{ (for zero mean)} \tag{9.51}$$

**Case 3:** Errors in Both Input and Output, $e_T$ (i.e. both variables $x_i$ and $y_i$ are noisy)

In this case, the error to be minimized is defined as the perpendicular distance to the line $y = a_T x$ as shown in Figure 9.11. This approach is called the *total least squares (TLS)* scheme, and from the figure the error can be written as

$$e_T^i = \frac{y_i - a_T x_i}{\sqrt{1 + a_T^2}} \tag{9.52}$$

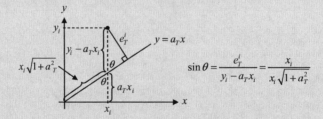

**Figure 9.11**   Representation of error $e_T$ normal to the line $y = a_T x$

Then the cost function $J_T$ is

$$J_T = \frac{1}{N} \sum_{i=1}^{N} \left(e_T^i\right)^2 = \frac{1}{N} \sum_{i=1}^{N} \frac{(y_i - a_T x_i)^2}{\left(1 + a_T^2\right)} \tag{9.53}$$

This is a non-quadratic function of $a_T$ and so there may be more than one extreme value. The necessary conditions for extrema are obtained from

$$\frac{dJ_T}{da_T} = \frac{1}{N}\sum_{i=1}^{N}\frac{2(y_i - a_T x_i)(-x_i)}{(1+a_T^2)} - \frac{1}{N}\sum_{i=1}^{N}\frac{(y_i - a_T x_i)^2(2a_T)}{(1+a_T^2)^2} = 0 \qquad (9.54)$$

This yields a quadratic equation in $a_T$ as

$$a_T^2\sum_{i=1}^{N}x_i y_i + a_T\left(\sum_{i=1}^{N}x_i^2 - \sum_{i=1}^{N}y_i^2\right) - \sum_{i=1}^{N}x_i y_i = 0 \qquad (9.55)$$

Thus, we note that the non-quadratic form of the cost function $J_T$ (9.53) results in two possible solutions for $a_T$, i.e. we need to find the correct $a_T$ that minimizes the cost function. We resolve this as follows. If we consider $N$ large then the cost function $J_T$ can be rewritten as

$$J_T = E\left[(e_T^i)^2\right] = E\left[\frac{(y_i - a_T x_i)^2}{1+a_T^2}\right]$$

$$= \frac{\sigma_y^2 + a_T^2\sigma_x^2 - 2a_T\sigma_{xy}}{1+a_T^2} \quad \text{(for zero mean)} \qquad (9.56)$$

Then equation (9.55) becomes

$$a_T^2\sigma_{xy} + a_T\left(\sigma_x^2 - \sigma_y^2\right) - \sigma_{xy} = 0 \qquad (9.57)$$

The solutions of this are given by

$$a_T = \frac{\left(\sigma_y^2 - \sigma_x^2\right) \pm \sqrt{\left(\sigma_x^2 - \sigma_y^2\right)^2 + 4\sigma_{xy}^2}}{2\sigma_{xy}} \qquad (9.58)$$

A typical form of the theoretical cost function (Equation (9.56)) may be drawn as in Figure 9.12 (Tan, 2005), where $+\sqrt{\ }$ and $-\sqrt{\ }$ denote the solutions (9.58) with positive and negative square root respectively.

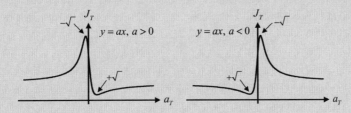

**Figure 9.12**   Theoretical cost function $J_T$ versus $a_T$

From this we see that the correct parameter $a_T$ (in a limiting form) for $y = a_T x$ that minimizes the cost function $J_T$ is given by the solution (9.58) which has the positive square root, i.e.

$$a_T = \frac{\left(\sigma_y^2 - \sigma_x^2\right) + \sqrt{\left(\sigma_x^2 - \sigma_y^2\right)^2 + 4\sigma_{xy}^2}}{2\sigma_{xy}} \qquad (9.59)$$

Accordingly, the correct parameter $a_T$ for Equation (9.55) is

$$a_T = \frac{\left(\sum\limits_{i=1}^{N} y_i^2 - \sum\limits_{i=1}^{N} x_i^2\right) + \sqrt{\left(\sum\limits_{i=1}^{N} x_i^2 - \sum\limits_{i=1}^{N} y_i^2\right)^2 + 4\left(\sum\limits_{i=1}^{N} x_i y_i\right)^2}}{2\sum\limits_{i=1}^{N} x_i y_i} \qquad (9.60)$$

### Frequency Response Identification

The relationship between $x(t)$ and $y(t)$ in the time domain is convolution (not a simple gain) - but it becomes a gain through the Fourier transform, i.e. $Y(f) = H(f)X(f)$, and the previous gain $a$ is now the complex-valued $H(f)$ which is frequency dependent.

With reference to Figure 9.8, i.e. considering measurement noise, suppose we have a series of measured results $X_{mi}(f)$ and $Y_{mi}(f)$. The index $i$ previously introduced now implies each sample realization that corresponds to each sample time history of length $T$. Accordingly, the form

$$\frac{1}{N}\sum_{i=1}^{N} x_i^2$$

used in the previous analysis can be replaced by

$$\frac{1}{N}\sum_{i=1}^{N} \frac{|X_{mi}(f)|^2}{T}$$

As we shall see in Chapter 10, this is an estimator for the power spectral density function of $x_m(t)$, i.e.

$$\tilde{S}_{x_m x_m}(f) = \frac{1}{N}\sum_{i=1}^{N} \frac{|X_{mi}(f)|^2}{T} \qquad (9.61)$$

Similarly, the cross-spectral density between $x_m(t)$ and $y_m(t)$ can be estimated by

$$\tilde{S}_{x_m y_m}(f) = \frac{1}{N}\sum_{i=1}^{N} \frac{X_{mi}^*(f)Y_{mi}(f)}{T} \qquad (9.62)$$

These results introduce three frequency response function estimators based on $a_1$, $a_2$ and $a_T$. A logical extension of the results to complex form yields the following:

1. **Estimator $H_1(f)$:** Based on Equation (9.46),

$$a_1 = \sum_{i=1}^{N} x_i y_i \bigg/ \sum_{i=1}^{N} x_i^2$$

the estimator $H_1(f)$ is defined as

$$H_1(f) = \frac{\tilde{S}_{x_m y_m}(f)}{\tilde{S}_{x_m x_m}(f)} \qquad (9.63)$$

We shall see later that this estimator is *unbiased with respect to the presence of output noise*, i.e. 'best' for errors (noise) on the output. Once again, the limiting (theoretical) version of this is

$$H_1(f) = \frac{S_{x_m y_m}(f)}{S_{x_m x_m}(f)} \qquad (9.64)$$

This estimator is probably the most widely used.

**2. Estimator $H_2(f)$:** Based on Equation (9.50),

$$a_2 = \sum_{i=1}^{N} y_i^2 \bigg/ \sum_{i=1}^{N} x_i y_i$$

the estimator $H_2(f)$ is defined as

$$H_2(f) = \frac{\tilde{S}_{y_m y_m}(f)}{\tilde{S}_{y_m x_m}(f)} \qquad (9.65)$$

This estimator is known to be 'best' for errors (noise) on the input, i.e. it is *unbiased with respect to the presence of input noise*. Note that the denominator is $\tilde{S}_{y_m x_m}(f)$ (not $\tilde{S}_{x_m y_m}(f)$). This is due to the location of the conjugate in the numerator of Equation (9.62), so $\tilde{S}_{y_m x_m}(f)$ must be used to satisfy the form $H(f) = Y(f)/X(f)$ (see Appendix E for a complex-valued least squares problem). Similar to the $H_1(f)$ estimator, the theoretical form of Equation (9.65) is

$$H_2(f) = \frac{S_{y_m y_m}(f)}{S_{y_m x_m}(f)} \qquad (9.66)$$

**3. Estimator $H_W(f)$** (also known as $H_s(f)$ or $H_v(f)$): This estimator is sometimes called the total least squares estimator. It has various derivations with slightly different forms – sometimes it is referred to as the $H_v(f)$ estimator (Leuridan *et al.*, 1986; Allemang and Brown, 2002) and as the $H_s(f)$ estimator (Wicks and Vold, 1986). Recently, White *et al.* (2006) generalized this estimator as a maximum likelihood (ML) estimator. We denote this ML estimator $H_W(f)$, which is

$$H_W(f)$$
$$= \frac{\tilde{S}_{y_m y_m}(f) - \kappa(f)\tilde{S}_{x_m x_m}(f) + \sqrt{\left[\tilde{S}_{x_m x_m}(f)\kappa(f) - \tilde{S}_{y_m y_m}(f)\right]^2 + 4\left|\tilde{S}_{x_m y_m}(f)\right|^2 \kappa(f)}}{2\tilde{S}_{y_m x_m}(f)}$$

$$(9.67)$$

where $\kappa(f)$ is the ratio of the spectra of the measurement noises, i.e. $\kappa(f) = S_{n_y n_y}(f)/S_{n_x n_x}(f)$.

This estimator is 'best' for errors (noise) on both input and output, i.e. it is *unbiased with respect to the presence of both input and output noise* provided that the ratio of noise spectra is known. Note that if $\kappa(f) = 0$ then $H_W(f) = H_2(f)$, and if $\kappa(f) \to \infty$

then $H_W(f) \rightarrow H_1(f)$ (see Appendix F for the proof). In practice, it may be difficult to know $\kappa(f)$. In this case, $\kappa(f) = 1$ may be a logical assumption, i.e. the noise power in the input signal is the same as in the output signal. If so, the estimator $H_W(f)$ becomes the solution of the TLS method which is often referred to as the $H_v(f)$ estimator (we use the notation $H_T(f)$ where the subscript $T$ denotes 'total'), i.e.

$$H_T(f) = \frac{\tilde{S}_{y_m y_m}(f) - \tilde{S}_{x_m x_m}(f) + \sqrt{\left[\tilde{S}_{x_m x_m}(f) - \tilde{S}_{y_m y_m}(f)\right]^2 + 4\left|\tilde{S}_{x_m y_m}(f)\right|^2}}{2\tilde{S}_{y_m x_m}(f)} \qquad (9.68)$$

Note that this is analogous to $a_T$ defined in Equation (9.60). The theoretical form of this is

$$H_T(f) = \frac{S_{y_m y_m}(f) - S_{x_m x_m}(f) + \sqrt{\left[S_{x_m x_m}(f) - S_{y_m y_m}(f)\right]^2 + 4\left|S_{x_m y_m}(f)\right|^2}}{2S_{y_m x_m}(f)} \qquad (9.69)$$

### The Biasing Effect of Noise on the Frequency Response Function Estimators $H_1(f)$ and $H_2(f)$

First, consider the effect of *output noise* only as described in Figure 9.6. The $H_1(f)$ estimator is

$$H_1(f) = \frac{S_{xy_m}(f)}{S_{xx}(f)} = \frac{S_{xy}(f) + S_{xn_y}(f)}{S_{xx}(f)} = \frac{S_{xy}(f)}{S_{xx}(f)} = H(f) \qquad (9.70)$$

Thus, $H_1(f)$ is *unbiased* if the noise is present on the *output only*. We assume that appropriate averaging and limiting operations are applied for this expression, i.e. *theoretical* spectral density functions are used. Now, consider the $H_2(f)$ estimator which becomes

$$H_2(f) = \frac{S_{y_m y_m}(f)}{S_{y_m x}(f)} = \frac{S_{yy}(f) + S_{n_y n_y}(f)}{S_{yx}(f)} = H(f)\left(1 + \frac{S_{n_y n_y}(f)}{S_{yy}(f)}\right) \qquad (9.71)$$

Note that this estimator is biased and overestimates $H(f)$ if the output noise is present, depending on the signal-to-noise ratio of the output signal (it may be different for each frequency). If the input is white noise, then the input power spectral density function $S_{xx}(f)$ is constant over the entire frequency range while the output power spectral density $S_{yy}(f)$ varies as the frequency changes, depending on the frequency response characteristics.

Now consider the case when only the *input noise* is present as shown in Figure 9.7. The $H_1(f)$ and $H_2(f)$ estimators are

$$H_1(f) = \frac{S_{x_m y}(f)}{S_{x_m x_m}(f)} = \frac{S_{xy}}{S_{xx}(f) + S_{n_x n_x}} = \frac{H(f)}{1 + S_{n_x n_x}/S_{xx}(f)} \qquad (9.72)$$

$$H_2(f) = \frac{S_{yy}(f)}{S_{yx_m}(f)} = \frac{S_{yy}(f)}{S_{yx}(f) + S_{yn_x}(f)} = \frac{S_{yy}(f)}{S_{yx}(f)} = H(f) \qquad (9.73)$$

Thus, it is shown that $H_2(f)$ is *unbiased* with respect to *input noise* while $H_1(f)$ is biased and underestimates $H(f)$ if the input noise is present. Note that the bias of the $H_1(f)$ estimator depends on the ratio $S_{n_x n_x}(f)/S_{xx}(f)$. If both noise and input signal are white noise, then the ratio is constant for all frequencies, i.e. the $H_1(f)$ estimator becomes simply a scaled version of the true frequency response function $H(f)$ (see MATLAB Example 9.4, Case (b)).

*Example*[M9.4]

Consider a system (with reference to Figure 9.8) that displays resonant and anti-resonant behaviour, i.e. as shown in Figure 9.13.

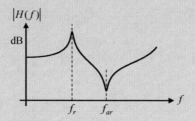

**Figure 9.13** A system with resonant and anti-resonant behaviour

Assume that both input and response are noise contaminated. The input and output signal-to-noise ratios (SNRs) are $S_{xx}(f)/S_{n_x n_x}(f)$ and $S_{yy}(f)/S_{n_y n_y}(f)$. Also, assume the noises are white.

Whilst the input SNR is unaffected by the system response, the output SNR is largest at resonance ($f_r$) and smallest at anti-resonance ($f_{ar}$). Accordingly the 'errors' at the output are (relatively) more significant at $f_{ar}$ than $f_r$, so estimator $H_1(f)$ is more appropriate than $H_2(f)$ for this frequency. Conversely, at frequency $f_r$ the output SNR is high, and so errors on input may be more significant and therefore $H_2(f)$ may be more appropriate.

Thus, $H_1(f)$ usually underestimates the frequency response function at resonances of the structure but gives better estimates at anti-resonances than $H_2(f)$. On the other hand, as mentioned earlier, $H_2(f)$ is relatively unbiased at resonances but significantly overestimates near the anti-resonances (see MATLAB Example 9.4, Case (a)). Thus, when both input and output noise are present the TLS estimator $H_T(f)$ (or $H_W(f)$ if $\kappa(f)$ can be measured) may be preferably used (see MATLAB Example 9.4, Case (c)). Alternatively, a combination of frequency response function estimates $H_1(f)$, $H_2(f)$ and $H_T(f)$ may also be used for different frequency regions appropriately.

Note that the biasing effect of noise on the estimators $H_1(f)$, $H_2(f)$ and $H_W(f)$ is limited to the magnitude spectrum only, i.e. the phase spectrum is unaffected by uncorrelated noise and is not biased. This can be easily verified from Equations (9.71) and (9.72), where $S_{n_y n_y}(f)$, $S_{yy}(f)$, $S_{n_x n_x}(f)$ and $S_{xx}(f)$ are all real valued. Thus, it follows that

$$\arg S_{x_m y_m}(f) = \arg\left(\frac{1}{S_{y_m x_m}(f)}\right) = \arg H_1(f) = \arg H_2(f) = \arg H_W(f) = \arg H(f)$$

(9.74)

This result indicates that the phase spectrum is less sensitive to noise. However, note that this is a theoretical result only. In practice we only have an estimate $\tilde{S}_{x_m y_m}(f)$. Thus, the phase spectrum also has errors as we shall see in Chapter 10.

### The Effect of Feedback

In some situations, there may be feedback in a dynamical system, as shown for example in Figure 9.14. The figure might depict a structure with frequency response function $H(f)$, with $x(t)$ the force and $y(t)$ the response (e.g. acceleration). The excitation is assumed to come from an electrodynamic shaker with input signal $r(t)$. The force applied depends on this excitation but is also affected by the back emf (electromotive force) effect due to the motion. This is modelled as the feedback path $G(f)$. A second input (uncorrelated with $r(t)$) to the system is modelled by the signal $n(t)$. This could come from another (unwanted) excitation.

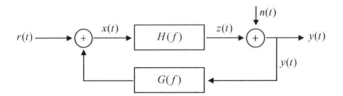

**Figure 9.14**   A system with feedback

The objective is to determine $H(f)$, i.e. the forward path frequency response function using the measured signals $x(t)$ (force) and $y(t)$ (acceleration). Simply using the $H_1(f)$ estimator turns out not to be helpful as the following demonstrates.

In this case, $X(f)$ and $Y(f)$ can be written as

$$X(f) = R(f) + G(f)Y(f) = \frac{R(f) + G(f)N(f)}{1 - H(f)G(f)}$$

$$Y(f) = H(f)X(f) + N(f) = \frac{N(f) + H(f)R(f)}{1 - H(f)G(f)}$$

(9.75)

Thus, the $H_1(f)$ estimator based on the measured signals $x(t)$ and $y(t)$ gives

$$H_1(f) = \frac{S_{xy}(f)}{S_{xx}(f)} = \frac{H(f)S_{rr}(f) + G^*(f)S_{nn}(f)}{S_{rr}(f) + |G(f)|^2 S_{nn}(f)}$$

(9.76)

which is not the required $H(f)$. Rather than determining $H(f)$, note that as the noise power gets large, $H_1(f)$ estimates the inverse of $G(f)$, i.e.

$$H_1(f) = \frac{S_{xy}(f)}{S_{xx}(f)} \approx \frac{1}{G(f)} \quad \text{if} \quad \frac{S_{rr}(f)}{S_{nn}(f)} \to 0$$

(9.77)

It is clear, however, that in the absence of disturbance $n(t)$, $H_1(f)$ does indeed result in $H(f)$ even in the presence of feedback.

From this we see that (if the additional input $n(t)$ is present) we need another approach and this was provided by Wellstead (1981), who proposed using a third signal, namely the excitation to the shaker $r(t)$. In essence he proposed an estimator referred to here as

$$H_3(f) = \frac{S_{ry}(f)}{S_{rx}(f)} \tag{9.78}$$

i.e. the ratio of two cross-spectral density functions. Equation (9.75) can be rearranged as

$$
\begin{aligned}
X(f)[1 - G(f)H(f)] &= R(f) + G(f)N(f) \\
Y(f)[1 - G(f)H(f)] &= H(f)R(f) + N(f)
\end{aligned}
\tag{9.79}
$$

Then, the cross-spectral density functions are

$$S_{rx}(f) = \frac{S_{rr}(f)}{1 - G(f)H(f)} \quad \text{and} \quad S_{ry}(f) = \frac{H(f)S_{rr}(f)}{1 - G(f)H(f)} \tag{9.80}$$

So $H_3(f) = H(f)$ even in the presence of disturbance $n(t)$ and feedback.

## 9.4 BRIEF SUMMARY

1. The input–output relationship in the time domain for a stationary random process $x(t)$ is

$$R_{yy}(\tau) = \int_0^\infty \int_0^\infty h(\tau_1)h(\tau_2)R_{xx}(\tau + \tau_1 - \tau_2)d\tau_1 d\tau_2 \quad \text{and}$$

$$R_{xy}(\tau) = \int_0^\infty h(\tau_1)R_{xx}(\tau - \tau_1)d\tau_1$$

and the corresponding frequency domain expressions are

$$S_{yy}(f) = |H(f)|^2 S_{xx}(f) \quad \text{and} \quad S_{xy}(f) = H(f)S_{xx}(f)$$

2. If the input $x(t)$ is white noise, then (for zero mean values)

$$R_{yy}(\tau) = \sigma_x^2 R_{hh}(\tau) \quad \text{and} \quad R_{xy}(\tau) = \sigma_x^2 h(\tau)$$

3. The ordinary coherence function between input $x(t)$ and output $y(t)$ is defined as

$$\gamma_{xy}^2(f) = \frac{|S_{xy}(f)|^2}{S_{xx}(f)S_{yy}(f)} \quad 0 \le \gamma_{xy}^2(f) \le 1$$

which measures the degree of linearity between $x(t)$ and $y(t)$.

4. When the effect of measurement noise on the *output* is considered, the coherent output power is defined as

$$S_{yy}(f) = \gamma_{xy_m}^2(f)S_{y_m y_m}(f)$$

and the *noise power* (or uncoherent output power) is defined as

$$S_{n_y n_y}(f) = \left[1 - \gamma_{x y_m}^2(f)\right] S_{y_m y_m}(f)$$

5. Power spectral and cross-spectral density functions can be estimated by

$$\tilde{S}_{x_m x_m}(f) = \frac{1}{N} \sum_{i=1}^{N} \frac{|X_{mi}(f)|^2}{T} \quad \text{and} \quad \tilde{S}_{x_m y_m}(f) = \frac{1}{N} \sum_{i=1}^{N} \frac{X_{mi}^*(f) Y_{mi}(f)}{T}$$

6. The frequency response function is estimated by:

(a) $H_1(f) = \dfrac{\tilde{S}_{x_m y_m}(f)}{\tilde{S}_{x_m x_m}(f)}$

which is unbiased with respect to the output noise;

(b) $H_2(f) = \dfrac{\tilde{S}_{y_m y_m}(f)}{\tilde{S}_{y_m x_m}(f)}$

which is unbiased with respect to the input noise;

(c) $H_W(f) = \dfrac{\tilde{S}_{y_m y_m}(f) - \kappa(f)\tilde{S}_{x_m x_m}(f) + \sqrt{\left[\tilde{S}_{x_m x_m}(f)\kappa(f) - \tilde{S}_{y_m y_m}(f)\right]^2 + 4\left|\tilde{S}_{x_m y_m}(f)\right|^2 \kappa(f)}}{2\tilde{S}_{y_m x_m}(f)}$

where $\kappa(f) = S_{n_y n_y}(f)/S_{n_x n_x}(f)$. This is unbiased with respect to both input and output noise. If $\kappa(f)$ is unknown, $\kappa(f) = 1$ may be used.

## 9.5  MATLAB EXAMPLES

---

*Example 9.1:* **System identification using spectral density functions: a first-order system**

Consider the following first-order system (see Equation (9.15))

$$T\dot{y}(t) + y(t) = x(t)$$

where the impulse response function is $h(t) = (1/T)e^{-t/T}$ and the frequency response function is $H(f) = 1/(1 + j2\pi f T)$. In this example, we use the band-limited white noise as an input $x(t)$; then the output $y(t)$ is obtained by the convolution, i.e. $y(t) = h(t) * x(t)$. The spectral density functions $S_{xx}(f)$ and $S_{xy}(f)$ are estimated using Welch's method (see Chapter 10 for details, and also see Comments 2 in MATLAB Example 8.10).

Then, we shall estimate the frequency response function based on Equation (9.12), $S_{xy}(f) = H(f)S_{xx}(f)$, i.e. $H_1(f) = S_{xy}(f)/S_{xx}(f)$. This estimate will be compared with the DFT of $h(t)$ (we mean here the DFT of the sampled, truncated impulse response function).

In this MATLAB example, we do not consider measurement noise. So, we note that $H_1(f) = H_2(f) = H_T(f)$.

| Line | MATLAB code | Comments |
|---|---|---|
| 1 | clear all | Define sampling rate and time variables t1 |
| 2 | fs=500; T1=1; T2=40; t1=0:1/fs:T1; | (for the impulse response function $h(t)$) and |
|  | t2=0:1/fs:T2-1/fs; | t2 (for the band-limited white noise input |
| 3 | T=0.1; | $x(t)$). Then, generate the impulse response |
| 4 | h=1/T*exp(-t1/T); | sequence accordingly which is truncated at |
|  |  | 1 second. |
| 5 | randn('state',0); | Generate the band-limited white noise input |
| 6 | x=randn(1,T2*fs); | signal $x(t)$. The cut-off frequency is set to |
| 7 | fc=30; [b, a] = butter(9,fc/(fs/2)); | 30 Hz for this example. The input sequence |
| 8 | x=filter(b,a,x); | has a zero mean value, and the variance is |
| 9 | x=x-mean(x); x=x/std(x); | one. |
| 10 | y=conv(h,x); y=y(1:end-length(h)+1); | Then, obtain the output sequence by a |
|  | % or y=filter(h,1,x); | convolution operation. Note that the output |
| 11 | y=y/fs; | sequence y is scaled by the sampling rate in |
|  |  | order to match its corresponding continuous |
|  |  | function. |
| 12 | N=4*fs; % N=10*fs; | Calculate the spectral density functions |
| 13 | Sxx=cpsd(x,x, hanning(N),N/2, N, fs, | using Welch's method; we use a Hann |
|  | 'twosided'); | window and 50 % overlap. The length of |
| 14 | Syy=cpsd(y,y, hanning(N),N/2, N, fs, | segment is defined by N, and is 4 seconds |
|  | 'twosided'); | long in this case. |
| 15 | Sxy=cpsd(x,y, hanning(N),N/2, N, fs, | Note that we defined both negative and |
|  | 'twosided'); | positive frequencies (Line 17), thus the |
| 16 | Sxx=fftshift(Sxx); Syy=fftshift(Syy); | MATLAB function 'fftshift' is used to shift |
|  | Sxy=fftshift(Sxy); | the zero-frequency component to the centre |
| 17 | f=fs*(-N/2:N/2-1)/N; | of spectrum. |
| 18 | H1=Sxy./Sxx; | Calculate $H_1(f) = S_{xy}(f)/S_{xx}(f)$, and also |
| 19 | H=fftshift(fft(h,N))/fs; | calculate $H(f)$ using the DFT of the impulse |
| 20 | Gamma=abs(Sxy).^2./(Sxx.*Syy); | response sequence. Also, compute the |
|  |  | coherence function. |
| 21 | figure(1) | Plot the 'calculated (estimated)' power |
| 22 | plot(f, 10*log10(Sxx)) | spectral density function and the magnitude |
| 23 | xlabel('Frequency (Hz)'); | spectrum of cross-spectral density function, |
|  | ylabel('\itS_x_x(\itf\rm) (dB)') | for the frequency range −30 to 30 Hz. |
| 24 | axis([-30 30 -35 -5]) | Note that these functions are only estimates |
| 25 | figure(2) | of true spectral density functions, i.e. they |
| 26 | plot(f, 10*log10(abs(Sxy))) | are $\tilde{S}_{xx}(f)$ and $\tilde{S}_{xy}(f)$. So, we may see some |
| 27 | xlabel('Frequency (Hz)'); | variability as shown in the figures. Note that |
|  | ylabel('|\itS_x_y(\itf\rm)| (dB)') | we use '10*log10(Sxx)' and |
| 28 | axis([-30 30 -35 -15]) | '10*log10(abs(Sxy))' for dB scale, since the |
|  |  | quantities are already power-like. |
| 29 | figure(3) | Plot the magnitude spectrum of both $H_1(f)$ |
| 30 | plot(f, 20*log10(abs(H1 ))); hold on | and $H(f)$ for the frequency range −30 to |
| 31 | xlabel('Frequency (Hz)'); | 30 Hz. |
|  | ylabel('|\itH\rm_1(\itf\rm)| (dB)') |  |
| 32 | plot(f, 20*log10(abs(H)), 'r:'); hold off |  |
| 33 | axis([-30 30 -30 5]) |  |

| 34 | figure(4) | Plot the phase spectrum of both $H_1(f)$ and |
| 35 | plot(f, unwrap(angle(H1))); hold on | $H(f)$ for the frequency range $-30$ to $30$ Hz. |
| 36 | xlabel('Frequency (Hz)'); | |
| | ylabel('arg\itH\rm_1(\itf\rm) (rad)') | |
| 37 | plot(f, unwrap(angle(H)), 'r:'); hold off | |
| 38 | axis([-30 30 -1.6 1.6]) | |
| 39 | figure(5) | Plot the coherence function. |
| 40 | plot(f, Gamma) | |
| 41 | xlabel('Frequency (Hz)'); | |
| | ylabel('Coherence function') | |
| 42 | axis([-150 150 0 1.1]) | |

## Results

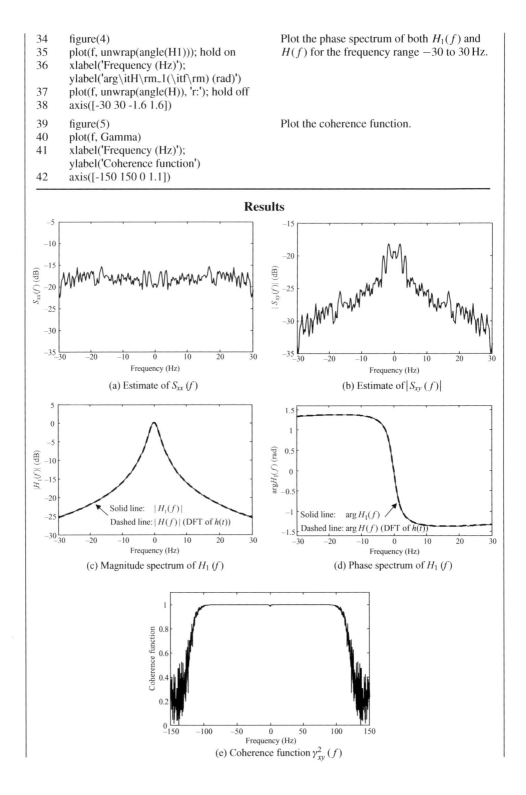

(a) Estimate of $S_{xx}(f)$

(b) Estimate of $|S_{xy}(f)|$

Solid line: $|H_1(f)|$
Dashed line: $|H(f)|$ (DFT of $h(t)$)

(c) Magnitude spectrum of $H_1(f)$

Solid line:   $\arg H_1(f)$
Dashed line: $\arg H(f)$ (DFT of $h(t)$)

(d) Phase spectrum of $H_1(f)$

(e) Coherence function $\gamma_{xy}^2(f)$

**Comments:**

1. Note that the coherence function $\gamma_{xy}^2(f) \approx 1$ within the frequency band of interest, except at the peak (i.e. at zero frequency). The drop of coherence function at $f = 0$ is due to the bias error. This bias error can be reduced by improving the resolution (see Chapter 10 for details). To improve the resolution, the length of the segment must be increased (but note that this reduces the number of averages). For example, replace the number 4 with 10 in Line 12 of the MATLAB code. This increases the window length in the time domain, thus increasing the frequency resolution. The result is shown in Figure (f), where the coherence function is almost unity including the value at $f = 0$.

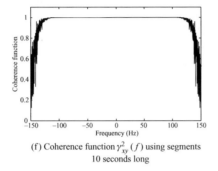

(f) Coherence function $\gamma_{xy}^2(f)$ using segments
10 seconds long

2. From Figures (c) and (d), we see that we have an almost perfect estimate for the frequency response function from $H_1(f)=\tilde{S}_{xy}(f)/\tilde{S}_{xx}(f)$. However, the individual spectral density function estimates show a large variability as shown in Figures (a) and (b). Note that, in theory, $S_{xx}(f)$ is constant and $S_{xy}(f)$ is a scaled version of $H(f)$ if the input is white noise.

It is emphasized that the errors are not due to the noise (we did not consider measurement noise in this example). In fact, these are the statistical errors inherent in the estimation processes (see Chapter 10). By comparing Figures (a)–(c), we may see that the estimate of $H(f)$ is less sensitive to the statistical errors than the estimates of spectral density functions. This will be discussed in Chapter 10.

Note that, even if there is no noise, the estimate of $H(f)$ may have large statistical errors if the number of averages (for the segment averaging method) is small. To demonstrate this, change the length of time (T2) in Line 2 of the MATLAB code, i.e. let T2 = 6. The result is shown in Figure (g), where we see relatively large random errors near the peak.

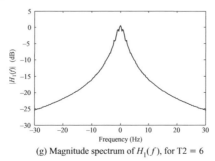

(g) Magnitude spectrum of $H_1(f)$, for T2 = 6

*Example 9.2:* **A first-order continuous system with a sinusoidal input**

Consider the same first-order system as in MATLAB Example 9.1. Now, the input is a sine function, i.e.

$$x(t) = A \sin(2\pi f_0 t) \quad \text{and} \quad R_{xx}(\tau) = \frac{A^2}{2} \cos(2\pi f_0 \tau)$$

Then, the output $y(t)$ can be written as

$$y(t) = \frac{A}{\sqrt{1 + (2\pi f_0 T)^2}} \sin(2\pi f_0 t + \theta) \quad \theta = \tan^{-1}(-2\pi f_0 T)$$

In section 9.1, we have seen that the autocorrelation and cross-correlation functions are (see Equations (9.23) and (9.25))

$$R_{yy}(\tau) = \frac{A^2}{2} \frac{1}{1 + (2\pi f_0 T)^2} \cos(2\pi f_0 \tau) \quad \text{and} \quad R_{xy}(\tau) = \frac{A^2}{2\sqrt{1 + (2\pi f_0 T)^2}} \sin(2\pi f_0 \tau + \phi)$$

where

$$\phi = \tan^{-1}\left(\frac{1}{2\pi f_0 T}\right)$$

In this example, we shall verify this.

| Line | MATLAB code | Comments |
|---|---|---|
| 1 | clear all | Same as in MATLAB Example 9.1, |
| 2 | fs=500; T1=1; T2=40; t1=0:1/fs:T1; | except that the input is now a 1 Hz sine |
|  | t2=0:1/fs:T2-1/fs; | function. |
| 3 | T=0.1; |  |
| 4 | h=1/T*exp(-t1/T); |  |
| 5 | A=2; f=1; w=2*pi*f; |  |
| 6 | x=A*sin(w*t2); |  |
| 7 | y=filter(h,1,x)/fs; |  |
| 8 | maxlag=2*fs; | Define the maximum lag and calculate |
| 9 | [Ryy, tau]=xcorr(y,y,maxlag, 'unbiased'); | the correlation functions. |
| 10 | [Rxy, tau]=xcorr(y,x,maxlag, 'unbiased'); |  |
| 11 | tau=tau/fs; |  |
| 12 | phi=atan(1/(w*T)); | Calculate the true $R_{yy}(\tau)$ and $R_{xy}(\tau)$ |
| 13 | Ryy_a=(A^2/2)*(1./(1+(w*T).^2)).*cos(w*tau); | using Equations (9.23) and (9.25). |
| 14 | Rxy_a=(A^2/2)*(1./sqrt(1+(w*T).^2)).*sin(w*tau+phi); |  |
| 15 | figure(1) | Plot both estimated and true |
| 16 | plot(tau,Ryy,tau,Ryy_a, 'r:') | autocorrelation functions (Ryy and |
| 17 | xlabel('Lag (\it\tau)'); | Ryy_a, respectively). |
|  | ylabel('\itR_y_y\rm(\it\tau\rm)') |  |
| 18 | figure(2) | Plot both estimated and true |
| 19 | plot(tau,Rxy,tau,Rxy_a, 'r:') | cross-correlation functions (Rxy and |
| 20 | xlabel('Lag (\it\tau)'); | Rxy_a, respectively). |
|  | ylabel('\itR_x_y\rm(\it\tau\rm)') |  |

**Results**

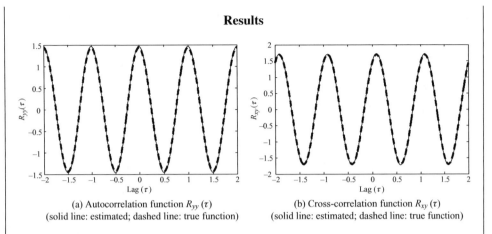

(a) Autocorrelation function $R_{yy}(\tau)$
(solid line: estimated; dashed line: true function)

(b) Cross-correlation function $R_{xy}(\tau)$
(solid line: estimated; dashed line: true function)

**Comments:** As mentioned in Section 9.1, this example has shown that the response to a sinusoidal input is the same sinusoid with scaled amplitude and shifted phase.

---

*Example 9.3:* **Transmission path identification**

We consider the simple acoustic problem as shown in Figure 9.15. Then, we may model the measured signal as

$$\text{Mic. } A = x(t) = as(t)$$

$$\text{Mic. } B = y(t) = bs(t - \Delta_1) + cs(t - \Delta_2)$$

$$(9.81)$$

where $\Delta_1$ and $\Delta_2$ are time delays.

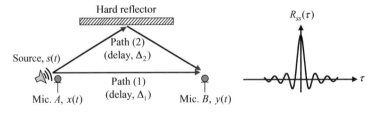

**Figure 9.15**  A simple acoustic example: transmission path identification

If the source signal $s(t)$ is broadband, then the autocorrelation function $R_{ss}(\tau)$ is narrow as depicted in the figure. By treating $x(t)$ as an input and $y(t)$ as an output, i.e. $y(t) = h(t) * x(t)$, as we have seen in Chapter 4, the impulse response function and frequency response function are given by

$$h(t) = \frac{b}{a}\delta(t - \Delta_1) + \frac{c}{a}\delta(t - \Delta_2)$$

$$(9.82)$$

$$H(f) = \frac{b}{a}e^{-j2\pi f \Delta_1}\left[1 + \frac{c}{b}e^{-j2\pi f(\Delta_2 - \Delta_1)}\right]$$

$$(9.83)$$

We now establish the delays and the relative importance of paths by forming the cross-correlation between $x(t)$ and $y(t)$ as

$$R_{xy}(\tau) = E[x(t)y(t+\tau)] = ab R_{ss}(\tau - \Delta_1) + ac R_{ss}(\tau - \Delta_2) \qquad (9.84)$$

This may be drawn as in Figure 9.16.

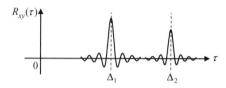

**Figure 9.16**   Cross-correlation function between $x(t)$ and $y(t)$

Note that $\Delta_1$ and $\Delta_2$ are identified if $R_{ss}(\tau)$ is 'narrow' compared with $\Delta_2 - \Delta_1$, and the relative magnitudes yield $b/c$. If the source signal has a bandwidth of $B$ as shown in Figure 9.17, then the autocorrelation function of $s(t)$ can be written as

$$R_{ss}(\tau) = AB\frac{\sin(\pi B\tau)}{\pi B\tau}\cos(2\pi f_0\tau) \qquad (9.85)$$

Thus, in order to resolve the delays, it is required that (roughly) $\Delta_2 - \Delta_1 > 2/B$.

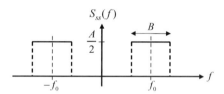

**Figure 9.17**   Power spectral density function of the band-limited signal $s(t)$

Time domain methods as outlined above are probably best – but we might also look at frequency domain methods. First, consider the cross-spectral density function for the simpler case of no reflector, i.e. as shown in Figure 9.18.

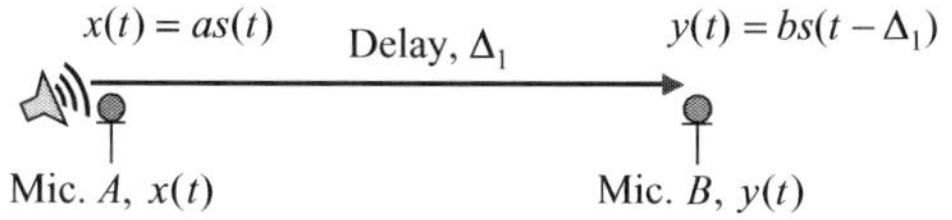

$x(t) = as(t)$          Delay, $\Delta_1$                    $y(t) = bs(t - \Delta_1)$

Mic. $A$, $x(t)$                                       Mic. $B$, $y(t)$

**Figure 9.18**   A simple acoustic example with no reflector

Then, $R_{xy}(\tau) = ab R_{ss}(\tau - \Delta_1)$ and $S_{xy}(f) = abe^{-j2\pi f \Delta_1} S_{ss}(f)$. So $\arg S_{xy}(f)$ gives the delay (see also MATLAB Example 8.9), but it turns out that the phase is more sensitive to other reflections (not the uncorrelated noise) than the correlation function.

Now reconsider the first problem (with a hard reflector). Suppose that $y(t)$ is noise contaminated, i.e. $y_m(t) = y(t) + n(t)$. If $n(t)$ is independent of $y(t)$ then $R_{xy_m}(\tau) = R_{xy}(\tau)$ and $S_{xy_m}(f) = S_{xy}(f)$. Thus, from Equation (9.84), the cross-spectral density function is

$$S_{xy_m}(f) = S_{xy}(f) = abe^{-j2\pi f \Delta_1} \left[ 1 + \frac{c}{b} e^{-j2\pi f (\Delta_2 - \Delta_1)} \right] S_{ss}(f) \qquad (9.86)$$

So the delay information is contained in the phase. However, unlike the single delay problem arg $S_{xy}(f)$ shows a mixture of two delay components. As will be seen later, although it is possible to identify both delays $\Delta_1$ and $\Delta_2$ from the cross-spectral density function, the frequency domain method is more difficult in this case. Also, consider the coherence function (see Equation (9.33)), which is

$$\gamma_{xy_m}^2(f) = \frac{|S_{xy_m}(f)|^2}{S_{xx}(f)S_{y_m y_m}(f)} = \frac{S_{yy}(f)}{S_{yy}(f) + S_{nn}(f)} = \frac{S_{yy}(f)}{S_{y_m y_m}(f)} \qquad (9.87)$$

For convenience, let $b = c$ and $\Delta_2 - \Delta_1 = \Delta$; then

$$S_{yy}(f) = 2b^2 [1 + \cos(2\pi f \Delta)] S_{ss}(f) \qquad (9.88)$$

So we see that $\gamma_{xy_m}^2(f) = 0$ at certain frequencies ($f = n/2\Delta$, $n = 1, 3, 5, \ldots$), i.e. the coherence collapses owing to destructive interference (i.e. the measurement SNR becomes very low).

In the above, we considered both individual transmission paths as non-dispersive. (Note that the two paths taken together are dispersive, i.e. the group delay is $-d\phi/d\omega \neq$ const.) In practical cases, we must first decide whether the paths are dispersive or non-dispersive. If dispersive, the propagation velocity varies with frequency. In such cases, broadband methods may not be successful since waves travel at different speeds. In order to suppress the dispersive effect the cross-correlation method is applied for *narrow* frequency bands, though this too has a smearing effect.

We now examine the transmission path identification problem described above, where the measured signal is

$$x(t) = as(t)$$

$$y_m(t) = y(t) + n(t) = bs(t - \Delta_1) + cs(t - \Delta_2) + n(t)$$

The cross-correlation function and the cross-spectral density function are

$$R_{xy_m}(\tau) = E[x(t)y_m(t + \tau)] = abR_{ss}(\tau - \Delta_1) + acR_{ss}(\tau - \Delta_2)$$

and

$$S_{xy_m}(f) = \left[ abe^{-j2\pi f \Delta_1} + ace^{-j2\pi f \Delta_2} \right] S_{ss}(f)$$

In this example, we shall compare the time domain method (using the cross-correlation function) and the frequency domain method (using the cross-spectral density function).

| Line | MATLAB code | Comments |
|------|-------------|----------|
| 1 | clear all | Define sampling rate and time variable. |
| 2 | fs=100; T=502; t=0:1/fs:T-1/fs; | Generate a band-limited white noise |
| 3 | randn('state',0); | signal, where the (full) bandwidth |
| 4 | s=randn(size(t)); | (equivalent to $B$ in Figure 9.17) is |
| 5 | fc=10; [b,a] = butter(9,fc/(fs/2)); | approximately 20 Hz ($-f_c$ to $f_c$). |
| 6 | s=filtfilt(b,a,s); | |

| | | |
|---|---|---|
| 7 | s=s-mean(s); s=s/std(s);<br>% Makes mean(s)=0 & std(s)=1; | |
| 8 | a=1; b=0.8; c=0.75; delta1=1; delta2=1.5;<br>% delta2=1.07; | Define parameters for signals $x(t)$ and<br>$y(t)$. |
| 9 | N1=2*fs; N2=T*fs-N1; | Also, define time delays, $\Delta_1 = 1$ and |
| 10 | x=a*s(N1+1:N1+N2); | $\Delta_2 = 1.5$. Note that $\Delta_2 - \Delta_1 = 0.5$. |
| 11 | y1=b*s(N1-(delta1*fs)+1:N1-<br>(delta1*fs)+N2); | Later, use $\Delta_2 = 1.07$, and compare the<br>cross-correlation functions. |
| 12 | y2=c*s(N1-(delta2*fs)+1:N1-<br>(delta2*fs)+N2); | Generate signals $x(t)$ and $y(t)$. Also, add<br>some noise to the signal $y(t)$. |
| 13 | y=y1+y2; | |
| 14 | randn('state',10); | |
| 15 | n=randn(size(y))*0.1; | |
| 16 | y=y+n; | |
| | | |
| 17 | maxlag=2*fs; | Calculate the cross-correlation function. |
| 18 | [Rxy, tau]=xcorr(y,x,maxlag, 'unbiased'); | |
| 19 | tau=tau/fs; | |
| | | |
| 20 | T1=50; | Calculate the (one-sided) spectral density |
| 21 | [Gxx, f]=cpsd(x,x, hanning(T1*fs),T1*fs/2,<br>T1*fs, fs); | functions and the coherence function. |
| 22 | [Gyy, f]=cpsd(y,y, hanning(T1*fs),T1*fs/2,<br>T1*fs, fs); | |
| 23 | [Gxy, f]=cpsd(x,y, hanning(T1*fs),T1*fs/2,<br>T1*fs, fs); | |
| 24 | Gamma=abs(Gxy).^2./(Gxx.*Gyy); | |
| | | |
| 25 | figure(1) | Plot the cross-correlation function. As |
| 26 | plot(tau(maxlag+1:end),Rxy(maxlag+1:end)) | shown in Figure (a), $\Delta_1 = 1$ and |
| 27 | xlabel('Lag (\it\tau)') | $\Delta_2 = 1.5$ are clearly identified. However, |
| 28 | ylabel('Cross-correlation') | if $\Delta_2 = 1.07$ is used, it is not possible to |
| 29 | axis([0 2 -0.2 0.8]) | detect the delays as shown in Figure (d).<br>Note that the bandwidth of the signal $s(t)$<br>is approximately 20 Hz, thus it is required<br>that $\Delta_2 - \Delta_1 > 0.1$ for this method to be<br>applicable. |
| | | |
| 30 | figure(2) | Plot the phase spectrum of the |
| 31 | plot(f,unwrap(angle(Gxy))) | cross-spectral density function $G_{xy}(f)$. |
| 32 | xlabel('Frequency (Hz)') | As shown in Figure (b), the phase curve |
| 33 | ylabel('arg\itG_x_y\rm(\itf\rm) (rad)') | is no longer a straight line, but it has a |
| 34 | axis([0 15 -90 0]) | 'periodic' structure. In fact, the relative<br>delay $\Delta_2 - \Delta_1$ can be found by<br>observing this periodicity as described in<br>the figure, while $\Delta_1$ can be obtained from<br>the overall slope of the phase curve.<br>Compare this phase spectrum with that of<br>a single delay problem (see MATLAB<br>Example 8.9). |
| | | |
| 35 | figure(3) | Plot the coherence function. Note that the |
| 36 | plot(f,Gamma) | coherence drops owing to the interference |
| 37 | xlabel('Frequency (Hz)');<br>ylabel('Coherence function') | between two delay components (see<br>Equations (9.87) and (9.88)). |
| 38 | axis([0 15 0 1]) | |

## Results

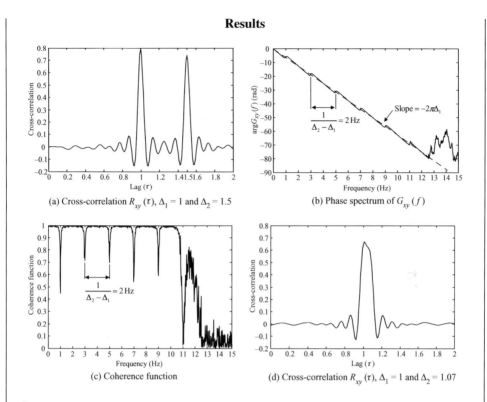

(a) Cross-correlation $R_{xy}(\tau)$, $\Delta_1 = 1$ and $\Delta_2 = 1.5$

(b) Phase spectrum of $G_{xy}(f)$

(c) Coherence function

(d) Cross-correlation $R_{xy}(\tau)$, $\Delta_1 = 1$ and $\Delta_2 = 1.07$

**Comments:** Note that the time domain method is much simpler and clearer. However, the signal must be wideband if the relative delay $\Delta_2 - \Delta_1$ is small, otherwise the time domain method may fail as shown in Figure (d).

---

*Example 9.4:* **Frequency response function estimators $H_1(f)$, $H_2(f)$ and $H_T(f)$**

Consider the following impulse response function of a two-degree-of-freedom system:

$$h(t) = \frac{A_1}{\omega_{d1}} e^{-\zeta_1 \omega_{n1} t} \sin \omega_{d1} t + \frac{A_2}{\omega_{d2}} e^{-\zeta_2 \omega_{n2} t} \sin \omega_{d2} t$$

In this example, we use the white noise as an input $x(t)$, and the output $y(t)$ is obtained by $y(t)=h(t) * x(t)$. We also consider the uncorrelated measurement noise.

Three FRF estimators, $H_1(f)$, $H_2(f)$, and $H_T(f)$, are compared for three different cases: Case (a), output noise only; Case (b), input noise only; and Case (c), both input and output noise.

Equations (9.63), (9.65) and (9.68) are used in this example, i.e.

$$H_1(f) = \frac{\tilde{S}_{x_m y_m}(f)}{\tilde{S}_{x_m x_m}(f)}, \quad H_2(f) = \frac{\tilde{S}_{y_m y_m}(f)}{\tilde{S}_{y_m x_m}(f)},$$

$$H_T(f) = \frac{\tilde{S}_{y_m y_m}(f) - \tilde{S}_{x_m x_m}(f) + \sqrt{\left[\tilde{S}_{x_m x_m}(f) - \tilde{S}_{y_m y_m}(f)\right]^2 + 4\left|\tilde{S}_{x_m y_m}(f)\right|^2}}{2\tilde{S}_{y_m x_m}(f)}$$

where the spectral density functions are estimated using the segment averaging method.

| Line | MATLAB code | Comments |
|---|---|---|
| 1 | clear all | Define parameters for the impulse response |
| 2 | A1=20; A2=30; f1=5; f2=15; wn1=2*pi*f1; wn2=2*pi*f2; | function $h(t)$, and generate the sequence accordingly. The sampling rate is chosen as |
| 3 | zeta1=0.05; zeta2=0.03; | 50 Hz, and the length of the impulse |
| 4 | wd1=sqrt(1-zeta1^2)*wn1; wd2=sqrt(1-zeta2^2)*wn2; | response function is 10 seconds. |
| 5 | fs=50; T1=10; t1=[0:1/fs:T1-1/fs]; | |
| 6 | h=(A1/wd1)*exp(- zeta1*wn1*t1).*sin(wd1*t1) + (A2/wd2)*exp(- zeta2*wn2*t1).*sin(wd2*t1); | |
| 7 | T= 50000; | Define the length of input signal, and |
| 8 | randn('state',0); | generate input white noise sequence 'x'. |
| 9 | x=randn(1,T*fs); | Then obtain the output sequence 'y'. |
| 10 | y=filter(h,1,x); % we do not scale for convenience | Note that we define very long sequences to minimize random errors on the estimation of the spectral density functions. This will be discussed in Chapter 10. |
| 11 | randn('state',10); | Generate the uncorrelated input |
| 12 | nx=0.5*randn(size(x)); % nx=0 for Case (a) | measurement noise and output measurement noise. Note that we define the noise such that |
| 13 | randn('state',20); | the variances of the input noise and the |
| 14 | ny=0.5*randn(size(y)); % ny=0 for Case (b) | output noise are the same, i.e. $\kappa(f)=1$. Add these noises to the input and output |
| 15 | x=x+nx; y=y+ny; | appropriately. Then clear the variables 'nx' |
| 16 | clear nx ny | and 'ny' (to save computer memory). This script is for Case (c). Replace Line 12 with 'nx=0' for Case (a), and replace Line 14 with 'ny=0' for Case (b). |
| 17 | [Gxx, f]=cpsd(x(1:T*fs),x(1:T*fs), hanning(T1*fs),T1*fs/2, T1*fs, fs); | Calculate the (one-sided) spectral density functions using the segment averaging |
| 18 | [Gyy, f]=cpsd(y(1:T*fs),y(1:T*fs), hanning(T1*fs),T1*fs/2, T1*fs, fs); | method. Then calculate the frequency response |
| 19 | [Gxy, f]=cpsd(x(1:T*fs),y(1:T*fs), hanning(T1*fs),T1*fs/2, T1*fs, fs); | function estimates $H_1(f)$, $H_2(f)$ and $H_T(f)$. Note that $H_T(f) = H_W(f)$ since $\kappa(f) = 1$. |
| 20 | [Gyx, f]=cpsd(y(1:T*fs),x(1:T*fs), hanning(T1*fs),T1*fs/2, T1*fs, fs); | Also calculate $H(f)$ by the DFT of the impulse response sequence. Then compare |
| 21 | H1=Gxy./Gxx; | the results. |
| 22 | H2=Gyy./Gyx; | |
| 23 | HT=(Gyy-Gxx + sqrt((Gxx-Gyy).^2 + 4*abs(Gxy).^2))./(2*Gyx); | |
| 24 | H=fft(h); | |

| | | |
|---|---|---|
| 25 | figure (1) | Plot the magnitude spectrum of both |
| 26 | plot(f,20*log10(abs(H1)), | $H_1(f)$ and $H(f)$. |
| | f,20*log10(abs(H(1:length(f)))), 'r:') | |
| 27 | xlabel('Frequency (Hz)'); | |
| | ylabel('|\itH\rm_1(\itf\rm)| (dB)') | |
| 28 | axis([0 25 -35 25]) | |
| | | |
| 29 | figure(2) | Plot the magnitude spectrum of both |
| 30 | plot(f,20*log10(abs(H2)), | $H_2(f)$ and $H(f)$. |
| | f,20*log10(abs(H(1:length(f)))), 'r:') | |
| 31 | xlabel('Frequency (Hz)'); | |
| | ylabel('|\itH\rm_2(\itf\rm)| (dB)') | |
| 32 | axis([0 25 -35 25]) | |
| | | |
| 33 | figure(3) | Plot the magnitude spectrum of both |
| 34 | plot(f,20*log10(abs(HT)), | $H_T(f)$ and $H(f)$. |
| | f,20*log10(abs(H(1:length(f)))), 'r:') | |
| 35 | xlabel('Frequency (Hz)'); | |
| | ylabel('|\itH_T(\itf\rm)| (dB)') | |
| 36 | axis([0 25 -35 25]) | |

**Results: Case (a) output noise only** (Replace Line 12 with 'nx=0'.)

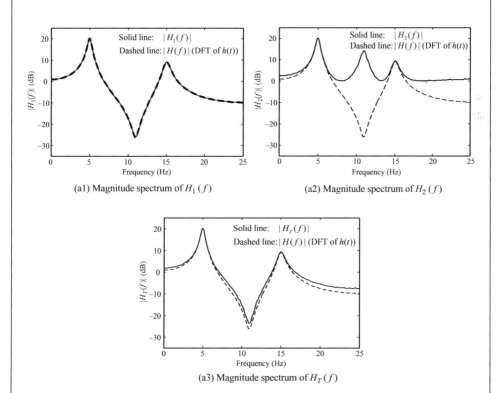

(a1) Magnitude spectrum of $H_1(f)$

(a2) Magnitude spectrum of $H_2(f)$

(a3) Magnitude spectrum of $H_T(f)$

**Results: Case (b) input noise only** (Replace Line 14 with 'ny=0'.)

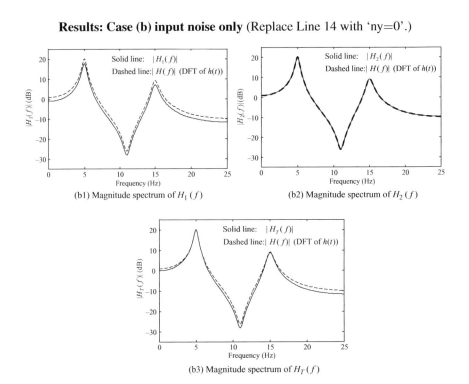

(b1) Magnitude spectrum of $H_1(f)$

(b2) Magnitude spectrum of $H_2(f)$

(b3) Magnitude spectrum of $H_T(f)$

**Results: Case (c) both input and output noise**

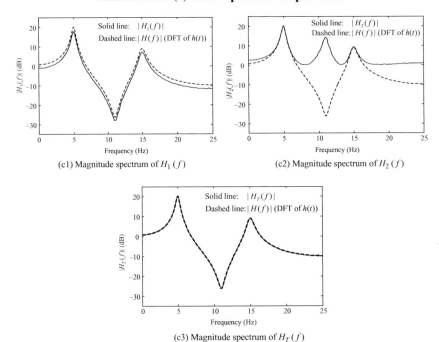

(c1) Magnitude spectrum of $H_1(f)$

(c2) Magnitude spectrum of $H_2(f)$

(c3) Magnitude spectrum of $H_T(f)$

**Comments:**

1. We have demonstrated that $H_1(f)$ is unbiased with respect to the output noise (see Case (a)), $H_2(f)$ is unbiased with respect to the input noise (see Case (b)), and $H_T(f)$ is unbiased with respect to both input and output noise if $\kappa(f) = 1$ (see Case (c)). Note that, for all different cases, the $H_T(f)$ estimator gives more consistent estimates of the frequency response function. Thus, the TLS estimator $H_T(f)$ (or $H_W(f)$ if $\kappa(f)$ is measurable) is highly recommended. However, in practical applications, it is always wise to compare all three estimators and choose the 'best' estimator based on some prior knowledge.

   As described in Equation (9.74) in Section 9.3, note that the phase spectrum of all three estimators is the same. To see this, type the following script in the MATLAB command window. The results are shown in Figure (d).

   > figure(4)
   > plot(f,unwrap(angle(H1)), f,unwrap(angle(H2)), f,unwrap(angle(HT)),
   >     f,unwrap(angle(H(1:length(f))))), 'k:')
   > xlabel('Frequency (Hz)'); ylabel('Phase spectrum (rad)')

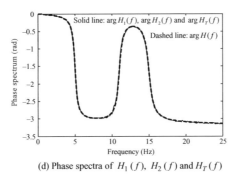

(d) Phase spectra of $H_1(f)$, $H_2(f)$ and $H_T(f)$

2. We note that the inverse DFT of the FRF estimate gives the corresponding estimated impulse response sequence. As mentioned in Chapter 6, this impulse response sequence can be regarded as an MA system (i.e. an FIR filter). In this MATLAB example, it has 500 MA coefficients. In real-time signal processing (such as active control), it may be useful if the number of coefficients can be reduced, especially for the case of a large number of filter coefficients. One approach to this is by curve fitting the estimated FRF data to a reduced order ARMA model (see Equation (6.12)). The basic procedure of the curve fitting algorithm can be found in Levi (1959). In MATLAB, the function called 'invfreqz' finds the coefficients for the ARMA model based on the estimated frequency response function. Type the following script to find the reduced order ARMA model (we use the ARMA(4,4) model, which has 10 coefficients in total):

   > [b,a]=invfreqz(HT, 2*pi*f/fs, 4,4, [], 30);
   > Hz=freqz(b,a,length(f),fs);
   > figure(5)
   > plot(f,20*log10(abs(Hz)),f,20*log10(abs(H(1:length(f))))), 'r:')
   > xlabel('Frequency (Hz)'); ylabel('Magnitude spectrum (dB)')
   > axis([0 25 -35 25])

```
figure(6)
plot(f,unwrap(angle(Hz)), f,unwrap(angle(H(1:length(f)))), 'r:')
xlabel('Frequency (Hz)'); ylabel('Phase spectrum (rad)')
```

The first line of the MATLAB script finds the coefficients for the ARMA(4,4) model based on the estimated FRF (we use the results of $H_T(f)$ in this example), and the second line evaluates the frequency response $H_Z(f)$ based on the coefficient vectors 'a' and 'b' obtained from the first line. Then, plot both magnitude and phase spectra of $H_Z(f)$ and compare with those of $H(f)$. The results are as in Figures (e) and (f).

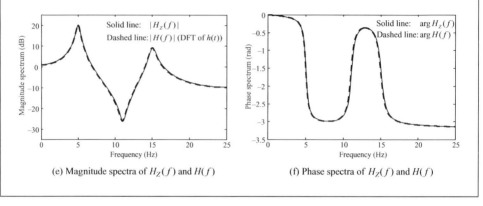

(e) Magnitude spectra of $H_Z(f)$ and $H(f)$ 　　　　　 (f) Phase spectra of $H_Z(f)$ and $H(f)$

***Example 9.5:* A practical example of system identification**

We consider the same experimental setup used in MATLAB Example 6.7 (impact testing of a structure), except that we use a band-limited white noise for the input signal as shown in Figure (a). In this experiment, the frequency band of the signal is set at 5 to 90 Hz and the sampling rate is chosen as $f_s = 256$ Hz.

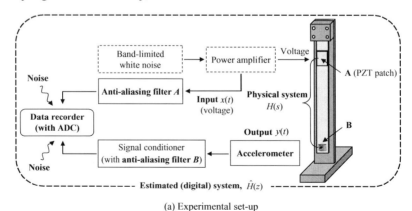

(a) Experimental set-up

In this example, we will compare the results of three different FRF estimators, $H_1(f)$, $H_2(f)$ and $H_T(f)$, using the measured data stored in the file 'beam_experiment.mat'.[1]

---

[1] The data files can be downloaded from the Companion Website (www.wiley.com/go/shin_hammond).

| Line | MATLAB code | Comments |
|------|-------------|----------|
| 1 | clear all | Load the measured data (x and y) which |
| 2 | load beam_experiment | are recorded for 20 seconds. Define the |
| 3 | fs=256; T=4; | sampling rate and the length of segment |
| 4 | [Gxx, f]=cpsd(x,x, hanning(T*fs),T*fs/2, T*fs, fs); | (4 seconds). |
| | | Calculate the (one-sided) spectral density |
| 5 | [Gyy, f]=cpsd(y,y, hanning(T*fs),T*fs/2, T*fs, fs); | functions. We use a Hann window and |
| | | 50 % overlap – this gives nine averages |
| 6 | [Gxy, f]=cpsd(x,y, hanning(T*fs),T*fs/2, T*fs, fs); | for each estimate. |
| | | Then calculate the frequency response |
| 7 | [Gyx, f]=cpsd(y,x, hanning(T*fs),T*fs/2, T*fs, fs); | function estimates $H_1(f)$, $H_2(f)$ and |
| | | $H_T(f)$. |
| 8 | H1=Gxy./Gxx; | |
| 9 | H2=Gyy./Gyx; | |
| 10 | HT=(Gyy – Gxx + sqrt((Gxx-Gyy).^2 + 4*abs(Gxy).^2))./(2*Gyx); | |
| 11 | figure (1) | Plot the magnitude spectra of $H_1(f)$, |
| 12 | plot(f,20*log10(abs(H1))) | $H_2(f)$ and $H_T(f)$ for the frequency range |
| 13 | xlabel('Frequency (Hz)'); ylabel('|\itH\rm_1(\itf\rm)| (dB)') | 5 to 90 Hz. |
| 14 | axis([5 90 -45 25]) | |
| 15 | figure (2) | |
| 16 | plot(f,20*log10(abs(H2))) | |
| 17 | xlabel('Frequency (Hz)'); ylabel('|\itH\rm_2(\itf\rm)| (dB)') | |
| 18 | axis([5 90 -45 25]) | |
| 19 | figure (3) | |
| 20 | plot(f,20*log10(abs(HT))) | |
| 21 | xlabel('Frequency (Hz)'); ylabel('|\itH_T(\itf\rm)| (dB)') | |
| 22 | axis([5 90 -45 25]) | |
| 23 | figure(4) | Plot the phase spectra of $H_1(f)$, $H_2(f)$ and |
| 24 | plot(f(21:361), unwrap(angle(H1(21:361))), f(21:361), unwrap(angle(H2(21:361))), f(21:361), unwrap(angle(HT(21:361)))); | $H_T(f)$ for the frequency range 5 to 90 Hz. Note that they are almost identical. |
| 25 | xlabel('Frequency (Hz)'); ylabel('Phase spectrum (rad)') | |
| 26 | axis([5 90 -7 0.5]) | |

**Results**

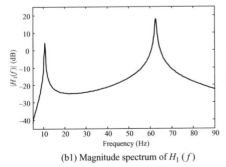

(b1) Magnitude spectrum of $H_1(f)$

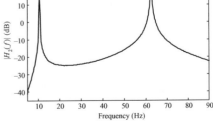

(b2) Magnitude spectrum of $H_2(f)$

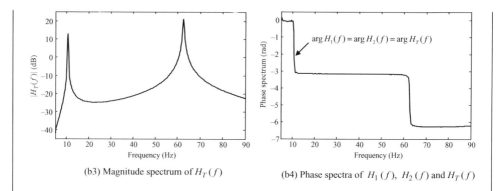

(b3) Magnitude spectrum of $H_T(f)$          (b4) Phase spectra of $H_1(f)$, $H_2(f)$ and $H_T(f)$

## Comments:

1. As shown in Figure (b4), the phase spectra of $H_1(f)$, $H_2(f)$ and $H_T(f)$ are the same. However, the results of magnitude spectra show that $H_1(f)$ considerably underestimates at resonances compared with other estimates $H_2(f)$ and $H_T(f)$. Figure (c) shows the differences in detail.

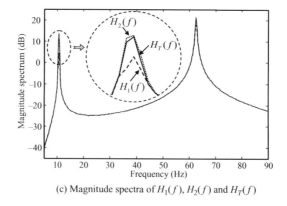

(c) Magnitude spectra of $H_1(f)$, $H_2(f)$ and $H_T(f)$

Also, plot the coherence function by typing the following script. The result is shown in Figure (d).

```
figure(5)
Gamma=abs(Gxy).^2./(Gxx.*Gyy);
plot(f, Gamma); axis([5 90 0 1.1])
xlabel('Frequency (Hz)'); ylabel('Coherence function')
```

Note that the value of the coherence function drops at resonances due to the bias error, which will be discussed in Chapter 10 (see also Comments 1 in MATLAB Example 9.1).

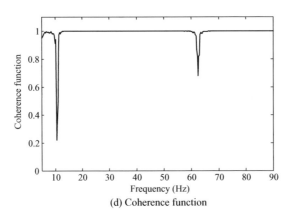

(d) Coherence function

2. With reference to the experimental setup in Figure (a), there is another practical aspect to be considered. It is often the case that the anti-aliasing filters $A$ and $B$ introduce different delays. For example, if the filter $B$ introduces more delay than the filter $A$, the phase spectrum becomes as shown in Figure (e). Thus, it is recommended that the same type of filter is used for both input and output.

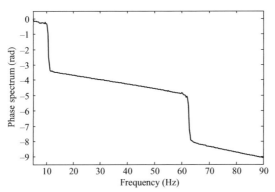

(e) Phase spectrum of FRF (delay in the filter $B$ > delay in the filter $A$)

# 10

# Estimation Methods and Statistical Considerations

## Introduction

So far, we have discussed random processes in terms of ideal quantities: probability density functions, correlation functions and spectral density functions. The results in Chapter 9 used these theoretical concepts, and the MATLAB examples used estimation methods that anticipated what is presented in this chapter. In this chapter, we introduce statistical estimation methods for random signals based on a single realization (record) of the process, and show how the theoretical quantities may be estimated and the accuracy of these estimates.

While omitting many mathematical justifications, the details of results quoted in this chapter may be found in many of the references, especially in Jenkins and Watts (1968) and Bendat and Piersol (2000). Readers who wish to know the details of statistical properties of random processes should refer to these two excellent texts.

## 10.1 ESTIMATOR ERRORS AND ACCURACY

Suppose we only have a single time record $x(t)$ with a length of $T$, taken from a stochastic process. If we want to know the mean value $\mu_x$ of the process, then a logical estimate (denoted by $\bar{x}$) of $\mu_x$ is

$$\bar{x} = \frac{1}{T} \int_0^T x(t)dt \qquad (10.1)$$

*Fundamentals of Signal Processing for Sound and Vibration Engineers*
K. Shin and J. K. Hammond.    © 2008 John Wiley & Sons, Ltd

The value obtained by Equation (10.1) is a sample value of a random variable, say $\bar{X}$, which has its own probability distribution (this is called the *sample distribution*). And $\bar{x}$ is a single realization of the random variable $\bar{X}$. Each time we compute a value $\bar{x}$ from a different length of record we get a different value. If our estimation procedure is 'satisfactory', we may expect that:

 (i) the scatter of values of $\bar{x}$ is not too great and they lie close to the true mean value $\mu_x$;
 (ii) the more data we use (i.e. the larger $T$), the better the estimate.

We now formalize these ideas. Let $\phi$ be the parameter we wish to estimate (i.e. $\phi$ is the theoretical quantity, e.g. $\mu_x$ above) and let $\hat{\Phi}$ be an estimator for $\phi$. Then $\hat{\Phi}$ is a random variable with its own probability distribution, e.g. as shown in Figure 10.1, where $\hat{\phi}$ is the value (or estimate) of the random variable $\hat{\Phi}$.

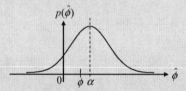

**Figure 10.1**　Probability density function of $\hat{\phi}$

We see here that the estimates $\hat{\phi}$ we would obtain can take a whole range of values but would predominantly take values near $\alpha$. It is often difficult to obtain the sampling distribution $p(\hat{\phi})$, and so we shall settle for a few summarizing properties.

### Bias

The bias of an estimator is defined as

$$b(\hat{\Phi}) = E[\hat{\Phi}] - \phi \tag{10.2}$$

i.e. the difference between the average of the estimator and the true value. Note that $E[\hat{\Phi}]$ is $\alpha$ in the cases shown in Figure 10.1. Thus, the bias is a measure of the average offset of the estimator. If $b(\hat{\Phi}) = 0$, then the estimator $\hat{\Phi}$ is 'unbiased'. Although it seems desirable to use an unbiased estimator, we may need to allow some bias of the estimator if the variability of the estimate can be reduced (relative to that of an unbiased estimator).

### Variance

The variance of an estimator is defined as

$$\text{Var}(\hat{\Phi}) = E\left[(\hat{\Phi} - E[\hat{\Phi}])^2\right] = E[\hat{\Phi}^2] - E^2[\hat{\Phi}] \tag{10.3}$$

This is a measure of the dispersion or spread of values of $\hat{\Phi}$ about its own mean value (see Section 7.3). Note that the square root of the variance is the standard deviation $\sigma(\hat{\Phi})$ of the estimator. In general, it is desirable to have a small variance, i.e. the probability

density function should be 'peaky'. This requirement often results in an increase of the
bias error.

## Mean Square Error

The mean square error (mse) of an estimator is a measure of the spread of values of $\hat{\Phi}$
about the *true* (*theoretical*) value $\phi$, i.e.

$$\text{mse}(\hat{\Phi}) = E\left[(\hat{\Phi} - \phi)^2\right] \qquad (10.4)$$

Since $E[(\hat{\Phi} - \phi)^2] = E[(\hat{\Phi} - E[\hat{\Phi}] + E[\hat{\Phi}] - \phi)^2] = E[(\hat{\Phi} - E[\hat{\Phi}])^2] + E[(E[\hat{\Phi}] - \phi)^2]$,
the above equation can be rewritten as

$$\text{mse}(\hat{\Phi}) = \text{Var}(\hat{\Phi}) + b^2(\hat{\Phi}) \qquad (10.5)$$

which shows that the mean square error reflects *both* variance and bias. Thus, the mean
square error is often used as a measure of the relative importance of bias and variance.
For example, if an estimator has the property that its mean square error is less than any
other estimators, it is said to be more *efficient* than other estimators.

    If the mean square error decreases as the sample size (amount of data) used to
compute the estimate increases, then the estimator is *consistent*. Sometimes the errors
are non-dimensionalized (normalized) by dividing them by the quantity being estimated
(for $\phi \neq 0$), e.g. as

$$\text{Bias error: } \varepsilon_b = \frac{b(\hat{\Phi})}{\phi} \qquad (10.6)$$

$$\text{Random error: } \varepsilon_r = \frac{\sigma(\hat{\Phi})}{\phi} \qquad (10.7)$$

$$\text{RMS error: } \varepsilon = \frac{\sqrt{\text{mse}(\hat{\Phi})}}{\phi} \qquad (10.8)$$

## Confidence Intervals

The estimate $\hat{\phi}$ we have discussed so far is a *point estimate*, i.e. a single value. It is
often desirable to define a certain *interval of values* in which the parameter is likely to
fall. For example, if we estimate a mean value $\bar{x}$ as 50, then perhaps it is 'likely' that
$\mu_x$ lies in the interval 45 to 55. This estimate is an *interval estimate*, and is called the
*confidence interval* when we attach a number describing the likelihood of the parameter
falling within the interval.

    For example, if we say 'a 95 % confidence interval for $\mu_x$ is (45, 55)', then this means
we are 95 % confident that $\mu_x$ lies in the range (45, 55). Note that this does not mean that
the probability that $\mu_x$ lies in the interval (45, 55) is 0.95, because $\mu_x$ is not a random
variable and so we cannot assign probabilities to it. Instead, we mean that if we could
realize a large number of samples and find a confidence interval for $\mu_x$ for *each* sample,

then approximately 95 % of these intervals would contain the true value $\mu_x$. In order to calculate confidence intervals, we need to know the sampling distribution of the estimator. We shall return to this problem when we discuss the spectral density function estimates.

In the following sections, we shall give a summary of: (i) definitions of commonly used estimators; (ii) some statistical properties of the estimators; and (iii) some computational aspects for calculating the estimates. Unless otherwise stated, we shall assume we are dealing with realizations of a continuous, stationary random process. Also, if the data are sampled we assume that the sampling rate is sufficiently high so that there is no aliasing.

## 10.2 MEAN VALUE AND MEAN SQUARE VALUE

### The Mean Value of x(t)

For a stationary stochastic process $x(t)$, the mean value (from data with length $T$) is estimated as

$$\hat{\mu}_x = \frac{1}{T} \int_0^T x(t)dt \tag{10.9}$$

where the true mean value is $\mu_x$. Note that we have changed our notation for sample mean from $\bar{x}$ to $\hat{\mu}_x$ to use the circumflex notation for an estimate. The average of this estimate is

$$E\left[\hat{\mu}_x\right] = \frac{1}{T} \int_0^T E[x(t)]dt = \frac{1}{T} \int_0^T \mu_x dt = \mu_x \tag{10.10}$$

i.e. $\hat{\mu}_x$ is unbiased. Now consider the mean square error which is

$$\text{mse}(\hat{\mu}_x) = E\left[(\hat{\mu}_x - \mu_x)^2\right] = E\left[\frac{1}{T^2} \int_0^T \int_0^T (x(t_1) - \mu_x)(x(t_2) - \mu_x)dt_1 dt_2\right]$$

$$= \frac{1}{T^2} \int_0^T \int_0^T C_{xx}(t_2 - t_1)dt_1 dt_2 = \frac{1}{T^2} \int_0^T \int_{-t_1}^{T-t_1} C_{xx}(\tau)d\tau dt_1 \tag{10.11}$$

where $\tau = t_2 - t_1$ and $C_{xx}(\tau)$ is the autocovariance function. By reversing the integration order and changing the limits of integration appropriately, this equation can be written as

$$\text{mse}(\hat{\mu}_x) = \frac{1}{T^2} \int_{-T}^{0} \int_{-\tau}^{T} C_{xx}(\tau)dt_1 d\tau + \frac{1}{T^2} \int_0^T \int_0^{T-\tau} C_{xx}(\tau)dt_1 d\tau$$

$$= \frac{1}{T} \int_{-T}^{T} \left(1 - \frac{|\tau|}{T}\right) C_{xx}(\tau)d\tau \tag{10.12}$$

The integrand is a triangular weighted covariance function, so the integral is finite. Thus $\text{mse}(\hat{\mu}_x) \to 0$ as $T \to \infty$, i.e. this estimator is *consistent*.

For example, if $C_{xx}(\tau) = Ke^{-\lambda|\tau|}$, then the power spectral density function is

$$S_{xx}(f) = K \frac{2\lambda}{\lambda^2 + (2\pi f)^2} \qquad \text{(assuming zero mean value)}$$

and the 3 dB bandwidth is $B = \lambda/\pi$ Hz. The mean square error may be approximated as

$$\text{mse}(\hat{\mu}_x) \approx \frac{1}{T} \int_{-\infty}^{\infty} C_{xx}(\tau)d\tau = \frac{1}{T} \int_{-\infty}^{\infty} Ke^{-\lambda|\tau|}d\tau = \frac{2K}{T\lambda} = \frac{2K}{\pi BT} \qquad (10.13)$$

i.e. it is inversely proportional to the bandwidth–time ($BT$) product of the data.

To perform the calculation using digitized data (sampled at every $\Delta$ seconds), the mean value can be estimated by (as Equation (8.39))

$$\hat{\mu}_x = \frac{1}{N} \sum_{n=0}^{N-1} x(n\Delta) \qquad (10.14)$$

### The Mean Square Value of x(t)

As for the mean value, the mean square value is estimated as

$$\hat{\psi}_x^2 = \frac{1}{T} \int_0^T x^2(t)dt \qquad (10.15)$$

The mean of this estimate is

$$E\left[\hat{\psi}_x^2\right] = \frac{1}{T} \int_0^T E[x^2(t)]dt = \psi_x^2 \qquad (10.16)$$

where the true mean square value is $\psi_x^2$. Thus, the estimator is unbiased. The variance of the estimate is

$$\text{Var}(\hat{\psi}_x^2) = E\left[(\hat{\psi}_x^2 - \psi_x^2)^2\right] = E\left[(\hat{\psi}_x^2)^2\right] - (\psi_x^2)^2$$

$$= \frac{1}{T^2} \int_0^T \int_0^T \left(E[x^2(t_1)x^2(t_2)] - (\psi_x^2)^2\right)dt_1 dt_2 \qquad (10.17)$$

If we assume that $x(t)$ is Gaussian we can use the following result to simplify the integrand. If the random variables $X_1$, $X_2$, $X_3$ and $X_4$ are jointly Gaussian, it can be shown that (Papoulis, 1991)

$$E[X_1 X_2 X_3 X_4] = E[X_1 X_2]E[X_3 X_4] + E[X_1 X_3]E[X_2 X_4]$$
$$+ E[X_1 X_4]E[X_2 X_3] - 2E[X_1]E[X_2]E[X_3]E[X_4] \qquad (10.18)$$

Using this result, it follows that $E[x^2(t_1)x^2(t_2)] = \psi_x^4 + 2R_{xx}^2(t_2 - t_1) - 2\mu_x^4$. After some manipulation by letting $\tau = t_2 - t_1$, Equation (10.17) becomes (Bendat and Piersol, 2000)

$$\text{Var}(\hat{\psi}_x^2) = \frac{2}{T} \int_{-T}^{T} \left(1 - \frac{|\tau|}{T}\right) \left(R_{xx}^2(\tau) - \mu_x^4\right) d\tau \tag{10.19}$$

Thus, for large $T$, if $R_{xx}(\tau)$ dies out 'quickly' compared with $T$, then

$$\text{Var}(\hat{\psi}_x^2) \approx \frac{2}{T} \int_{-\infty}^{\infty} \left(R_{xx}^2(\tau) - \mu_x^4\right) d\tau \tag{10.20}$$

For example, if $\mu_x = 0$ and $R_{xx}(\tau) = Ke^{-\lambda|\tau|}$, then

$$\text{Var}(\hat{\psi}_x^2) \approx \frac{2K^2}{T\lambda} = \frac{2K^2}{\pi BT} \tag{10.21}$$

Since $\hat{\psi}_x^2$ is an unbiased estimator, $\text{mse}(\hat{\psi}_x^2) = \text{Var}(\hat{\psi}_x^2)$, which is also inversely proportional to the bandwidth–time product of the data. Note that the normalized rms error is

$$\varepsilon = \frac{\sqrt{\text{mse}(\hat{\psi}_x^2)}}{\psi_x^2} = \frac{\sqrt{\text{Var}(\hat{\psi}_x^2)}}{\psi_x^2} \approx \frac{\sqrt{2}K}{K\sqrt{\pi BT}} = \sqrt{\frac{2}{\pi BT}} \tag{10.22}$$

where

$$\psi_x^2 = \frac{1}{T} \int_0^T R_{xx}(0) dt = K \quad \text{(from Equation (10.16))}$$

In practice, we often calculate the variance of the signal by subtracting the mean first from the data. In digital form, we might estimate the variance of $x(n\Delta)$ by

$$\text{Var}(x) \approx \hat{\sigma}_x^2 = \frac{1}{N} \sum_{n=0}^{N-1} (x(n\Delta) - \hat{\mu}_x)^2 \tag{10.23}$$

However, if the observations are independent the above can be shown to be a biased estimate. Thus, the divisor $N - 1$ is frequently used, i.e. the unbiased estimate is

$$\hat{\sigma}_x^2 = \frac{1}{N-1} \sum_{n=0}^{N-1} (x(n\Delta) - \hat{\mu}_x)^2 \tag{10.24}$$

## 10.3  CORRELATION AND COVARIANCE FUNCTIONS

### The Autocorrelation (Autocovariance) Function

If a signal $x(t)$ is defined for $0 \le t \le T$, then there are two commonly used estimates of the theoretical autocovariance function $C_{xx}(\tau)$. They may be written as $\hat{C}^b_{xx}(\tau)$ and $\hat{C}_{xx}(\tau)$, where

$$\hat{C}^b_{xx}(\tau) = \frac{1}{T} \int_0^{T-|\tau|} (x(t) - \hat{\mu}_x)(x(t+|\tau|) - \hat{\mu}_x)dt \quad 0 \le |\tau| < T \tag{10.25}$$
$$= 0 \hspace{6.5cm} |\tau| > T$$

and

$$\hat{C}_{xx}(\tau) = \frac{1}{T - |\tau|} \int_0^{T-|\tau|} (x(t) - \hat{\mu}_x)(x(t+|\tau|) - \hat{\mu}_x)dt \quad 0 \le |\tau| < T \tag{10.26}$$
$$= 0 \hspace{7cm} |\tau| > T$$

The superscript $b$ in Equation (10.25) denotes a biased estimate. Often the latter expression is used since it is unbiased. However, both these estimators are used because they have intuitive appeal and should be compared on the basis of some criterion (e.g. the mean square error) to choose between them. The estimates for the theoretical autocorrelation function $R_{xx}(\tau)$ may be expressed as $\hat{R}^b_{xx}(\tau)$ and $\hat{R}_{xx}(\tau)$ in the same way as above by omitting $\hat{\mu}_x$ in the equations.

For convenience, suppose that the process $x(t)$ has zero mean (so that $C_{xx}(\tau) = R_{xx}(\tau)$). Then, as in Jenkins and Watts (1968), we can calculate the bias and variance:

1. *Bias:* Since

$$E\left[\hat{R}^b_{xx}(\tau)\right] = \frac{1}{T} \int_0^{T-|\tau|} E\left[x(t)x(t+|\tau|)\right]dt \quad 0 \le |\tau| < T$$

the expected value of the biased estimator is

$$E\left[\hat{R}^b_{xx}(\tau)\right] = R_{xx}(\tau)\left(1 - \frac{|\tau|}{T}\right) \quad 0 \le |\tau| < T \tag{10.27}$$
$$= 0 \hspace{5cm} |\tau| > T$$

and the expected value of the unbiased estimator is

$$E\left[\hat{R}_{xx}(\tau)\right] = R_{xx}(\tau) \quad 0 \le |\tau| < T \tag{10.28}$$
$$= 0 \hspace{3.5cm} |\tau| > T$$

That is, for $\tau$ in the range $0 \le |\tau| < T$, $\hat{R}_{xx}(\tau)$ is an unbiased estimator, whilst $\hat{R}^b_{xx}(\tau)$ is biased but the bias is small when $|\tau|/T \ll 1$ (i.e. asymptotically unbiased).

2. *Variance:* The variances of biased and unbiased estimators are

$$\text{Var}\big(\hat{R}^b_{xx}(\tau)\big) = \frac{1}{T^2} \int_{-(T-\tau)}^{T-\tau} (T - \tau - |r|)\big(R^2_{xx}(r) + R_{xx}(r + \tau)R_{xx}(r - \tau)\big)dr \qquad (10.29)$$

$$\text{Var}\big(\hat{R}_{xx}(\tau)\big) = \frac{1}{(T - |\tau|)^2} \int_{-(T-\tau)}^{T-\tau} (T - \tau - |r|)\big(R^2_{xx}(r) + R_{xx}(r + \tau)R_{xx}(r - \tau)\big)dr$$

$$\qquad (10.30)$$

When $T$ is large compared with $\tau$, a useful approximation for both equations is

$$\text{Var}\big(\hat{R}_{xx}(\tau)\big) \approx \text{Var}\big(\hat{R}^b_{xx}(\tau)\big) \approx \frac{1}{T} \int_{-\infty}^{\infty} \big(R^2_{xx}(r) + R_{xx}(r + \tau)R_{xx}(r - \tau)\big)dr \qquad (10.31)$$

Note that the variance of the estimates is inversely proportional to the length of data, i.e. $\text{Var}\big(\hat{R}_{xx}(\tau)\big) \propto 1/T$. This shows that both estimators are consistent. Thus, the autocorrelation function may be estimated with diminishing error as the length of the data increases (this also shows that the estimate of the autocorrelation function is ergodic).

## Comparison of the Two Estimators, $\hat{R}^b_{xx}(\tau)$ and $\hat{R}_{xx}(\tau)$

For $\tau \ll T$, there is little difference between the two estimators, i.e. $\hat{R}_{xx}(\tau) \approx \hat{R}^b_{xx}(\tau)$. However, Jenkins and Watts conjecture that $\text{mse}(\hat{R}_{xx}(\tau)) > \text{mse}(\hat{R}^b_{xx}(\tau))$. In fact, by considering the divisor in Equation (10.30), it is easy to see that as $\tau \to T$ the variance of the unbiased estimator $\hat{R}_{xx}(\tau)$ tends to infinity (i.e. diverges). It is this behaviour that makes the unbiased estimator unsatisfactory. However, we note that the unbiased estimator is often used in practical engineering despite the relatively larger mean square error. As a rough guide to using $\hat{R}_{xx}(\tau)$, the ratio of the maximum lag to the total data length, $\tau_{max}/T$, should not exceed 0.1.

Another important feature of the estimators is that adjacent autocorrelation function estimates will have (in general) strong correlations, and so the sample autocorrelation function $\hat{R}_{xx}(\tau)$ (and $\hat{R}^b_{xx}(\tau)$) gives more strongly correlated results than the original time series $x(t)$, i.e. the estimate may not decay as rapidly as might be expected to (Jenkins and Watts, 1968).

## The Cross-correlation (Cross-covariance) Function

If $x(t)$ and $y(t)$ are random signals defined for $0 \le t \le T$, the sample cross-covariance function is defined as

$$\hat{C}_{xy}(\tau) = \frac{1}{T - \tau} \int_0^{T-\tau} (x(t) - \hat{\mu}_x)(y(t + \tau) - \hat{\mu}_y)dt \qquad 0 \le \tau < T$$

$$= \frac{1}{T - |\tau|} \int_{|\tau|}^{T} (x(t) - \hat{\mu}_x)(y(t + \tau) - \hat{\mu}_y)dt \qquad -T < \tau \le 0 \qquad (10.32)$$

$$= 0 \qquad |\tau| > T$$

This is an unbiased estimator. The same integral but with divisor $T$ can also be used, which is then the biased estimator $\hat{C}_{xy}^b(\tau)$. Similarly the sample cross-correlation function may be expressed as $\hat{R}_{xy}(\tau)$ or $\hat{R}_{xy}^b(\tau)$ without subtracting mean values.

The estimators for the cross-correlation (cross-covariance) function have statistical properties that are very similar to those of the autocorrelation (autocovariance) function: $\hat{R}_{xy}(\tau)$ is unbiased whilst $\hat{R}_{xy}^b(\tau)$ is asymptotically unbiased, and $\hat{R}_{xy}^b(\tau)$ has a smaller mean square error, i.e. $\mathrm{mse}(\hat{R}_{xy}(\tau)) > \mathrm{mse}(\hat{R}_{xy}^b(\tau))$.

## Methods of Calculation Using Sampled Data

From sampled data, the autocovariance and cross-covariance functions are evaluated from

$$\hat{C}_{xx}(m\Delta) = \frac{1}{N-m} \sum_{n=0}^{N-m-1} (x(n\Delta) - \hat{\mu}_x)(x((n+m)\Delta) - \hat{\mu}_x) \quad 0 \le m \le N-1$$

(10.33)

and

$$\hat{C}_{xy}(m\Delta) = \frac{1}{N-m} \sum_{n=0}^{N-m-1} (x(n\Delta) - \hat{\mu}_x)(y((n+m)\Delta) - \hat{\mu}_y) \quad 0 \le m \le N-1$$

(10.34)

where both $x(n\Delta)$ and $y(n\Delta)$ are $N$-point sequences, i.e. they are defined for $n = 0, 1, \ldots, N-1$, and $m$ is the lag that may take values $0 \le m \le N-1$. Note that these are unbiased estimators, and the divisor is $N$ for the biased estimators $\hat{C}_{xx}^b(m\Delta)$ and $\hat{C}_{xy}^b(m\Delta)$. The same expressions are applied for the computation of autocorrelation and cross-correlation functions, e.g. $\hat{R}_{xx}(m\Delta)$ and $\hat{R}_{xy}(m\Delta)$ are obtained without subtracting mean values in Equations (10.33) and (10.34).

We note that the autocorrelation function is even, i.e. $\hat{R}_{xx}(-m\Delta) = \hat{R}_{xx}(m\Delta)$, and the cross-correlation function has a property that

$$\hat{R}_{xy}(-m\Delta) = \hat{R}_{yx}(m\Delta) = \frac{1}{N-m} \sum_{n=0}^{N-m-1} y(n\Delta)x((n+m)\Delta) \quad 0 \le m \le N-1$$

(10.35)

The above expressions are the so-called 'mean lagged product' formulae and are evaluated directly if there are not too many multiply and add operations. However, it turns out that it is quicker to use FFT methods to evaluate these indirectly.

### *Autocorrelation via FFT*[1]

The basis of this method lies in our earlier discussions in Chapter 6 on the convolution of two sequences (i.e. the convolution sum). Recall that if $y(n) = \sum_{m=0}^{N-1} h(m)x(n-m)$, and $h(n)$ and $x(n)$ are $N$-point sequences, then $y(n)$ is a $(2N-1)$-point sequence. Since the DFT of the convolution of two sequences is the product of the DFTs of two sequences, *as long as the sequences are padded out with zeros to avoid circular convolution*, the sequence $y(n)$ can be obtained by $y(n) = \mathrm{IDFT}[H(k)X(k)]$.

---

[1] See Bendat and Piersol (2000) for more details.

The correlation calculation is very similar. In fact $\sum_{n=0}^{N-1} x(n)y(n+m)$ is a convolution of $x(-n)$ with $y(n)$ and the DFT of this is

$$\text{DFT}\left[\sum_{n=0}^{N-1} x(n)y(n+m)\right] = \sum_{m=0}^{N-1}\sum_{n=0}^{N-1} x(n)y(n+m)e^{-j(2\pi/N)mk} = X^*(k)Y(k) \qquad (10.36)$$

Thus, the required correlation function $\hat{R}_{xy}(m\Delta)$ is the IDFT $[X^*(k)Y(k)]$ and then *scaled* by $1/(N-m)$. Note that we must ensure that *circular effects are removed by adding zeros*.

Pictorially, the computation of the autocorrelation function can be illustrated as in Figure 10.2. Note that the sequence is effectively *periodic* when the DFT is used.

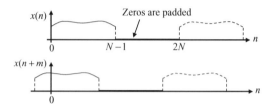

**Figure 10.2**   Pictorial description of the computation of the autocorrelation function

We can see that the correlation of these *periodic* sequences is the same as the *linear* correlation if there are as many zeros appended as data points. So the autocorrelation function (without explicitly noting the sampling interval $\Delta$)

$$\hat{R}_{xx}(m) = \frac{1}{N-m}\sum_{n=0}^{N-m-1} x(n)x(n+m)$$

is obtained by:

1. Take $x(n)$ ($N$ points) and add $N$ zeros to it.
2. Form $X(k)$ ($2N$-point DFT).
3. Form IDFT $[X^*(k)X(k)]$ and then scale appropriately by $1/(N-m)$.

The result will have the appearance as shown in Figure 10.3. This basic idea can be generalized to cross-correlation functions.

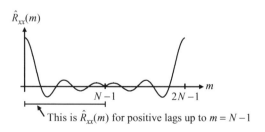

**Figure 10.3**   Results of autocorrelation computation using the DFT

## 10.4 POWER SPECTRAL DENSITY FUNCTION

There are two main approaches to estimating the power spectral density function, namely *parametric* (more recent) and non-parametric (*traditional*), as shown in Figure 10.4.

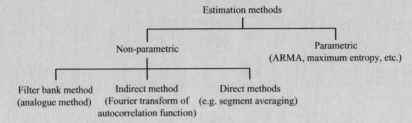

**Figure 10.4**   Classification of the estimation methods for the power spectral density function

Estimation methods for the power spectral density function considered here will relate to the 'traditional' methods rather than 'parametric' methods. We shall outline three methods for the estimation of the power spectral density function:

- Method (1): 'Analogue' method (filter bank method)
- Method (2): Fourier transform of the autocorrelation function (indirect method)
- Method (3): Direct methods.

Note that Method (3) is the most widely used since with the advent of the FFT it is the quickest.

### Method (1): 'Analogue' Method (Filter Bank Method)

The word 'analogue' is in quotes because this method can also be implemented digitally, but it is convenient to refer to continuous signals. The basic scheme is indicated in Figure 10.5.

The basis of this method is that the variance of the signal is the area under the power spectral density function curve (assuming zero mean value), i.e.

$$\text{Var}\,(x(t)) = \sigma_x^2 = \int_0^\infty G_{xx}(f)df \tag{10.37}$$

where $G_{xx}(f)$ is regarded as a measure of the distribution of the power of the process over frequency. So if we wish to know the power in the signal over some frequency band $f_c \pm B/2$,

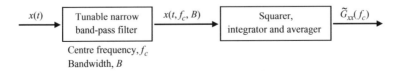

**Figure 10.5**   Concept of the filter bank method

then we pass the signal through a filter with that passband, square (to get the power), average to reduce the fluctuations and divide by the bandwidth to obtain the 'density', i.e.

$$\tilde{G}_{xx}(f_c) = \frac{1}{BT} \int_0^T x^2(t, f_c, B)dt \tag{10.38}$$

The key elements in any spectral estimation scheme are: (i) a procedure to 'home in' on a narrow band, i.e. *good resolution* (low bias); (ii) the subsequent *smoothing* of the squared estimate (i.e. low variance).

Let us assume an ideal band-pass filter and discuss the bias and variance of this estimate. The frequency response function of an ideal band-pass filter is shown in Figure 10.6.

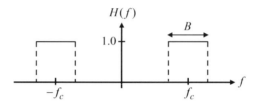

**Figure 10.6** Frequency response function of an ideal band-pass filter

### Bias

The bias of the smoothed estimator is obtained from averaging $\tilde{G}_{xx}(f_c)$, i.e.

$$E\left[\tilde{G}_{xx}(f_c)\right] = \frac{1}{BT} \int_0^T E\left[x^2(t, f_c, B)\right]dt = \frac{E\left[x^2(t, f_c, B)\right]}{B} \tag{10.39}$$

Note that $E\left[x^2(t, f_c, B)\right]$ is the variance of the output of the filter, i.e.

$$E\left[x^2(t, f_c, B)\right] = \int_{f_c-B/2}^{f_c+B/2} G_{xx}(f)df$$

Thus,

$$E\left[\tilde{G}_{xx}(f_c)\right] = \frac{1}{B} \int_{f_c-B/2}^{f_c+B/2} G_{xx}(f)df \tag{10.40}$$

So, in general, $E[\tilde{G}_{xx}(f_c)] \neq G_{xx}(f_c)$, i.e. the estimate is biased. Expanding $G_{xx}(f)$ in a Taylor series about the point $f = f_c$ gives

$$G_{xx}(f) \approx G_{xx}(f_c) + (f - f_c)G'_{xx}(f_c) + \frac{(f - f_c)^2}{2!}G''_{xx}(f_c) \tag{10.41}$$

Substituting this into the integral, we get (Bendat and Piersol, 2000)

$$E\left[\tilde{G}_{xx}(f_c)\right] \approx G_{xx}(f_c) + \underbrace{\frac{B^2}{24} G''_{xx}(f_c)}_{bias} \tag{10.42}$$

Note that at a *peak*, $G''_{xx}(f) < 0$, so the power spectral density is *underestimated* (on average); at a *trough*, $G''_{xx}(f) > 0$, so we have an *overestimate*, i.e. the dynamic range is reduced as illustrated in Figure 10.7. Note also that poor resolution (large $B$) introduces more bias error.

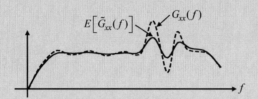

**Figure 10.7**   Demonstration of the bias error of $\tilde{G}_{xx}(f)$

As can be seen from Equation (10.42), bias depends on the resolution $B$ (i.e. the filter bandwidth) relative to the fine structure of the spectrum. As an example, consider a simple oscillator (with a damping ratio of $\zeta$ and a resonance at $f_r$) excited by white noise. The output power spectral density may be as shown in Figure 10.8, where the half-power point bandwidth is given by $B_r \approx 2\zeta f_r$. For a given ideal filter with a bandwidth of $B$ centred at frequency $f_r$, the normalized bias error at $f_r$ can be shown to be (Bendat and Piersol, 2000)

$$\varepsilon_b = \frac{b\left(\tilde{G}_{xx}(f_r)\right)}{G_{xx}(f_r)} \approx -\frac{1}{3}\left(\frac{B}{B_r}\right)^2 \tag{10.43}$$

Note that if $B = B_r$, the normalized bias error is $-33.3\%$, whereas the bias error may be negligible if $B < B_r/4$ (where $|\varepsilon_b| < 2.1\%$).

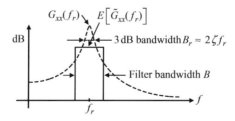

**Figure 10.8**   Illustration of 3 dB bandwidth and the filter bandwidth

### Variance and the Mean Square Error

Assuming that the process is Gaussian and $G_{xx}(f)$ is constant over the bandwidth of the filter, the variance of the estimate may be shown to be (Newland, 1984; Bendat and Piersol, 2000)

$$\text{Var}\left(\tilde{G}_{xx}(f)\right) \approx \frac{G_{xx}^2(f)}{BT} \tag{10.44}$$

The mean square error is the sum of the variance and the square of bias, and we normalize this to give

$$\varepsilon^2 = \frac{\text{Var}\left(\tilde{G}_{xx}(f)\right) + b^2\left(\tilde{G}_{xx}(f)\right)}{G_{xx}^2(f)} \approx \frac{1}{BT} + \frac{B^4}{576}\left(\frac{G_{xx}''(f)}{G_{xx}(f)}\right)^2 \tag{10.45}$$

Note the *conflict* – to suppress bias the filter bandwidth $B$ must be small (i.e. fine resolution), but to reduce the variance the product $BT$ must be large. Note also that the product $BT$ relates to *controllable* parameters, i.e. $B$ is the filter bandwidth (not the data bandwidth), and the averaging time $T$ obviously affects the variance. While maintaining small filter bandwidth, the only way to reduce the mean square error is by increasing the averaging time $T$.

### Comments on the Choice of Filter Bandwidth[2]

The basic choice is between the constant (*absolute*) bandwidth and the constant (*relative*) percentage (%) bandwidth. The *constant bandwidth* gives uniform resolution on a linear frequency scale, as shown in Figure 10.9.

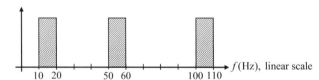

**Figure 10.9**    Constant bandwidth (10 Hz) filter

For constant bandwidth, the centre frequency of an ideal filter is defined as

$$f_c = \frac{f_u + f_l}{2} \tag{10.46}$$

where $f_u$ and $f_l$ are defined as in Figure 10.10. The centre frequency is simply the arithmetic mean of the upper and the lower frequencies.

Constant bandwidth is useful if the signal has harmonically related components, i.e. for detecting a harmonic pattern. However, note that if the bandwidth is satisfactory at high frequencies, it is 'coarse' at the one below and swamps the next lowest on a logarithmic scale (see Figure 10.11).

---

[2] See Randall (1987) for more details.

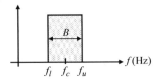

**Figure 10.10**  Centre frequency of the constant bandwidth filter

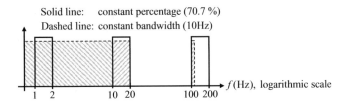

**Figure 10.11**  Constant percentage (70.7 %) bandwidth filter

Since in many cases it is natural and efficient to analyse spectra using a constant percentage bandwidth, e.g. structural response, where if each mode has roughly the same damping, then the 3 dB bandwidth increases with frequency.

The *constant percentage bandwidth* gives uniform resolution on a *logarithmic* frequency scale, as shown in Figure 10.11.

For the constant percentage bandwidth, the centre frequency of an ideal filter is defined as (see Figure 10.12)

$$\log f_c = \frac{\log f_u + \log f_l}{2} \tag{10.47}$$

or

$$f_c = \sqrt{f_u \cdot f_l} \tag{10.48}$$

We consider two special cases: *octave* and *third octave filters*. The octave filters have a passband such that $f_u = 2f_l$, so $f_c = \sqrt{2} f_l$ and the relative bandwidth is

$$\text{Relative bandwidth} = \frac{f_u - f_l}{f_c} = \frac{f_l}{f_c} = \frac{1}{\sqrt{2}} \approx 70.7\,\% \text{ (for octave filters)} \tag{10.49}$$

i.e. a constant percentage of 70.7 %. Starting with a reference centre frequency of 1000 Hz, it is possible to cover three decades in frequency with 10 octave bands ranging from 22.5 Hz

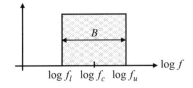

**Figure 10.12**  Centre frequency of the constant percentage bandwidth filter

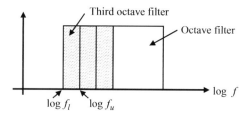

**Figure 10.13**   Comparison of the third octave filter and the octave filter

(lower frequency for a centre frequency of 31.5 Hz) to 22.5 kHz (upper frequency for a centre frequency of 16 kHz).

Third octave filters (1/3 octave filters) are obtained as illustrated in Figure 10.13, i.e. each octave band is divided into three geometrically equal subsections.

As shown in the figure, $\log(f_u/f_l) = \log 2$ for the octave filter, so $\log(f_u/f_l) = \frac{1}{3}\log 2$ for the third octave filter, i.e. $f_u = 2^{1/3} f_l$. (Note that this is approximately 1/10 of a decade, i.e. $\frac{1}{3}\log 2 \approx 0.1 = \frac{1}{10}\log 10$.) The centre frequency is $f_c = \sqrt{f_u \cdot f_l} = \sqrt{2^{1/3} f_l^2} = 2^{1/6} f_l$, and the bandwidth for the third octave filter is

$$\text{Bandwidth} = f_u - f_l = (2^{1/3} - 1)f_l \tag{10.50}$$

and the relative bandwidth is

$$\text{Relative bandwidth} = \frac{f_u - f_l}{f_c} = \frac{2^{1/3} - 1}{2^{1/6}} \approx 23.1\,\% \tag{10.51}$$

Note that, similar to the third octave filter, an $m$ octave filter may be defined so that $f_u = 2^{1/m} f_l$.

The above considerations relate to 'ideal' filters. Various other definitions of bandwidth exist, e.g. 3 dB bandwidth and noise bandwidth as shown in Figure 10.14. As mentioned in Section 4.6, the (effective) noise bandwidth is defined as the width of an ideal rectangular filter that would accumulate the same noise power from a white noise source as the practical filter with the same reference transmission level. The 3 dB bandwidth is the width of the practical filter at the 3 dB points. Although the noise bandwidth and the 3 dB bandwidth are close to each other, the 3 dB bandwidth may be more useful when describing structural responses and

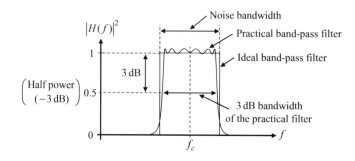

**Figure 10.14**   Noise bandwidth and 3 dB bandwidth of a practical filter

is often preferably used since it is easier to measure. Later in this section, we shall define another bandwidth for spectral windows, which is effectively the resolution bandwidth.

As a final comment, note that the phase characteristics of the filters are unimportant for power spectra measurements.

## Method (2): Fourier Transform of the Autocorrelation Function (Indirect Method)[3]

Consider a sample record $x(t)$, where $|t| < T/2$, and the corresponding Fourier transform given by $X_T(f) = \int_{-T/2}^{T/2} x(t)e^{-j2\pi ft}dt$. Then, the raw (or sample) power spectral density function is

$$\hat{S}_{xx}(f) = \frac{|X_T(f)|^2}{T} = \frac{1}{T}\int\limits_{-T/2}^{T/2}\int\limits_{-T/2}^{T/2} x(t)e^{-j2\pi ft}x(t_1)e^{j2\pi ft_1}dt\,dt_1 \qquad (10.52)$$

Now, transforming the double integral by setting $u = t - t_1$ and $v = t_1$, as shown in Figure 10.15, then Equation (10.52) may be rewritten as

$$\hat{S}_{xx}(f) = \int\limits_0^T \left[ \frac{1}{T}\int\limits_{-T/2}^{T/2-u} x(u+v)x(v)dv \right] e^{-j2\pi fu}du$$

$$+ \int\limits_{-T}^0 \left[ \frac{1}{T}\int\limits_{-T/2-u}^{T/2} x(u+v)x(v)dv \right] e^{-j2\pi fu}du \qquad (10.53)$$

By definition, the term in the first square bracket is $\hat{R}_{xx}^b(u)$ for $0 \le u < T$, and the term in the second square bracket is $\hat{R}_{xx}^b(u)$ for $-T < u \le 0$, i.e. $\hat{S}_{xx}(f) = \int_{-T}^T \hat{R}_{xx}^b(u)e^{-j2\pi fu}du$.

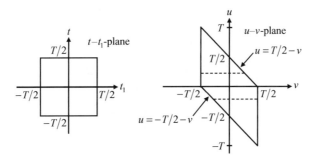

**Figure 10.15**   Transformation of the regions of integration

---

[3] See Jenkins and Watts (1968) for more details.

Thus, an estimate of the power spectral density from a length of data $T$ (we assume that $x(t)$ is defined for $|t| < T/2$ can be written as

$$\hat{S}_{xx}(f) = \int_{-T}^{T} \hat{R}_{xx}^{b}(\tau)e^{-j2\pi f\tau}d\tau \qquad (10.54)$$

Note that the sample power spectral density $\hat{S}_{xx}(f)$ is related to the sample autocorrelation function $\hat{R}_{xx}^{b}(\tau)$ (biased estimator with divisor $T$). This relationship suggests that we might estimate the power spectral density by first forming the sample autocorrelation function and Fourier transforming this. (However, we do not presume the validity of the Wiener–Khinchin theorem – which will follow shortly.) Note that $\hat{R}_{xx}^{b}(\tau) = 0$ for $|\tau| > T$, thus

$$\hat{S}_{xx}(f) = \int_{-\infty}^{\infty} \hat{R}_{xx}^{b}(\tau)e^{-j2\pi f\tau}d\tau = F\{\hat{R}_{xx}^{b}(\tau)\} \qquad (10.55)$$

However, as mentioned in Chapter 8, this is termed the 'raw' power spectral density since it turns out that the variability of this estimator is independent of the data length $T$ as we shall see soon. Now, first consider the bias of this estimator.

### Bias of $\hat{S}_{xx}(f)$

Averaging $\hat{S}_{xx}(f)$ gives

$$E\left[\hat{S}_{xx}(f)\right] = \int_{-T}^{T} E\left[\hat{R}_{xx}^{b}(\tau)\right]e^{-j2\pi f\tau}d\tau \qquad (10.56)$$

and using Equation (10.27), this becomes

$$E\left[\hat{S}_{xx}(f)\right] = \int_{-T}^{T} R_{xx}(\tau)\left(1 - \frac{|\tau|}{T}\right)e^{-j2\pi f\tau}d\tau \quad 0 \leq |\tau| < T \qquad (10.57)$$

Thus, if $T$ is sufficiently large, i.e. $T \to \infty$, then

$$\lim_{T \to \infty} E\left[\hat{S}_{xx}(f)\right] = \lim_{T \to \infty} \frac{E\left[|X_T(f)|^2\right]}{T} = S_{xx}(f) = \int_{-\infty}^{\infty} R_{xx}(\tau)e^{-j2\pi f\tau}d\tau \qquad (10.58)$$

So $\hat{S}_{xx}(f)$ is an *asymptotically unbiased* estimator. The above result (10.58) proves the Wiener–Khinchin theorem introduced in Chapter 8.

### Variance of $\hat{S}_{xx}(f)$

As a prerequisite we need to discuss the properties of a random variable having the so-called chi-squared distribution. We first outline some important results on this (Jenkins and Watts, 1968).

### The Chi-squared Distribution

Let $X_1, X_2, \ldots, X_n$ be $n$ independent random variables, *each of which has a normal distribution with zero mean and unit standard deviation*, and define a new random variable

$$\chi_n^2 = X_1^2 + X_2^2 + \cdots + X_n^2 \tag{10.59}$$

The distribution of $\chi_n^2$ is called the *chi-squared distribution* with '$n$ degrees of freedom', where the number of degrees of freedom represents the number of independent random variables $X_i$. The general form of the $\chi_\nu^2$ probability density function with $\nu$ degrees of freedom is

$$p_{\chi_\nu^2}(x) = \frac{1}{2^{\nu/2}\Gamma(\nu/2)}x^{(\nu/2)-1}e^{-x/2} \quad 0 \le x \le \infty \tag{10.60}$$

where $\Gamma(\frac{\nu}{2}) = \int_0^\infty e^{-t}t^{(\nu/2)-1}dt$ is the gamma function. For some values of $\nu$, $p_{\chi_\nu^2}(x)$ are shown in Figure 10.16.

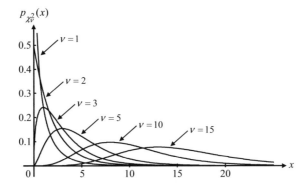

**Figure 10.16** Chi-squared probability density functions

For a small value of $\nu$, the distribution is non-symmetrical, but as $\nu$ increases the chi-squared distribution tends to Gaussian, as predicted by the central limit theorem. The first two moments of the $\chi_\nu^2$ random variable are

$$E\left[\chi_\nu^2\right] = \nu \tag{10.61}$$

$$\text{Var}(\chi_\nu^2) = 2\nu \tag{10.62}$$

We now summarize two important properties of the chi-squared distribution: (i) the decomposition theorem for chi-squared random variables; (ii) approximation by a chi-squared distribution. The first property states that if a random variable $\chi_\nu^2$ is decomposed into $k$ random variables according to $\chi_\nu^2 = \chi_{\nu_1}^2 + \chi_{\nu_2}^2 + \cdots + \chi_{\nu_k}^2$ and if $\nu_1 + \nu_2 + \cdots + \nu_k = \nu$, then the random variables $\chi_{\nu_i}^2$ are mutually independent. Conversely, if $k$ independent random variables $\chi_{\nu_i}^2$ are added together, then the sum is $\chi_\nu^2$, where

$$\nu = \nu_1 + \nu_2 + \cdots + \nu_k \tag{10.63}$$

The second property states that: suppose we have a *positive-valued* random variable $Y$, and we wish to approximate its distribution by $a\chi_\nu^2$ where $a$ and $\nu$ are unknown, but we may know the mean and variance of $Y$, i.e. $\mu_y$ and $\sigma_y^2$ are known. Then in this case,

$E[Y] = \mu_y = E[a\chi_v^2] = av$ and $\sigma_y^2 = \text{Var}(a\chi_v^2) = a^2\text{Var}(\chi_v^2) = 2a^2v = 2\mu_y^2/v$. Thus $a$ and $v$ are found from

$$a = \frac{\mu_y}{v} \tag{10.64}$$

$$v = \frac{2\mu_y^2}{\sigma_y^2} \tag{10.65}$$

**Variance Considerations**

We now return to study the variance of $\hat{S}_{xx}(f)$. The sample power spectral density function can be written as ($x(t)$ is defined for $|t| < T/2$)

$$\hat{S}_{xx}(f) = \frac{|X_T(f)|^2}{T} = \frac{1}{T}\left|\int_{-T/2}^{T/2} x(t)e^{-j2\pi ft}dt\right|^2 = \frac{1}{T}\left|\int_{-\infty}^{\infty} x(t)e^{-j2\pi ft}dt\right|^2$$

$$= \frac{1}{T}\left\{\left[\int_{-\infty}^{\infty} x(t)\cos(2\pi ft)dt\right]^2 + \left[\int_{-\infty}^{\infty} x(t)\sin(2\pi ft)dt\right]^2\right\}$$

$$= \frac{1}{T}[X_c^2(f) + X_s^2(f)] \tag{10.66}$$

where $X_c(f)$ and $X_s(f)$ are Fourier cosine and sine transforms of $x(t)$.

Let us assume that $x(t)$ is a Gaussian process with zero mean value; then $X_c(f)$ and $X_s(f)$ are also Gaussian and have zero mean values. Furthermore, it can be shown that $X_c(f)$ and $X_s(f)$ are uncorrelated and have approximately equal variances (Jenkins and Watts, 1968). Now if the variances were unity, we could use the properties of the chi-squared distribution to say $\hat{S}_{xx}(f)$ is related to a $\chi_2^2$ distribution (note that $\hat{S}_{xx}(f)$ is a squared quantity and so positive valued). We do this as follows. Let

$$\frac{1}{T}E[X_c^2(f)] = \frac{1}{T}E[X_s^2(f)] = \sigma^2 \text{ (say, for each frequency } f) \tag{10.67}$$

and

$$E[\hat{S}_{xx}(f)] \approx S_{xx}(f) = 2\sigma^2 \tag{10.68}$$

Then, it can be shown that

$$\frac{2\hat{S}_{xx}(f)}{S_{xx}(f)} = \frac{X_c^2(f)}{T\sigma^2} + \frac{X_s^2(f)}{T\sigma^2} \tag{10.69}$$

which is the sum of two squared Gaussian random variables with unit variances (note that $X_c(f)$ and $X_s(f)$ are jointly normally distributed (see Appendix G for justification) and uncorrelated, so they are independent). Therefore the random variable $2\hat{S}_{xx}(f)/S_{xx}(f)$

is distributed as a chi-squared random variable with two degrees of freedom, i.e. $\chi_2^2$ (for all values of sample length $T$). Using Equation (10.62), the variance is

$$\text{Var}\left(\frac{2\hat{S}_{xx}(f)}{S_{xx}(f)}\right) = 4$$

and so

$$\frac{4\text{Var}\left(\hat{S}_{xx}(f)\right)}{S_{xx}^2(f)} = 4 \tag{10.70}$$

i.e.

$$\text{Var}(\hat{S}_{xx}(f)) = S_{xx}^2(f) \tag{10.71}$$

or

$$\sigma(\hat{S}_{xx}(f)) = S_{xx}(f) \tag{10.72}$$

This important result states that the estimator $\hat{S}_{xx}(f)$ has a variance that is *independent of sample length T*, i.e. $\hat{S}_{xx}(f)$ is an *inconsistent* estimate of $S_{xx}(f)$. Furthermore, the random error of the estimate is substantial, i.e. *the standard deviation of the estimate is as great as the quantity being estimated*. These undesirable features lead to the estimate $\hat{S}_{xx}(f)$ being referred to as the 'raw' spectrum estimate or 'raw periodogram'. As it stands, $\hat{S}_{xx}(f)$ is not a useful estimator and we must reduce the random error. This may be accomplished by 'smoothing' as indicated below. However, as we shall see, the penalty for this is the degradation of accuracy due to bias error.

### Smoothed Spectral Density Estimators

As we have already discussed, the averaged $\hat{S}_{xx}(f)$ is given by Equation (10.57), where the integrand is $R_{xx}(\tau)(1 - |\tau|/T)$. This motivates us to study the effect of introducing a *lag window*, $w(\tau)$, i.e. the estimate may be smoothed in the frequency domain by Fourier transforming the product of the autocorrelation function estimate and $w(\tau)$ which has a Fourier transform $W(f)$ (called a 'spectral window'). This defines the smoothed spectral estimator $\tilde{S}_{xx}(f)$ as

$$\tilde{S}_{xx}(f) = \int_{-\infty}^{\infty} \hat{R}_{xx}^b(\tau)w(\tau)e^{-j2\pi f\tau}d\tau \tag{10.73}$$

Recall that

$$\int_{-\infty}^{\infty} x(t)w(t)e^{-j2\pi ft}dt = \int_{-\infty}^{\infty} X(g)W(f-g)dg$$

i.e. $\tilde{S}_{xx}(f) = \hat{S}_{xx}(f) * W(f)$, which is the convolution of the raw spectral density with a spectral window. Thus the right hand side of Equation (10.73) has an alternative form

$$\tilde{S}_{xx}(f) = \int\limits_{-\infty}^{\infty} \hat{S}_{xx}(g)W(f-g)dg \qquad (10.74)$$

This estimation procedure may be viewed as a smoothing operation in the frequency domain. Thus the lag window $w(\tau)$ results in the spectral window $W(f)$ 'smoothing' the raw periodogram $\hat{S}_{xx}(f)$ through the convolution operation. In fact, the above method is the basis of the correlation method of estimating spectral density functions. In the time domain, the lag window can be regarded as reducing the 'importance' of values of $\hat{R}^b_{xx}(\tau)$ as $\tau$ increases.

It is necessary to start all over again and study the bias and variance properties of this new estimator $\tilde{S}_{xx}(f)$ where clearly the window function $w(\tau)$ will now play an important role. Jenkins and Watts (1968) and Priestley (1981) give a detailed discussion of this problem. We shall only quote the main results here.

Some commonly used window functions are listed in Table 10.1. The rectangular window is included for completeness, and other window functions may also be used. Note that the discussions on window functions given in Chapter 4 are directly related to this case.

The lag windows $w(\tau)$ are shown in Figure 10.17, where $w(0) = 1$ for all windows, and the spectral windows $W(f)$ are shown in Figure 10.18. Note that the spectral windows which take negative values might give rise to *negative* spectral density estimates in this approach.

**Table 10.1** Commonly used lag and spectral windows

| Window name | Lag window, $w(\tau)$ | | Spectral window, $W(f)$ |
|---|---|---|---|
| Rectangular | $w(\tau) = 1$ $= 0$ | $\|\tau\| \le T_w$ $\|\tau\| > T_w$ | $W(f) = 2T_w\left(\dfrac{\sin(2\pi f T_w)}{2\pi f T_w}\right)$ |
| Bartlett | $w(\tau) = 1 - \dfrac{\|\tau\|}{T_w}$ $= 0$ | $\|\tau\| \le T_w$ $\|\tau\| > T_w$ | $W(f) = T_w\left(\dfrac{\sin(\pi f T_w)}{\pi f T_w}\right)^2$ |
| Hann(ing) | $w(\tau) = \dfrac{1}{2}\left(1 + \cos\dfrac{\pi\tau}{T_w}\right)$ $= 0$ | $\|\tau\| \le T_w$ $\|\tau\| > T_w$ | $W(f) = T_w\left(\dfrac{\sin(2\pi f T_w)}{2\pi f T_w}\right)\dfrac{1}{1-(2fT_w)^2}$ |
| Hamming | $w(\tau) = 0.54 + 0.46\cos\dfrac{\pi\tau}{T_w}$ $= 0$ | $\|\tau\| \le T_w$ $\|\tau\| > T_w$ | $W(f) = \dfrac{\left[0.54\pi^2 - 0.08(2\pi f T_w)^2\right]\sin(2\pi f T_w)}{2\pi f T_w\left[\pi^2 - (2\pi f T_w)^2\right]}$ |
| Parzen | $w(\tau) = 1 - 6\left(\dfrac{\tau}{T_w}\right)^2 + 6\left(\dfrac{\|\tau\|}{T_w}\right)^3$ $= 2\left(1 - \dfrac{\|\tau\|}{T_w}\right)^3$ $= 0$ | $\|\tau\| \le \dfrac{T_w}{2}$ $\dfrac{T_w}{2} < \|\tau\| \le T_w$ $\|\tau\| > T_w$ | $W(f) = \dfrac{3}{4}T_w\left(\dfrac{\sin(\pi f T_w/2)}{\pi f T_w/2}\right)^4$ |

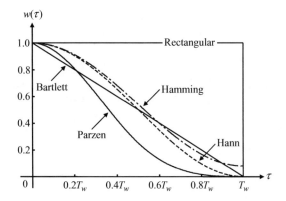

**Figure 10.17**   Lag windows (for $\tau \geq 0$)

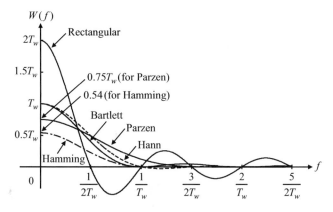

**Figure 10.18**   Spectral windows (for $f \geq 0$)

**Bias Considerations**

From Equation (10.74), the average of the smoothed power spectral density function is

$$E\left[\tilde{S}_{xx}(f)\right] = \int_{-\infty}^{\infty} S_{xx}(g)W(f - g)dg \qquad (10.75)$$

So, $\tilde{S}_{xx}(f)$ is biased. Note that this equation indicates how the estimate may be distorted by the smearing and leakage effect of the spectral windows. In fact, for large $T$, bias is shown to be

$$b\left(\tilde{S}_{xx}(f)\right) = E\left[\tilde{S}_{xx}(f)\right] - S_{xx}(f) \approx \int_{-\infty}^{\infty} [w(\tau) - 1]\, R_{xx}(\tau)e^{-j2\pi f\tau}\, d\tau \qquad (10.76)$$

Note that the bias is different for each lag window. The details are given in Jenkins and Watts (1968). We may comment broadly that the general effect is to reduce the dynamic range of the spectra as in the filter bank method of the spectral estimation, i.e. peaks are underestimated and troughs are overestimated.

As can be seen from Equation (10.75), the bias is reduced as the spectral window gets narrower, i.e. as the width of the lag window $w(\tau)$ gets larger, the spectral window becomes $W(f) \to \delta(f)$. So, $\tilde{S}_{xx}(f)$ is asymptotically unbiased (for $T_w \to \infty$). However, the spectral windows cannot be made too narrow since then there is little smoothing and the random errors increase, so we need $W(f)$ to have some width to do the smoothing. Consequently, once again we need a compromise between bias and variance, i.e. a trade-off between the resolution and the random error. Note that, sometimes, the bias problem is referred to under 'bandwidth considerations' since small bias is associated with small bandwidth of the window function.

**Variance Considerations**

From Equation (10.74), $\tilde{S}_{xx}(f)$ can be considered as a weighted sum of values of $\hat{S}_{xx}(f)$. Thus, it may be argued that $n\tilde{S}_{xx}(f)/S_{xx}(f)$ is approximately distributed as a chi-squared random variable with $n$ degrees of freedom, $\chi_n^2$, where the number of degrees of freedom is defined as

$$n = \frac{2T}{\int_{-\infty}^{\infty} w^2(\tau)d\tau} = \frac{2T}{I} \tag{10.77}$$

which depends on the window and data length (Jenkins and Watts, 1968).

Also, since $\mathrm{Var}(n\tilde{S}_{xx}(f)/S_{xx}(f)) = 2n$, it can be shown that

$$\mathrm{Var}(\tilde{S}_{xx}(f)) = \frac{S_{xx}^2(f)}{n/2} = \frac{S_{xx}^2(f)}{T/I} \tag{10.78}$$

$$\sigma(\tilde{S}_{xx}(f)) = \frac{S_{xx}(f)}{\sqrt{n/2}} = \frac{S_{xx}(f)}{\sqrt{T/I}} \tag{10.79}$$

Now, $1/I$ can be argued to be a measure of the resolution bandwidth $B$ of the window (see below for justification), so the number of degrees of freedom is $n = 2BT$. Thus, the above equations can be rewritten as

$$\frac{\mathrm{Var}(\tilde{S}_{xx}(f))}{S_{xx}^2(f)} = \frac{1}{BT} \tag{10.80}$$

$$\frac{\sigma(\tilde{S}_{xx}(f))}{S_{xx}(f)} = \frac{1}{\sqrt{BT}} \tag{10.81}$$

To justify that $1/I$ is a measure of bandwidth, consider an ideal filter shown in Figure 10.19, where

$$I = \int_{-\infty}^{\infty} w^2(\tau)d\tau = \int_{-\infty}^{\infty} W^2(f)df \tag{10.82}$$

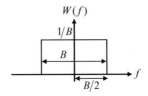

**Figure 10.19** An ideal filter with a resolution bandwidth of $B$

Since

$$\int_{-\infty}^{\infty} W^2(f)df = \frac{1}{B}$$

it can be shown that

$$1/I = \frac{1}{\int_{-\infty}^{\infty} w^2(\tau)d\tau} = \frac{1}{\int_{-\infty}^{\infty} W^2(f)df} = B \qquad (10.83)$$

When the 'non-ideal' filters (Bartlett, Hann, Parzen, etc.) are used, then $1/\int_{-\infty}^{\infty} w^2(\tau)d\tau$ defines a generalized bandwidth. This is the effective resolution of the spectral window.

A summary of bandwidths, biases and variance ratios of some window functions is given in Table 10.2 (Jenkins and Watts, 1968; Priestley, 1981).

**Table 10.2** Properties of some spectral windows

| Window name | Bandwidth $1/I = B$ | DOF | Variance ratio, $\dfrac{\mathrm{Var}\left(\tilde{S}_{xx}(f)\right)}{S_{xx}^2(f)}$ | Approximate bias |
|---|---|---|---|---|
| Rectangular | $\dfrac{0.5}{T_w}$ | $\dfrac{T}{T_w}$ | $\dfrac{2T_w}{T}$ | N/A |
| Barlett | $\dfrac{1.5}{T_w}$ | $\dfrac{3T}{T_w}$ | $\dfrac{0.667T_w}{T}$ | $\approx \dfrac{1}{T_w}\displaystyle\int_{-\infty}^{\infty} -|\tau|R_{xx}(\tau)e^{-j2\pi f\tau}d\tau$ |
| Hann(ing) | $\dfrac{1.333}{T_w}$ | $\dfrac{2.667T}{T_w}$ | $\dfrac{0.75T_w}{T}$ | $\approx \dfrac{0.063}{T_w^2}S_{xx}''(f)$ |
| Hamming | $\dfrac{1.26}{T_w}$ | $\dfrac{2.52T}{T_w}$ | $\dfrac{0.795T_w}{T}$ | $\approx \dfrac{0.058}{T_w^2}S_{xx}''(f)$ |
| Parzen | $\dfrac{1.86}{T_w}$ | $\dfrac{3.71T}{T_w}$ | $\dfrac{0.539T_w}{T}$ | $\approx \dfrac{0.152}{T_w^2}S_{xx}''(f)$ |

*Note:* $S_{xx}''(f)$ is the second derivative of the spectrum at frequency $f$, $T$ is the total data length and $T_w$ is as defined in Table 10.1 (i.e. half of the lag window length).

*General Comments on Window Functions and* $\tilde{S}_{xx}(f)$

Although the rectangular window is included in Table 10.2, it is rarely used since its spectral window side lobes cause large 'leakage' effects. However, in Chapter 8, the rectangular window function is applied to estimate the spectral density functions in MATLAB Examples 8.8–8.10 (note that we have used a very long data length $T$).

Bias of the Bartlett window is of order $1/T_w$ and so will in general be greater than other windows (order $1/T_w^2$). Among the Hann, Hamming and Parzen windows, the Parzen window has the smallest variance but has the greatest bias. Since the bias of these windows depends on $S_{xx}''(f)$, larger bias occurs at sharp peaks and troughs than other frequency regions. From the table, we can easily see that the bias is reduced as $T_w$ increases, but the variance is increased. The variance can then be reduced by increasing the total data length $T$.

We note that: (i) when the bias is small, $\tilde{S}_{xx}(f)$ is said to reproduce $S_{xx}(f)$ with high *fidelity*; (ii) when the variance is small, the estimator is said to have high *stability* (Jenkins and Watts, 1968). The choice of window functions should depend on whether the concern is for statistical stability (low variance) or high fidelity (small bias), although in general we must consider both. For example, if the spectral density function has narrow peaks of importance we may willingly tolerate some loss of stability to resolve the peak properly, while if the spectral density function is smooth then bias errors are not likely to be so important. Thus, we may state that the estimator $\tilde{S}_{xx}(f)$ is approximately unbiased and has a low variance only if the sufficiently narrow resolution bandwidth and yet long enough data are used.

When spectra are estimated via the autocorrelation function many texts give only approximate values for the degrees of freedom and the resolution bandwidth. For example, with digital analysis with $N$ data points (sampled at every $\Delta$ seconds) and a maximum correlation lag $M$, the number of degrees of freedom is usually quoted as $2N/M$ and the resolution bandwidth as $1/(M\Delta)$ (this may corresponds to the rectangular window function).

Finally, instead of multiplying the sample autocorrelation function by the lag window and Fourier transforming the weighted sample autocorrelation function, an alternative procedure is to do the smoothing in the frequency domain, i.e. form the raw periodogram and perform the frequency convolution. This 'frequency smoothing' will be referred to again soon.

## Method (3): Direct Methods

We shall now discuss the basis for forming smoothed power spectral density estimates without first forming the autocorrelation functions. These methods are probably the most widely used because of computational considerations. There are two methods, although they can be combined if required, namely (i) segment averaging; (ii) frequency smoothing.

*Segment Averaging (Welch's Method)*[M10.1]

The segment averaging method has become very popular mainly because of its fast speed of computation. The method is discussed in Jenkins and Watts (1968) as Bartlett's smoothing procedure and in Welch (1967) in some detail. The basic procedure is outlined with reference to Figure 10.20.

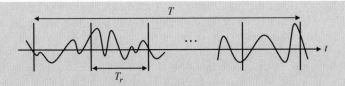

**Figure 10.20** The basis of the segment averaging method

Consider the data (length $T$) which is segmented into $q$ separate time slices each of length $T_r$ such that $qT_r = T$ (in this case non-overlapping). Now we form the raw periodogram for each slice as

$$\hat{S}_{xx_i}(f) = \frac{1}{T_r}\left|X_{T_{ri}}(f)\right|^2 \quad \text{for } i = 1, 2, \ldots, q \tag{10.84}$$

We saw earlier that this is distributed as a chi-squared random variable with two degrees of freedom. We might expect that by averaging successive raw spectra the underlying behaviour would reinforce and the variability would reduce, i.e. form

$$\tilde{S}_{xx}(f) = \frac{1}{q}\sum_{i=1}^{q} \hat{S}_{xx_i}(f) \tag{10.85}$$

We can estimate the variance reduction by the following argument. Note that for each segment $2\hat{S}_{xx_i}(f)/S_{xx}(f)$ is a $\chi_2^2$ random variable. From Equation (10.85),

$$\frac{2\tilde{S}_{xx}(f)\cdot q}{S_{xx}(f)} = \sum_{i=1}^{q} \frac{2\hat{S}_{xx_i}(f)}{S_{xx}(f)}$$

Thus, $(2\tilde{S}_{xx}(f)\cdot q)/S_{xx}(f)$ is the sum of $q\chi_2^2$ random variables and so assuming that these are essentially independent of each other then this is approximated as $\chi_{2q}^2$. From Equation (10.61),

$$E\left[\frac{2\tilde{S}_{xx}(f)\cdot q}{S_{xx}(f)}\right] \approx 2q \tag{10.86}$$

from which $E[\tilde{S}_{xx}(f)] \approx S_{xx}(f)$ (i.e. $\tilde{S}_{xx}(f)$ is approximately unbiased). From Equation (10.62),

$$\text{Var}\left(\frac{2\tilde{S}_{xx}(f)\cdot q}{S_{xx}(f)}\right) \approx 4q \tag{10.87}$$

Thus,

$$\frac{4q^2}{S_{xx}^2(f)}\text{Var}\left(\tilde{S}_{xx}(f)\right) \approx 4q$$

i.e. it follows that

$$\frac{\text{Var}\left(\tilde{S}_{xx}(f)\right)}{S_{xx}^2(f)} \approx \frac{1}{q} \tag{10.88}$$

$$\frac{\sigma\left(\tilde{S}_{xx}(f)\right)}{S_{xx}(f)} \approx \frac{1}{\sqrt{q}} \tag{10.89}$$

This can be expressed differently. For example, the resolution bandwidth of the rectangular data window is $B = 1/T_r = q/T$. Thus, Equation (10.88) can be written as

$$\frac{\text{Var}\left(\tilde{S}_{xx}(f)\right)}{S_{xx}^2(f)} \approx \frac{1}{BT} \tag{10.90}$$

which is the same as Equation (10.80). Note that by segmenting the data, the resolution bandwidth becomes wider since $T_r < T$. We must be aware that the underlying assumption of the above results is that each segment of the data must be independent (see the comments in MATLAB Example 10.1). Clearly this is generally not the case, particularly if the segments in the segment averaging overlap. This is commented on in the following paragraphs.

To summarize the segment averaging method:

1. Resolution bandwidth: $B \approx \frac{1}{T_r} = \frac{q}{T}$
2. Degrees of freedom: $n = 2q = 2BT$
3. Variance ratio: $\frac{\text{Var}(\tilde{S}_{xx}(f))}{S_{xx}^2(f)} \approx \frac{1}{BT} = \frac{1}{q}$

While the above description summarizes the essential features of the method, Welch (1967) and Bingham et al. (1967) give more elaborate procedures and insight. Since the use of a rectangular data window introduces leakage, the basic method above is usually modified by using other data windows. This is often called linear tapering. The word 'linear' here does not refer to the window shape but to the fact that it operates on the data directly and not on the autocorrelation function (see Equation (10.73)) where it is sometimes called quadratic tapering.

The use of a data window on a segment before transforming reduces leakage. However, since the windows have tapering ends, the values obtained for $\hat{S}_{xx_i}(f)$ must be compensated for the 'power reduction' introduced by the window. This results in the calculation of 'modified' periodograms for each segment of the form

$$\hat{S}_{xx_i}(f) = \frac{\dfrac{1}{T_r} \left| \displaystyle\int\limits_{i\text{th interval}} x(t)w(t)e^{-j2\pi ft}\,dt \right|^2}{\dfrac{1}{T_r} \displaystyle\int\limits_{-T_r/2}^{T_r/2} w^2(t)\,dt} \tag{10.91}$$

where the denominator compensates for the power reduction, and is unity for the rectangular window and 3/8 for the Hann window.

Finally, we note that the use of a data window ignores some data because of the tapered shape of the window. Intuitively the overlapping of segments compensates for this in some way (Welch, 1967), though it is not easy to relate these results to the indirect method of Fourier transforming the autocorrelation function. We must remember that ideally segments should be independent to obtain the variance reduction – and with overlapping this is compromised. We simply quote the following results from Welch (1967) which is based on Gaussian white noise. If the segments overlap by one-half of their length (50 % overlap), and the total data length is $N$ points and each individual segment length is $L$ points, then

1. The number of degrees of freedom is $n \approx 2(\frac{2N}{L} - 1)$
2. The variance ratio is $\frac{\text{Var}(\tilde{S}_{xx}(f))}{S_{xx}^2(f)} \approx \frac{1}{n/2}$
3. The resolution is $1/(L\Delta) = f_s/L$, but note that this depends on the data window being used.

### Frequency Smoothing

This approach is based on the comments given in Method (2) (in the last paragraph). The method is as follows:

1. Form the raw periodogram from the data length $T$.
2. Average $l$ neighbouring estimates of this spectrum, i.e. form

$$\tilde{S}_{xx}(f_k) = \frac{1}{l} \sum_{i=1}^{l} \hat{S}_{xx}(f_i) \tag{10.92}$$

where the $f_i$ surround $f_k$.

As before, we can argue that $(2\tilde{S}_{xx}(f) \cdot l)/S_{xx}(f)$ is distributed as $\chi_{2l}^2$ and

$$\frac{\text{Var}(\tilde{S}_{xx}(f))}{S_{xx}^2(f)} \approx \frac{1}{l} \tag{10.93}$$

The resolution bandwidth before smoothing is approximately $1/T$, but after smoothing it is approximately $l/T$ since $l$ neighbouring values are averaged. This method is effectively the same as the indirect method (Method (2)).

Note that one might combine both segment averaging and frequency smoothing to get an estimate with $2lq$ degrees of freedom and then the resolution bandwidth is approximately $lq/T$.

## Confidence Intervals for Spectral Estimates

We now discuss the 'interval estimates' based on the point estimates for the smoothed spectral density function $\tilde{S}_{xx}(f)$. We have seen that $n\tilde{S}_{xx}(f)/S_{xx}(f)$ is distributed as a $\chi_n^2$ random variable where the probability density function for $\chi_n^2$ is of the form shown in Figure 10.21, i.e. the values taken by $n\tilde{S}_{xx}(f)/S_{xx}(f)$ are much more likely to fall within the hump than the tail or near $x = 0$.

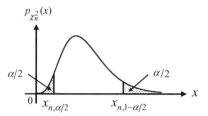

**Figure 10.21**  The creation of confidence intervals

If we choose a number $\alpha$ $(0 < \alpha < 1)$ such that sections of area $\alpha/2$ are as shown marked off by the points $x_{n,\alpha/2}$ and $x_{n,1-\alpha/2}$, then the following probability statement can be made:

$$P\left[x_{n,\alpha/2} \leq \frac{n\tilde{S}_{xx}(f)}{S_{xx}(f)} \leq x_{n,1-\alpha/2}\right] = 1 - \alpha \qquad (10.94)$$

The points $x_{n,\alpha/2}$ and $x_{n,1-\alpha/2}$ can be obtained from tables of $\chi_n^2$ for different values of $\alpha$. Now the inequality can be solved for the true spectral density $S_{xx}(f)$ from the following equivalent inequality:

$$\frac{n\tilde{S}_{xx}(f)}{x_{n,1-\alpha/2}} \leq S_{xx}(f) \leq \frac{n\tilde{S}_{xx}(f)}{x_{n,\alpha/2}} \qquad (10.95)$$

Thus, for a particular sample value $\tilde{S}_{xx}(f)$ (a point estimate), the $100(1-\alpha)\%$ confidence limits for $S_{xx}(f)$ are

$$\frac{n}{x_{n,1-\alpha/2}}\tilde{S}_{xx}(f) \quad \text{and} \quad \frac{n}{x_{n,\alpha/2}}\tilde{S}_{xx}(f) \qquad (10.96)$$

and the *confidence interval* is the difference between these two limits.

Note that on a linear scale the confidence interval depends on the estimate $\tilde{S}_{xx}(f)$, but on a log scale the confidence limits are

$$\log\left(\frac{n}{x_{n,1-\alpha/2}}\right) + \log\left(\tilde{S}_{xx}(f)\right) \quad \text{and} \quad \log\left(\frac{n}{x_{n,\alpha/2}}\right) + \log\left(\tilde{S}_{xx}(f)\right) \qquad (10.97)$$

and so the *interval* is $\log(n/x_{n,\alpha/2}) - \log(n/x_{n,1-\alpha/2})$ which is independent of $\tilde{S}_{xx}(f)$. Thus, if the spectral estimate $\tilde{S}_{xx}(f)$ is plotted on a logarithmic scale, then the confidence interval for the spectrum can be represented by a constant interval about the estimate. Figure 10.22 indicates the behaviour of $n/x_{n,\alpha/2}$ and $n/x_{n,1-\alpha/2}$ (Jenkins and Watts, 1968).

From Figure 10.22, we can clearly see that the confidence interval decreases as the number of degrees of freedom $n$ increases. For example, if $n = 100$ (approximately 50 averages for the segment averaging method), the 95 % confidence interval is about $[0.77\tilde{S}_{xx}(f), 1.35\tilde{S}_{xx}(f)]$. Sometimes, the number of degrees of freedom is referred to as the 'statistical degrees of freedom (Stat DOF)', and more than 120 degrees of freedom is often required in many random vibration testing standards (e.g. MIL-STD-810F and IEC 60068-2-64).

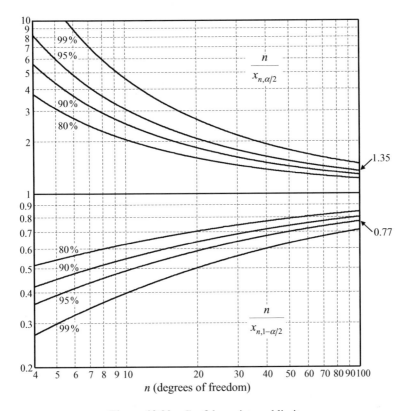

**Figure 10.22**  Confidence interval limits

## 10.5  CROSS-SPECTRAL DENSITY FUNCTION[M10.1,10.3]

The basic considerations given in Section 10.4 relate also to cross-spectral density function estimation together with some additional features, but we shall not go into any detail here. Detailed results can be found in Jenkins and Watts (1968). We shall merely summarize some important features.

The raw cross-spectral density function can be obtained by Fourier transforming the raw cross-correlation function, i.e.

$$\hat{S}_{xy}(f) = \int_{-T}^{T} \hat{R}_{xy}^{b}(\tau) e^{-j2\pi f \tau} d\tau \tag{10.98}$$

and this has the same unsatisfactory properties as the raw power spectral density function. Thus, as before, a lag window $w(\tau)$ is introduced to smooth the estimate, i.e.

$$\tilde{S}_{xy}(f) = \int_{-T}^{T} \hat{R}_{xy}^{b}(\tau) w(\tau) e^{-j2\pi f \tau} d\tau \tag{10.99}$$

Note that the unbiased estimator $\hat{R}_{xy}(\tau)$ may also be used in place of $\hat{R}^b_{xy}(\tau)$ provided that the maximum lag $\tau_{max}$ is relatively small compared with $T$. Alternatively, the smoothed estimate can be obtained by the segment averaging method or by frequency smoothing of the raw cross-spectral density. For example, if the segment averaging method is used the raw and smoothed cross-spectral density functions are

$$\hat{S}_{xy_i}(f) = \frac{1}{T_r}\left[X^*_{T_{ri}}(f)Y_{T_{ri}}(f)\right] \quad \text{for } i = 1, 2, \ldots, q \tag{10.100}$$

$$\tilde{S}_{xy}(f) = \frac{1}{q}\sum_{i=1}^{q}\hat{S}_{xy_i}(f) \tag{10.101}$$

The smoothed estimate $\tilde{S}_{xy}(f)$ may be written in the form

$$\tilde{S}_{xy}(f) = \left|\tilde{S}_{xy}(f)\right|e^{j\,\arg\,\tilde{S}_{xy}(f)} \tag{10.102}$$

Roughly speaking, one can show that the variances of the amplitude $\left|\tilde{S}_{xy}(f)\right|$ and the phase $\arg\tilde{S}_{xy}(f)$ are proportional to $1/BT$ where $B$ is the resolution bandwidth and $T$ is the data length.

Whilst the general effect of smoothing is much the same as for the power spectral density estimate, we note in addition, though, that the amplitude and phase estimators are also strongly dependent on the 'true' coherence function $\gamma^2_{xy}(f)$. So, as Jenkins and Watts (1968) observed, *the sampling properties of the amplitude and phase estimators may be dominated by the 'uncontrollable' influence of the coherence spectrum $\gamma^2_{xy}(f)$ rather than by the 'controllable' influence of the smoothing factor $1/BT$.* For example, the variance of the modulus and phase of $\tilde{S}_{xy}(f)$ are shown to be (Bendat and Piersol, 2000)

$$\frac{\text{Var}\left(\left|\tilde{S}_{xy}(f)\right|\right)}{\left|S_{xy}(f)\right|^2} \approx \frac{1}{\gamma^2_{xy}(f)}\cdot\frac{1}{BT} \tag{10.103}$$

$$\text{Var}\big(\arg\tilde{S}_{xy}(f)\big) \approx \frac{1-\gamma^2_{xy}(f)}{\gamma^2_{xy}(f)}\cdot\frac{1}{2BT} \tag{10.104}$$

Note the 'uncontrollable' influence of true coherence function $\gamma^2_{xy}(f)$ on the variability of the estimate. Note also that the variance of $\arg\tilde{S}_{xy}(f)$ is not normalized. Particularly, if $x(t)$ and $y(t)$ are fully linearly related, i.e. $\gamma^2_{xy}(f) = 1$, then $\text{Var}(\arg\tilde{S}_{xy}(f)) \approx 0$. Thus, we see that the random error of the phase estimator is much smaller than that of the amplitude estimator.

Similar to the power spectral density estimate, in general, the estimator $\tilde{S}_{xy}(f)$ is approximately unbiased when $T$ is sufficiently large and the resolution bandwidth is narrow. However, there is another important aspect: since $\tilde{R}_{xy}(\tau)$ is not (in general) symmetric, it is necessary to ensure that its maximum is well within the window $w(\tau)$ or serious bias errors result. For example, if $y(t) = x(t - \Delta)$, then it can be shown that

(Schmidt, 1985b) for rectangular windows

$$E\left[\tilde{S}_{xy}(f)\right] \approx \left(1 - \frac{\Delta}{T_r}\right) S_{xy}(f) \tag{10.105}$$

where $T_r$ is the length of window (or the length of segment). Note that the time delay between signals results in biased estimates (see MATLAB Example 10.3). This problem may be avoided by 'aligning' the two time series so that the cross-correlation function $\tilde{R}_{xy}(\tau)$ has a maximum at $\tau = 0$.

## 10.6 COHERENCE FUNCTION[M10.2]

The estimate for the coherence function is made up from the estimates of the smoothed power and cross-spectral density functions as

$$\tilde{\gamma}_{xy}^2(f) = \frac{\left|\tilde{S}_{xy}(f)\right|^2}{\tilde{S}_{xx}(f)\tilde{S}_{yy}(f)} \tag{10.106}$$

It should be noted that if 'raw' spectral density functions are used on the right hand side of the equation, it can be easily verified that the sample coherence function $\hat{\gamma}_{xy}^2(f)$ is always 'unity' for all frequencies for any signals $x$ and $y$ (even if they are unrelated).

Detailed calculations are given in Jenkins and Watts (1968) for the sampling properties of the smoothed coherence function $\tilde{\gamma}_{xy}^2(f)$, but roughly speaking, the variance of $\tilde{\gamma}_{xy}^2(f)$ is proportional to $1/BT$ (which is *known*, so a controllable parameter) and also depends on $\gamma_{xy}^2(f)$ (which is *unknown*, so is an uncontrollable parameter), where the variance of $\tilde{\gamma}_{xy}^2(f)$ is shown to be

$$\frac{\text{Var}\left(\tilde{\gamma}_{xy}^2(f)\right)}{\left(\gamma_{xy}^2(f)\right)^2} \approx \frac{2\left(1 - \gamma_{xy}^2(f)\right)^2}{\gamma_{xy}^2(f)} \cdot \frac{1}{BT} \tag{10.107}$$

This expression is sometimes used as an approximate guide after measurements have been made by replacing $\gamma_{xy}^2(f)$ with $\tilde{\gamma}_{xy}^2(f)$.

Jenkins and Watts (1968) show that the bias of this estimator is proportional to the square of the derivative of the phase spectrum $\arg S_{xy}(f)$. For example, if the Hann window is used the normalized bias error can be expressed by

$$\frac{b\left(\tilde{\gamma}_{xy}^2(f)\right)}{\gamma_{xy}^2(f)} \approx -\frac{0.126}{T_w^2}\left(\frac{d}{df}\left(\arg S_{xy}(f)\right)\right)^2 \tag{10.108}$$

for large $T$ (total data length), where $T_w$ is half of the lag window length as defined in Table 10.1. For the Parzen window, 0.126 is replaced by 0.304. The above equation means that the estimator is sensitive to delays between $x(t)$ and $y(t)$. Similar to the cross-spectral density function estimate, such bias can be reduced by realigning the processes, i.e. aligning the peak in the cross-correlation between $x(t)$ and $y(t)$ to occur at zero lag.

Note also that, if $x(t)$ and $y(t)$ are the input and output of a lightly damped oscillator, severe bias errors are likely to occur at resonant (and anti-resonant) frequencies where the phase changes rapidly. Since the resolution bandwidth is inversely proportional to the length of the lag window (e.g. for the Hann window $B = 1.333/T_w$ as shown in Table 10.2), the bias error can be reduced by improving the resolution of the estimate, i.e. the resolution bandwidth $B$ should be reduced (see MATLAB Example 10.2). In Figure 10.23, the coherence function is estimated for simulated input/output results for a lightly damped oscillator ($\zeta = 0.02$) with natural frequency at 1 Hz. The theoretical value of coherence is unity and the resolutions used are shown in the figure, where it can be seen that the bias is reduced as the resolution bandwidth $B$ decreases.

If the resolution bandwidth is chosen adequately, the bias of $\tilde{\gamma}_{xy}^2(f)$ may be approximated (Carter *et al.*, 1973) by

$$b\left(\tilde{\gamma}_{xy}^2(f)\right) \approx \frac{\left(1 - \gamma_{xy}^2(f)\right)^2}{BT} \tag{10.109}$$

This shows that the estimate $\tilde{\gamma}_{xy}^2(f)$ is asymptotically unbiased (i.e. for large $BT$).

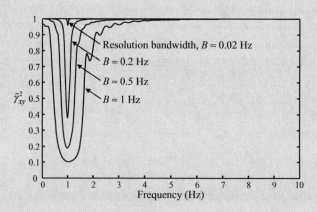

**Figure 10.23** Bias in the coherence function estimate

## 10.7 FREQUENCY RESPONSE FUNCTION

The frequency response function is estimated using smoothed spectral density functions. For example, the estimate of $H_1(f)$ can be obtained from

$$\tilde{H}_1(f) = \frac{\tilde{S}_{xy}(f)}{\tilde{S}_{xx}(f)} \tag{10.110}$$

Note that we use the notation $\tilde{H}_1(f)$ to distinguish it from the theoretical quantity $H_1(f)$, though it is not explicitly used in Chapter 9. The results for errors and confidence limits can be found in Bendat and Piersol (1980, 2000) and Otnes and Enochson (1978). A few results from Bendat and Piersol are quoted below.

Bias errors in the frequency response estimate $\tilde{H}_1(f)$ arise from:

(a) bias in the estimation procedure;
(b) nonlinear effects;
(c) bias in power spectral and cross-spectral density function estimators;
(d) measurement noise on input (note that uncorrelated output noise does not cause bias).

In connection with (a), we would get bias effects since

$$E\left[\tilde{H}_1(f)\right] = E\left[\frac{\tilde{S}_{xy}(f)}{\tilde{S}_{xx}(f)}\right] \neq \frac{E\left[\tilde{S}_{xy}(f)\right]}{E\left[\tilde{S}_{xx}(f)\right]}$$

i.e. $E[\tilde{H}_1(f)] \neq H_1(f)$. However, this effect is usually small if $BT$ is large. In connection with (b), use of Equation (10.110) produces the best linear approximation (in the least squares sense) for the frequency response function. In connection with (c), bias in the power spectral and cross-spectral density function may be significant at peaks and troughs. These are suppressed by having narrow resolution bandwidth. In connection with (d), we have already discussed this in Chapter 9 (i.e. various FRF estimators $H_1(f)$, $H_2(f)$ and $H_W(f)$ (or $H_T(f)$) are discussed to cope with the measurement noise).

Finally, the variances of the modulus and phase of $\tilde{H}_1(f)$ are

$$\frac{\text{Var}\left(|\tilde{H}_1(f)|\right)}{|H_1(f)|^2} \approx \frac{1 - \gamma_{xy}^2(f)}{\gamma_{xy}^2(f)} \cdot \frac{1}{2BT} \qquad (10.111)$$

$$\text{Var}\left(\arg \tilde{H}_1(f)\right) \approx \frac{1 - \gamma_{xy}^2(f)}{\gamma_{xy}^2(f)} \cdot \frac{1}{2BT} \qquad (10.112)$$

This shows that, similar to the estimates $\tilde{S}_{xy}(f)$ and $\tilde{\gamma}_{xy}^2(f)$, the variances depend on both the controllable parameter $BT$ and the uncontrollable parameter $\gamma_{xy}^2(f)$. Note that the right hand sides of Equations (10.111) and (10.112) are the same. Also, comparing with the results of the cross-spectral density estimate $\tilde{S}_{xy}(f)$ shown in Equations (10.103) and (10.104), we see that the normalized variance of $|\tilde{H}_1(f)|$ is smaller than that of $|\tilde{S}_{xy}(f)|$, while the variances of the phase estimators are the same, i.e. $\text{Var}(\arg \tilde{H}_1(f)) = \text{Var}(\arg \tilde{S}_{xy}(f))$. In practice, this implies that we may need shorter data length (or fewer number of averages) for the FRF estimate than the cross-spectral density estimate. Note that, if $\gamma_{xy}^2(f) = 1$, then both $\text{Var}(|\tilde{H}_1(f)|)$ and $\text{Var}(\arg \tilde{H}_1(f))$ approach zero.

The random errors of $\tilde{H}_2(f)$ may be similar to those of $\tilde{H}_1(f)$ since the $H_2(f)$ estimator can be thought of as reversing the role of input and output defined for $H_1(f)$ in the optimization scheme (discussed in Chapter 9). The random errors of $\tilde{H}_W(f)$ (or $\tilde{H}_T(f)$) are not as obvious as the others. However, if there is no measurement noise it can easily be seen that all three theoretical quantities are the same, i.e. $H_1(f) = H_2(f) = H_W(f)$. Thus, apart from the error due to the measurement noise, we may anticipate that the random errors are similar to the $H_1(f)$ estimator. Details of the statistical properties of $H_W(f)$ can be found in White *et al.* (2006).

We summarize the normalized random errors of various estimates in Table 10.3, where the factor $BT$ can be replaced by the number of averages $q$ for the segment averaging method (assuming that the data segments used are mutually uncorrelated).

**Table 10.3** Random errors for some smoothed estimators

| Estimator | Random error, $\varepsilon_r = \dfrac{\sigma(\hat{\Phi})}{\phi}$ |
|---|---|
| $\tilde{S}_{xy}(f)$ | $\varepsilon_r \approx \dfrac{1}{\sqrt{BT}}$ |
| $\left\|\tilde{S}_{xy}(f)\right\|$ | $\varepsilon_r \approx \dfrac{1}{\left\|\gamma_{xy}(f)\right\|\sqrt{BT}}$ |
| $\arg \tilde{S}_{xy}(f)$ | $\sigma\left(\arg\tilde{S}_{xy}(f)\right) \approx \dfrac{\left[1 - \gamma_{xy}^2(f)\right]^{1/2}}{\left\|\gamma_{xy}(f)\right\|\sqrt{2BT}}$ |
| $\left\|\tilde{H}_1(f)\right\|$ | $\varepsilon_r \approx \dfrac{\left[1 - \gamma_{xy}^2(f)\right]^{1/2}}{\left\|\gamma_{xy}(f)\right\|\sqrt{2BT}}$ |
| $\arg \tilde{H}_1(f)$ | $\sigma\left(\arg\tilde{H}_1(f)\right) \approx \dfrac{\left[1 - \gamma_{xy}^2(f)\right]^{1/2}}{\left\|\gamma_{xy}(f)\right\|\sqrt{2BT}}$ |
| $\tilde{\gamma}_{xy}^2(f)$ | $\varepsilon_r \approx \dfrac{\sqrt{2}\left[1 - \gamma_{xy}^2(f)\right]}{\left\|\gamma_{xy}(f)\right\|\sqrt{BT}}$ |

## 10.8 BRIEF SUMMARY

1. Estimator errors are defined by

$$\text{Bias: } b(\hat{\Phi}) = E[\hat{\Phi}] - \phi$$
$$\text{Variance: } \text{Var}(\hat{\Phi}) = E[\hat{\Phi}^2] - E^2[\hat{\Phi}]$$
$$\text{Mean square error: } \text{mse}(\hat{\Phi}) = \text{Var}(\hat{\Phi}) + b^2(\hat{\Phi})$$

The normalized errors are

$$\text{Bias error: } \varepsilon_b = b(\hat{\Phi})/\phi$$
$$\text{Random error: } \varepsilon_r = \sigma(\hat{\Phi})/\phi$$
$$\text{RMS error: } \varepsilon = \sqrt{\text{mse}(\hat{\Phi})}/\phi$$

2. $\hat{R}_{xx}(\tau)$ and $\hat{R}_{xy}(\tau)$ are the unbiased autocorrelation and cross-correlation function estimates; however, the biased (but asymptotically unbiased) estimators $\hat{R}_{xx}^b(\tau)$ and $\hat{R}_{xy}^b(\tau)$ have a smaller mean square error. When unbiased estimators are used, the ratio of the maximum lag to the total data length, $\tau_{max}/T$, should not exceed 0.1.
   Correlation functions may be estimated with arbitrarily small error if the length of the data is sufficiently long.
3. The 'raw' power spectral density function $\hat{S}_{xx}(f)$ is an asymptotically unbiased estimator; however, the variance of $\hat{S}_{xx}(f)$ is $\mathrm{Var}\left(\hat{S}_{xx}(f)\right) = S_{xx}^2(f)$.
4. The 'smoothed' power spectral density function $\tilde{S}_{xx}(f)$ can be obtained by

$$\tilde{S}_{xx}(f) = \int_{-\infty}^{\infty} \hat{R}_{xx}^b(\tau)w(\tau)e^{-j2\pi f\tau}d\tau$$

or

$$\tilde{S}_{xx}(f) = \frac{1}{q}\sum_{i=1}^{q} \hat{S}_{xx_i}(f), \quad \text{where } \hat{S}_{xx_i}(f) = \frac{1}{T_r}\left|X_{T_{ri}}(f)\right|^2$$

5. The bias error of $\tilde{S}_{xx}(f)$ is usually small if $S_{xx}(f)$ is smooth. However, the estimator $\tilde{S}_{xx}(f)$ usually underestimates the peaks and overestimates the troughs (i.e. dynamic range is reduced). The bias error can be reduced by improving the resolution bandwidth. The resolution bandwidths and approximate bias errors for various lag windows are shown in Table 10.2.
6. The variance of $\tilde{S}_{xx}(f)$ is given by

$$\frac{\mathrm{Var}\left(\tilde{S}_{xx}(f)\right)}{S_{xx}^2(f)} = \frac{1}{BT}$$

   where the $BT$ can be replaced by the number of averages $q$ for the segment averaging method (the number of degree is $n = 2BT$). The random error is reduced as the product $BT$ becomes large. However, in general, we need to trade-off between the resolution (bias error) and the variability (random error).
   While maintaining the good resolution (low bias) the only way to reduce the random error is by increasing the data length $T$.
7. The cross-spectral density function estimate $\tilde{S}_{xy}(f)$ has similar statistical properties to those of $\tilde{S}_{xx}(f)$. However, this estimator depends on the 'true' coherence function $\gamma_{xy}^2(f)$ which is an 'uncontrollable' parameter.
   Time delay between two signals $x(t)$ and $y(t)$ can introduce a severe bias error.
8. The statistical properties of the coherence function estimate $\tilde{\gamma}_{xy}^2(f)$ depend on both the true coherence function $\gamma_{xy}^2(f)$ and the product $BT$. The random error is reduced by increasing the product $BT$. However, significant bias error may occur if the resolution bandwidth is wide when $\arg S_{xy}(f)$ changes rapidly.
9. The estimator $\tilde{H}_1(f)$ also depends on both the controllable parameter $BT$ and the uncontrollable parameter $\gamma_{xy}^2(f)$.
   The random errors of various estimators are summarized in Table 10.3.

## 10.9  MATLAB EXAMPLES

---

### *Example 10.1:* **Statistical errors of power and cross-spectral density functions**

Consider the same example (2-DOF system) as in MATLAB Example 9.4, i.e.

$$h(t) = \frac{A_1}{\omega_{d1}} e^{-\zeta_1 \omega_{n1} t} \sin \omega_{d1} t + \frac{A_2}{\omega_{d2}} e^{-\zeta_2 \omega_{n2} t} \sin \omega_{d2} t$$

Again, we use the white noise as an input $x(t)$, and the output $y(t)$ is obtained by $y(t) = h(t) * x(t)$. However, we do not consider the measurement noise.

Since we use the white noise input (band-limited up to $f_s/2$, where $f_s$ is the sampling rate, i.e. $\sigma_x^2 = \int_{-f_s/2}^{f_s/2} S_{xx}(f) df = 1$), the theoretical spectral density functions are $S_{xx}(f) = \sigma_x^2/f_s$, $S_{yy}(f) = |H(f)|^2 \sigma_x^2/f_s$ and $S_{xy}(f) = H(f)\sigma_x^2/f_s$. These theoretical values are compared with the estimated spectral density functions.

The segment averaging method is used to obtain smoothed spectral density functions $\tilde{S}_{xx}(f)$, $\tilde{S}_{yy}(f)$ and $\tilde{S}_{xy}(f)$. Then, for a given data length $T$, we demonstrate how the bias error and the random error change depending on the resolution bandwidth $B \approx 1/T_r$, where $T_r$ is the length of the segment.

---

| Line | MATLAB code | Comments |
|------|-------------|----------|
| 1 | clear all | Same as MATLAB Example 9.4, except |
| 2 | A1=20; A2=30; f1=5; f2=15; wn1=2*pi*f1; wn2=2*pi*f2; | that the damping ratios are smaller, i.e. we use a more lightly damped system. |
| 3 | zeta1=0.02; zeta2=0.01; | |
| 4 | wd1=sqrt(1-zeta1^2)*wn1; wd2=sqrt(1-zeta2^2)*wn2; | |
| 5 | fs=100; T1=10; t1=[0:1/fs:T1-1/fs]; | |
| 6 | h=(A1/wd1)*exp(-zeta1*wn1*t1). *sin(wd1*t1) + (A2/wd2) *exp(-zeta2*wn2*t1).*sin(wd2*t1); | |
| 7 | T= 2000; % T=10000; | Define the data length T seconds. First, |
| 8 | randn('state',0); | use T = 2000, then compare the results |
| 9 | x=randn(1,T*fs); | with the cases of T = 10000 (when Tr = |
| 10 | y=filter(h,1,x)/fs; % scaled appropriately. | 20 is used at Line 11). Generate the white noise input sequence 'x' ($\sigma_x^2 = 1$), and then obtain the output sequence 'y' (scaled appropriately). |
| 11 | Tr=4; N=Tr*fs; % Tr=20; | Define the length of segment Tr |
| 12 | [Sxx, f]=cpsd(x,x, hanning(N),N/2, N, fs, 'twosided'); | seconds. First, use Tr = 4 (approximately 1000 averages), then |
| 13 | [Syy, f]=cpsd(y,y, hanning(N),N/2, N, fs, 'twosided'); | compare the results with the cases of Tr = 20 (approximately 200 averages). |

| 14 | [Sxy, f]=cpsd(x,y, hanning(N),N/2, N, fs, 'twosided'); | Obtain the spectral density estimates using the segment averaging method (Hann window with 50 % overlap is used). Also, calculate $H(f)$ by the DFT of the impulse response sequence (scaled appropriately). |
|----|---|---|
| 15 | H=fft(h)/fs; % scaled appropriately. | |
| 16 | f1=fs*(0:length(H)-1)/length(H); | |

| 17 | figure (1) | Plot the power spectral density estimate $\tilde{S}_{xx}(f)$, where 'fs' is multiplied. Also, plot the variance of the signal $\sigma_x^2 = \tilde{S}_{xx}(f) \cdot f_s = 1$ (0 dB) for comparison. |
|----|---|---|
| 18 | plot(f,10*log10(fs*Sxx), f, zeros(size(f)), 'r:') | |
| 19 | xlabel('Frequency (Hz)') | |
| 20 | ylabel('Estimate of \itS_x_x\rm(\itf\rm) (dB)') | |
| 21 | axis([0 30 -10 10]) | |

| 22 | figure(2) | Plot the power spectral density estimate $\tilde{S}_{yy}(f)$ and the theoretical power spectral density function $S_{yy}(f) = |H(f)|^2 \sigma_x^2 / f_s$. |
|----|---|---|
| 23 | plot(f,10*log10(Syy), f1,10*log10(abs(H).^2/fs), 'r:') | |
| 24 | xlabel('Frequency (Hz)') | |
| 25 | ylabel('Estimate of \itS_y_y\rm(\itf\rm) (dB)') | |
| 26 | axis([0 30 -100 -20]) | |

| 27 | figure(3) | Plot the magnitude spectra of $\tilde{S}_{xy}(f)$ and $S_{xy}(f) = H(f)\sigma_x^2 / f_s$. |
|----|---|---|
| 28 | plot(f,10*log10(abs(Sxy)), f1,10*log10(abs(H)/fs), 'r:') | |
| 29 | xlabel('Frequency (Hz)') | |
| 30 | ylabel('Estimate of |\itS_x_y\rm(\itf\rm)| (dB)') | |
| 31 | axis([0 30 -60 -20]) | |

| 32 | figure(4) | Plot the phase spectra of $\tilde{S}_{xy}(f)$ and $S_{xy}(f)$. |
|----|---|---|
| 33 | plot(f,unwrap(angle(Sxy)), f1,unwrap(angle(H)), 'r:') | |
| 34 | xlabel('Frequency (Hz)') | |
| 35 | ylabel('Estimate of arg\itS_x_y\rm(\itf\rm) (rad)') | |
| 36 | axis([0 30 -3.5 0]) | |

**Results: Case (a)** $T_r = 4$ seconds at Line 11 and $T = 2000$ at Line 7 (1000 averages)

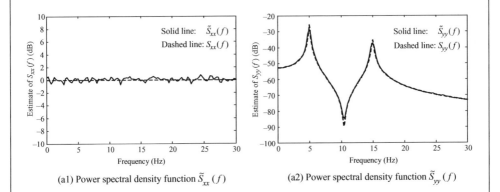

(a1) Power spectral density function $\tilde{S}_{xx}(f)$

(a2) Power spectral density function $\tilde{S}_{yy}(f)$

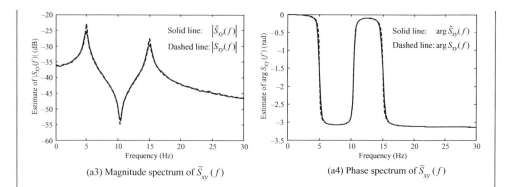

(a3) Magnitude spectrum of $\tilde{S}_{xy}(f)$          (a4) Phase spectrum of $\tilde{S}_{xy}(f)$

**Comments:** Since the Hann window is used, the resolution bandwidth is $B \approx 1.33/T_w \approx 0.67$ Hz, where $T_w \approx T_r/2$. Note that both $\tilde{S}_{yy}(f)$ and $|\tilde{S}_{xy}(f)|$ underestimate the peaks and overestimate the trough owing to the bias error.

**Results: Case (b)** $T_r = 20$ seconds at Line 11 and $T = 2000$ at Line 7 (200 averages)

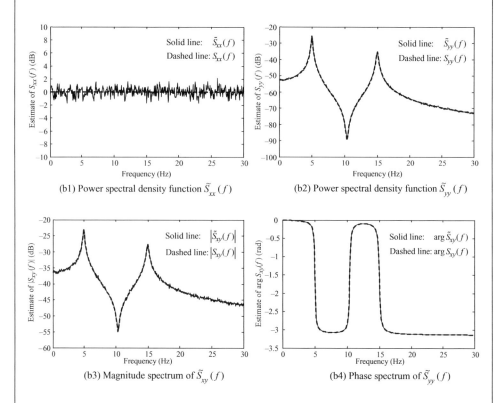

(b1) Power spectral density function $\tilde{S}_{xx}(f)$          (b2) Power spectral density function $\tilde{S}_{yy}(f)$

(b3) Magnitude spectrum of $\tilde{S}_{xy}(f)$          (b4) Phase spectrum of $\tilde{S}_{yy}(f)$

**Comments:** In this case, the resolution bandwidth is $B \approx 0.13$ Hz. It can be shown that the bias errors of spectral density estimates $\tilde{S}_{yy}(f)$ and $\tilde{S}_{xy}(f)$ are greatly reduced owing to the improvement of the resolution. However, the random error is increased since the number of averages is decreased. Note that $\arg \tilde{S}_{xy}(f)$ has much less random error than $|\tilde{S}_{xy}(f)|$ (almost no random error is present in this example since $\gamma_{xy}^2(f) = 1$).

**Results: Case (c)** $T_r = 20$ seconds at Line 11 and $T = 10\,000$ at Line 7 (1000 averages)

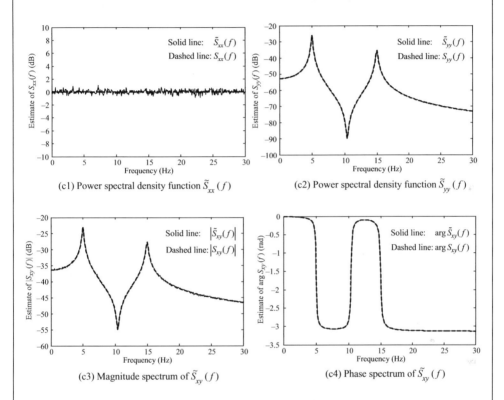

(c1) Power spectral density function $\tilde{S}_{xx}(f)$

(c2) Power spectral density function $\tilde{S}_{yy}(f)$

(c3) Magnitude spectrum of $\tilde{S}_{xy}(f)$

(c4) Phase spectrum of $\tilde{S}_{xy}(f)$

**Comments:** While maintaining the narrow resolution bandwidth, increasing the number of averages results in better estimates.

**Comments on the segment averaging method:** As mentioned in Section 10.4, the underlying assumption for the segment averaging method is that each segment of the data must be uncorrelated. If it is correlated, the random error will not reduce appreciably. To demonstrate this, use $T = 2000$ and $Tr = 20$ (i.e. Case (b)), and add the following script between Line 9 and Line 10. Then run this MATLAB program again and compare the result with Case (b).

$$x=[x\ 2*x\ 3*x\ 4*x\ 5*x];\ x=x-mean(x);\ x=x/std(x);$$

Now, the total data length is $5 \times 2000 = 10\,000$ seconds, so the number of averages is approximately 1000 which is the same as in Case (c). However, the random error will not reduce since correlated data are repeatedly used. For example, the results of $\tilde{S}_{xx}(f)$ and $|\tilde{S}_{xy}(f)|$ are shown in Figures (d).

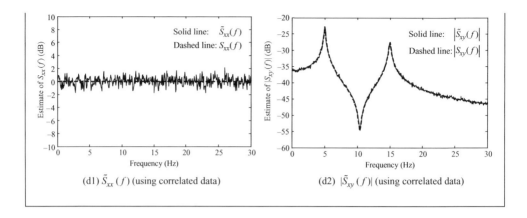

(d1) $\tilde{S}_{xx}(f)$ (using correlated data)          (d2) $|\tilde{S}_{xy}(f)|$ (using correlated data)

---

### *Example 10.2:* Bias error of the coherence function estimate

In the previous MATLAB example, we did not consider the coherence function estimator $\tilde{\gamma}_{xy}^2(f)$. We shall examine the bias error of $\tilde{\gamma}_{xy}^2(f)$, using the same system as in the previous example. Note that the half-power point bandwidths at resonances ($f_1 = 5$ and $f_2 = 15$) are $B_{r1} = 2\zeta_1 f_1 = 0.2\,\text{Hz}$ and $B_{r2} = 2\zeta_2 f_2 = 0.3\,\text{Hz}$

As mentioned in Section 10.6, considerable bias error may occur at resonances and anti-resonances where the phase of the cross-spectral density function changes rapidly, e.g. if the Hann window is used the normalized bias error is (i.e. Equation (10.108))

$$\frac{b\left(\tilde{\gamma}_{xy}^2(f)\right)}{\gamma_{xy}^2(f)} \approx -\frac{0.126}{T_w^2}\left(\frac{d}{df}\left(\arg S_{xy}(f)\right)\right)^2$$

In this example, various resolution bandwidths are used: $B_1 = 1\,\text{Hz}$, $B_2 = 0.5\,\text{Hz}$, $B_3 = 0.2\,\text{Hz}$ and $B_4 = 0.05\,\text{Hz}$. For each resolution bandwidth, approximately 1000 averages are used so that the random error is negligible. A Hann window with 50 % overlap is used. The length of each segment for the Hann window is obtained by $T_r = 2T_w \approx 2 \times 1.33/B$, where $B$ is the resolution bandwidth (see Table 10.2).

---

| Line | MATLAB code | Comments |
|------|-------------|----------|
| 1 | clear all | Same as MATLAB Example 10.1. |
| 2 | A1=20; A2=30; f1=5; f2=15; wn1=2*pi*f1; wn2=2*pi*f2; | |
| 3 | zeta1=0.02; zeta2=0.01; | |
| 4 | wd1=sqrt(1-zeta1^2)*wn1; wd2=sqrt(1-zeta2^2)*wn2; | |
| 5 | fs=100; T1=10; t1=[0:1/fs:T1-1/fs]; | |
| 6 | h=(A1/wd1)*exp(-zeta1*wn1*t1). *sin(wd1*t1) + (A2/wd2)*exp(-zeta2*wn2*t1). *sin(wd2*t1); | |

| 7 | B1=1; B2=0.5; B3=0.2; B4=0.05; | Define the resolution bandwidths: B1, |
|---|---|---|
| 8 | N1=fix(1.33*2/B1*fs); N2=fix(1.33*2/B2*fs); | B2, B3 and B4. Then, calculate the |
| 9 | N3=fix(1.33*2/B3*fs); N4=fix(1.33*2/B4*fs); | number of points of a segment for each |
| 10 | Ns=500; Nt=N4*Ns; | bandwidth. Define the total number of |
| 11 | randn('state',0); | segments Ns = 500 that results in |
| 12 | x=randn(1,Nt); | approximately 1000 averages if 50 % |
| 13 | y=filter(h,1,x); | overlap is used. |
| | % we do not scale for convenience. | Generate white noise input sequence |
| | | 'x' and the output sequence 'y'. |

| 14 | [Gamma_1, f] = mscohere(x(1:Ns*N1), | Calculate the coherence function |
|---|---|---|
| | y(1:Ns*N1), hanning(N1), [], N4, fs); | estimates $\tilde{\gamma}_{xy}^2(f)$ for each resolution |
| 15 | [Gamma_2, f] = mscohere(x(1:Ns*N2), | bandwidth using the MATLAB |
| | y(1:Ns*N2), hanning(N2), [], N4, fs); | function 'mscohere'. |
| 16 | [Gamma_3, f] = mscohere(x(1:Ns*N3), | Also, calculate $H(f)$ by the DFT of |
| | y(1:Ns*N3), hanning(N3), [], N4, fs); | the impulse response sequence. We |
| 17 | [Gamma_4,f] = mscohere(x(1:Ns*N4), | calculate this to compare $\tilde{\gamma}_{xy}^2(f)$ and |
| | y(1:Ns*N4), hanning(N4), [], N4, fs); | arg $H(f)$. Note that arg $H(f) =$ |
| 18 | H=fft(h, N4); | arg $S_{xy}(f)$. |
| | % we do not scale for convenience. | |

| 19 | figure (1) | Plot the coherence function estimates |
|---|---|---|
| 20 | plot(f, [Gamma_1 Gamma_2 Gamma_3 | $\tilde{\gamma}_{xy}^2(f)$ for each resolution bandwidth. |
| | Gamma_4]) | |
| 21 | xlabel('Frequency (Hz)') | |
| 22 | ylabel('Estimate of \it\gamma_x_y\rm^ | |
| | 2(\itf\rm)') | |
| 23 | axis([0 30 0 1]) | |

| 24 | figure(2) | Plot arg $H(f)$ which is the same as |
|---|---|---|
| 25 | plot(f,unwrap(angle(H(1:length(f))))) | arg $S_{xy}(f)$. |
| 26 | xlabel('Frequency (Hz)') | |
| 27 | ylabel('arg\itH\rm(\itf\rm) = | |
| | arg\itS_x_y\rm(\itf\rm) (rad)') | |
| 28 | axis([0 30 -3.5 0]) | |

## Results

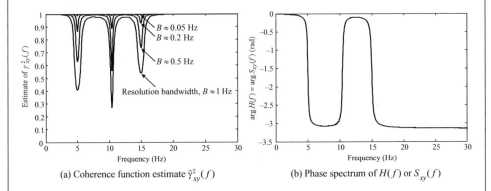

(a) Coherence function estimate $\tilde{\gamma}_{xy}^2(f)$      (b) Phase spectrum of $H(f)$ or $S_{xy}(f)$

**Comments:** Note the large bias error at the resonances and anti-resonance. Also note that the bias error decreases as the resolution bandwidth gets narrower.

Another point on the bias error in the coherence function is that it depends on the 'window function' used in the estimation (Schmidt, 1985a). For example, if we use a rectangular window, i.e. replace 'hanning' in Lines 14–17 with 'rectwin', then we may not see the drop of coherence at resonances as shown in Figure (c). Readers may care to try different window functions.

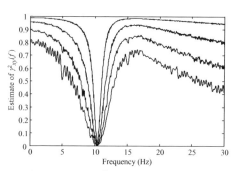

(c) Coherence function estimate $\tilde{\gamma}^2_{xy}(f)$ (rectangular window function is used)

---

***Example 10.3:* Bias error of the cross-spectral density function estimate** (time delay problem)

In Section 10.5, we mentioned that the cross-spectral density function estimate $\tilde{S}_{xy}(f)$ produces a biased result if time delay is present between two signals. For example if $y(t) = x(t - \Delta)$, then the average of $\tilde{S}_{xy}(f)$ is (i.e. Equation (10.105) for a rectangular window)

$$E\left[\tilde{S}_{xy}(f)\right] \approx \left(1 - \frac{\Delta}{T_r}\right) S_{xy}(f)$$

In this example, we use the white noise signal for $x(t)$ (band-limited up to $f_s/2$), and $y(t) = x(t - \Delta)$ where $\Delta = 1$ second. Since it is a pure delay problem, the cross-spectral density function is

$$S_{xy}(f) = e^{-j2\pi f \Delta} S_{xx}(f)$$

i.e. $|S_{xy}(f)| = S_{xx}(f) = \sigma_x^2/f_s$ (see MATLAB Example 10.1).

We shall examine the bias error of $\tilde{S}_{xx}(f)$ for various values of $T_r$. Note that the bias error can only be reduced by increasing the window length (in effect, improving the resolution) or by aligning two signals (Jenkins and Watts, 1968), e.g. $y(t)$ may be replaced by $y'(t) = y(t + \Delta)$ if $\Delta$ can be found from the cross-correlation function (arg $\tilde{S}_{xy}(f)$ must be compensated later).

| Line | MATLAB code | Comments |
|------|-------------|----------|
| 1 | clear all | Define the delay $\Delta = 1$ second and the |
| 2 | delta=1; fs=20; | window length $T_r$. We compare the |
| 3 | Tr=1.1; % Tr=1.1, 2, 5, 50; | results of four different window |
| 4 | N=Tr*fs; Nt=1000*N; | lengths $T_r = 1.1, 2, 5$ and 50 seconds. |
| | | 'N' is the number of points in the |
| | | segment and 'Nt' is the total data |
| | | length. |
| 5 | randn('state',0); | Generate white noise sequence 'x' and |
| 6 | x=randn(1,Nt+delta*fs); | the delayed sequence 'y'. (Note that |
| 7 | y=x(1:length(x)-delta*fs); | $\sigma_x^2 = \sigma_y^2 = 1$.) Then, calculate the |
| 8 | x=x(delta*fs+1:end); | cross-spectral density function |
| 9 | [Sxy, f]=cpsd(x,y, rectwin(N), 0, 1000, fs, 'twosided'); | estimate $\tilde{S}_{xy}(f)$. In this example, the rectangular window with no overlap is used. So, the number of averages is 1000. |
| 10 | figure (1) | Plot the magnitude spectrum of $\tilde{S}_{xy}(f)$ |
| 11 | plot(f,fs*abs(Sxy), f, ones(size(f)), 'r:') | (multiplied by the sampling rate) and |
| 12 | xlabel('Frequency (Hz)') | the theoretical value which is unity |
| 13 | ylabel('Estimate of \|\itS_x_y\rm(\itf\rm)\| (linear scale)') | (note that $\left\|S_{xy}(f)\right\| \cdot f_s = \sigma_x^2 = 1$). |
| 14 | axis([0 10 0 1.1]) | |
| 15 | figure(2) | Plot the phase spectrum of $\tilde{S}_{xy}(f)$ and |
| 16 | plot(f, unwrap(angle(Sxy)), [0 10], [0 -2*pi*10*delta], 'r:') | the theoretical value of arg $S_{xy}(f)$ which is $-2\pi f \Delta$. |
| 17 | xlabel('Frequency (Hz)') | Run this MATLAB program again |
| 18 | ylabel('Estimate of arg\itS_x_y\rm(\itf\rm) (rad)') | using different values of $T_r$. |
| 19 | axis([0 10 -65 0]) | |

## Results

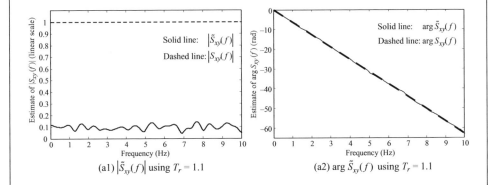

(a1) $\left|\tilde{S}_{xy}(f)\right|$ using $T_r = 1.1$

(a2) arg $\tilde{S}_{xy}(f)$ using $T_r = 1.1$

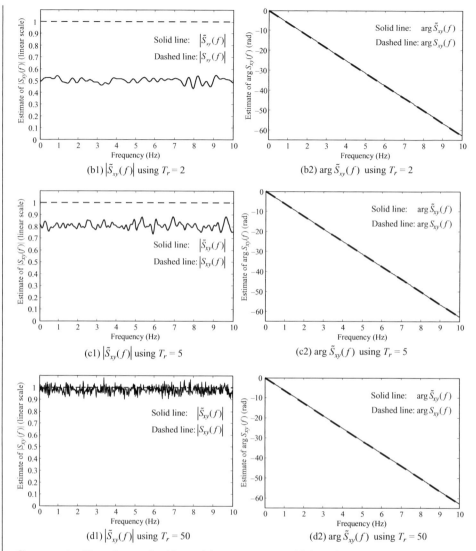

(b1) $\left|\tilde{S}_{xy}(f)\right|$ using $T_r = 2$

(b2) arg $\tilde{S}_{xy}(f)$ using $T_r = 2$

(c1) $\left|\tilde{S}_{xy}(f)\right|$ using $T_r = 5$

(c2) arg $\tilde{S}_{xy}(f)$ using $T_r = 5$

(d1) $\left|\tilde{S}_{xy}(f)\right|$ using $T_r = 50$

(d2) arg $\tilde{S}_{xy}(f)$ using $T_r = 50$

**Comments:** Note that a significant bias error occurs if the window length $T_r$ is short. However, it is interesting to see that arg $\tilde{S}_{xy}(f)$ is almost unaffected as long as $T_r > \Delta$, as one might expect from Equation (10.105).

# 11

# Multiple-Input/Response Systems

## Introduction

This chapter briefly introduces some additions to the work presented so far. The natural extension is to multiple-input and multiple-output systems. The concepts of residual spectra and partial and multiple coherence functions offer insight into the formal matrix solutions. Finally principal component analysis is summarized and related to the total least squares method of Chapter 9.

## 11.1 DESCRIPTION OF MULTIPLE-INPUT, MULTIPLE-OUTPUT (MIMO) SYSTEMS

Consider the multiple-input, multiple-output system depicted in Figure 11.1.

Assuming that the system is composed of linear elements, then any single output $y_j(t)$ (say) is

$$y_j(t) = \sum_{i=1}^{m} h_{ji}(t) * x_i(t) \tag{11.1}$$

where $h_{ji}(t)$ is the impulse response function relating the $i$th input to the $j$th output. Fourier transforming this yields

$$Y_j(f) = \sum_{i=1}^{m} H_{ji}(f) X_i(f) \tag{11.2}$$

where $H_{ji}(f)$ is the frequency response function relating the $i$th input to the $j$th response.

The Fourier transform of the set of all responses can be arranged as a vector as

$$\mathbf{Y}(f) = \mathbf{H}(f)\mathbf{X}(f) \tag{11.3}$$

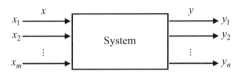

**Figure 11.1**    A multiple-input, multiple-output system

where $\mathbf{Y}(f)$ is an $n \times 1$ vector of responses, $\mathbf{X}(f)$ is an $m \times 1$ vector of inputs and $\mathbf{H}(f)$ is $n \times m$ matrix of frequency response functions. For simplicity of notation we write the transforms as $\mathbf{X}(f)$ rather than $\mathbf{X}_T(f)$, implying a data length $T$. Also, we imply below the proper limitation as $T \to \infty$ etc.

From this the $n \times n$ output spectral density matrix $S_{\mathbf{YY}}(f) = E[\mathbf{Y}^*(f)\mathbf{Y}^T(f)]$ is

$$S_{\mathbf{YY}}(f) = \mathbf{H}^*(f)S_{\mathbf{XX}}(f)\mathbf{H}^T(f) \qquad (11.4)$$

where $S_{\mathbf{XX}}(f)$ is the $m \times m$ input spectral density matrix. This expression generalizes Equation (9.8). Note that both these matrices $S_{\mathbf{XX}}(f)$ and $S_{\mathbf{YY}}(f)$ include cross-spectra relating the various inputs for $S_{\mathbf{XX}}(f)$ and outputs for $S_{\mathbf{YY}}(f)$.

Similarly, the input–output spectral density matrix may be expressed as $S_{\mathbf{XY}}(f) = E[\mathbf{X}^*(f)\mathbf{Y}^T(f)]$, which becomes

$$S_{\mathbf{XY}}(f) = S_{\mathbf{XX}}(f)\mathbf{H}^T(f) \qquad (11.5)$$

This is the generalization of Equation (9.12). It is tempting to use this as the basis for 'identification' of the matrix $\mathbf{H}^T(f)$ by forming

$$\mathbf{H}^T(f) = S_{\mathbf{XX}}^{-1}(f)S_{\mathbf{YY}}(f) \qquad (11.6)$$

Immediately apparent is the potential difficulty in that we need the inverse of $S_{\mathbf{XX}}(f)$, which might be singular. This arises if there is a linear dependency between inputs, i.e. if at least one input can be regarded as a linear combination of the others. Under these circumstances the determinant of $S_{\mathbf{XX}}(f)$ is zero and the rank of $S_{\mathbf{XX}}(f)$ is less than $m$. The pseudo-inverse of $S_{\mathbf{XX}}(f)$ may be employed but this is not followed up here.

## 11.2  RESIDUAL RANDOM VARIABLES, PARTIAL AND MULTIPLE COHERENCE FUNCTIONS

The matrix formulation in Equation (11.6) is a compact approach to dealing with multiple-input, multiple-output systems. However, there are other approaches aimed at revealing and interpreting the nature and relative importance of signals and transmission paths in systems. One such method is described below. This is demonstrated here by using a very simple example, namely a two-input, single-output system. This can easily be generalized to more inputs – and for more outputs each output can be taken in turn.

Let us start by saying that we measure three signals $(x_1(t), x_2(t), x_3(t))$ and wish to know how these signals may be related. An approach to this would be to 'strip out' progressively the 'effect' of first one signal on the other two, and then what is left of the next from the last remaining one (and so on if we had more signals). This 'stripping out' of one signal's effect on another yields what is called a 'residual' process. Comprehensive studies on residual processes can be found in Bendat (1976a, 1976b, 1978) and Bendat and Piersol (1980, 2000). We illustrate how this approach may be helpful by choosing any one of the three (say $x_3(t)$) and identifying it as an 'output' $y(t)$ arising from inputs $x_1(t)$ and $x_2(t)$.

So we consider a two-input, single-output system with some uncorrelated output measurement noise as shown in Figure 11.2.

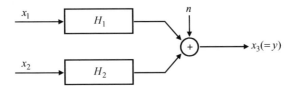

**Figure 11.2**    Two-input, single-output system

On the basis of the measurements taken, this is a three-component process, $x_1, x_2, x_3(= y)$, where we reiterate that it may be convenient (but not necessary) to regard $y$ as an output. Based on the assumed structure we might wish to quantify:

1. The relative magnitude of noise to 'linear effects', i.e. how much of $y$ is accounted for by linear operations on $x_1$ and $x_2$.
2. The relative importance of inputs $x_1$ and $x_2$, i.e. how much of $y$ comes from each of $x_1$ and $x_2$.
3. The frequency response functions $H_1$ and $H_2$ (i.e. estimate $H_1$ and $H_2$ from $x_1$, $x_2$ and $y$).

To start with, it is useful to remind ourselves of the concept and use of the ordinary coherence function. With reference to Figure 11.3, suppose we have two signals $x$, $y$ and we seek a linear 'link' between them. Then, Figure 11.3 may be redrawn as Figure 11.4.

**Figure 11.3**    A single-input, single-output system with measurement noise on the output

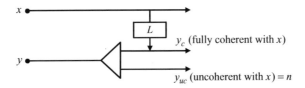

**Figure 11.4**    Alternative expression of Figure 11.3

If we try to estimate the 'best' linear operation (optimal filter $L$) on $x$ that minimizes $E[y_{uc}^2]$, then its frequency response function is given by

$$L(f) = \frac{S_{xy}(f)}{S_{xx}(f)} \tag{11.7}$$

Also, the coherent output power is $S_{y_c y_c}(f) = \gamma_{xy}^2(f)S_{yy}(f)$ and the uncoherent (noise) power is $S_{y_{uc} y_{uc}}(f) = [1 - \gamma_{xy}^2(f)]S_{yy}(f)$. In fact, $y_{uc}$ is the *residual* random variable resulting from $y$ after a linear prediction of $y$ based on $x$ has been subtracted. Note that, in Figure 11.4, the noise is interpreted as what is 'left' in the output after the linear effects of $x$ have been removed.

We now return to the problem of three processes $x_1$, $x_2$, $x_3(= y)$. Figure 11.2 can be decomposed into the two stages below, as shown in Figure 11.5.

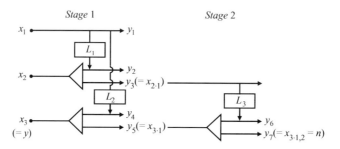

**Figure 11.5**   Alternative expression of Figure 11.2

We should emphasize that we assume that we can only use the three measured signals $x_1$, $x_2$ and $x_3(= y)$. Furthermore we restrict ourselves to second-order properties of stationary random processes. Accordingly, the only information we have available is the $3 \times 3$ spectral density matrix linking $x_1$, $x_2$ and $x_3(= y)$. All subsequent manipulations involve the elements of this (Hermitian) matrix.

### Stage 1

In Figure 11.5, Stage 1 depicts the 'stripping out' (in a least squares optimization sense) of the signal $x_1$ from $x_2$ and $x_3(= y)$. The signal denoted $x_{2\cdot1}$ is therefore what is left of $x_2$ when the linearly correlated part of $x_1$ has been removed. The notation $x_{2\cdot1}$ denotes the 'residual' random variable. Similarly, $x_{3\cdot1}$ denotes what is left of $x_3$ when the linearly related part of $x_1$ has been removed.

To put a physical interpretation on this – it is as though process $x_1$ is 'switched off' and $x_{2\cdot1}$ and $x_{3\cdot1}$ are what remains of $x_2$ and $x_3$ when this is done. (Once again we emphasize that this switching off is in a least squares sense. Thus it picks out the linear link between the signals.) The linear links between $x_1$ and $x_2$, $x_3$ are denoted $L_1$ and $L_2$. These and the following quantities can be expressed in terms of spectra relating the residual random variables. It should be noted that the filters $L_1$ and $L_2$ are mathematical ideal (generally non-causal) linear filters – not 'physical' filters (and so should not be identified with $H_1$ and $H_2$). This introduces the concept of residual spectral densities and partial coherence functions:

- *Residual spectral density functions* are 'usual' spectral density functions formed from residual variables.
- *Partial coherence functions* are ordinary coherence functions formed from residual variables.

First, for the pair $x_1$ and $x_2$, the 'optimal' linear filter linking $x_1$ and $x_2$ is

$$L_1(f) = \frac{S_{12}(f)}{S_{11}(f)} \tag{11.8}$$

where $S_{12}(f)$ is short for $S_{x_1 x_2}(f)$ etc. The power spectral density of that part of $x_2$ which is coherent with $x_1$ is

$$S_{y_2 y_2}(f) = \gamma_{12}^2(f) S_{22}(f) \tag{11.9}$$

The 'noise' output power is $S_{y_3 y_3}(f)$ which is written as $S_{22\cdot1}(f)$, i.e.

$$S_{y_3 y_3}(f) = S_{22\cdot1}(f) = \left[1 - \gamma_{12}^2(f)\right] S_{22}(f) \tag{11.10}$$

Similarly, for the pair $x_1$ and $x_3$, the optimal filter is

$$L_2(f) = \frac{S_{13}(f)}{S_{11}(f)} \tag{11.11}$$

The spectral density of $y(= x_3)$ is $S_{yy}(f) = S_{33}(f) = S_{y_4 y_4}(f) + S_{y_5 y_5}(f)$, where $S_{y_4 y_4}(f)$ is the power spectral density of that part of $y$ that is coherent with $x_1$ and $S_{y_5 y_5}(f) = S_{33\cdot1}(f)$ is the power spectral density of that part of $y$ that is uncoherent with $x_1$, i.e.

$$S_{y_4 y_4}(f) = \gamma_{13}^2(f) S_{33}(f) \tag{11.12}$$

$$S_{y_5 y_5}(f) = S_{33\cdot1}(f) = \left[1 - \gamma_{13}^2(f)\right] S_{33}(f) \tag{11.13}$$

From Equations (11.10) and (11.13), we see that the residual spectral density functions $S_{22\cdot1}(f)$ and $S_{33\cdot1}(f)$ are computed from the 'usual' spectral density functions and ordinary coherence functions. Similarly, the residual spectral density function $S_{23\cdot1}(f)$ which is the cross-spectral density between $x_{2\cdot1}$ and $x_{3\cdot1}$ can also be expressed in terms of the spectral density functions formed from the measured signal $x_1$, $x_2$ and $x_3$. We do this as follows.

From the definition of the cross-spectral density function, $S_{23\cdot1}(f)$ is

$$S_{23\cdot1}(f) = \lim_{T \to \infty} \frac{E[X_{2\cdot1}^*(f) X_{3\cdot1}(f)]}{T} \tag{11.14}$$

Since $X_2 = Y_2 + X_{2\cdot1} = L_1 X_1 + X_{2\cdot1}$ and $X_3 = Y_4 + X_{3\cdot1} = L_2 X_1 + X_{3\cdot1}$, and using Equations (11.8) and (11.11), it can be shown that

$$S_{23\cdot1}(f) = S_{23}(f) - \frac{S_{21}(f) S_{13}(f)}{S_{11}(f)} \tag{11.15}$$

So the residual spectral density $S_{23\cdot1}(f)$ can be computed in terms of the usual spectral density functions.

We now introduce the concept of the partial coherence function. This is the 'ordinary' coherence function but linking residual random variables. The partial coherence function between $x_{2\cdot1}$ and $x_{3\cdot1}$ can be computed using the above results, and is

$$\gamma_{23\cdot1}^2(f) = \frac{|S_{23\cdot1}(f)|^2}{S_{22\cdot1}(f) S_{33\cdot1}(f)} \tag{11.16}$$

### Stage 2

In Figure 11.5, Stage 2 depicts the 'removal' of $x_{2\cdot1}$ from $x_{3\cdot1}$. As before, for the pair $x_{2\cdot1}$ and $x_{3\cdot1}$, the optimal filter is

$$L_3(f) = \frac{S_{23\cdot1}(f)}{S_{22\cdot1}(f)} \qquad (11.17)$$

In the figure, $x_{3\cdot1,2}$ denotes the residual variable arising when the linear effects of both $x_1$ and $x_2$ are removed from $x_3$. The output powers of uncoherent and coherent components with $x_{2\cdot1}$ are

$$S_{y_7 y_7}(f) = S_{nn}(f) = S_{33\cdot1,2}(f) = \left[1 - \gamma_{23\cdot1}^2(f)\right] S_{33\cdot1}(f) \qquad (11.18)$$

$$S_{y_6 y_6}(f) = \gamma_{23\cdot1}^2(f) S_{33\cdot1}(f) \qquad (11.19)$$

Note that $S_{33\cdot1,2}(f)$ is the power spectral density of that part of $y$ unaccounted for by linear operations on $x_1$ and $x_2$, i.e. the uncorrelated noise power. Now, combining the above results and using Figure 11.5, the power spectral density of $y$ can be decomposed into

$$
\begin{aligned}
S_{yy}(f) = S_{33}(f) &= S_{y_4 y_4}(f) + S_{y_6 y_6}(f) + S_{y_7 y_7}(f) \\
&= \gamma_{13}^2(f) S_{33}(f) \qquad && \text{part fully coherent with } x_1 \\
&+ \gamma_{23\cdot1}^2(f) S_{33\cdot1}(f) \qquad && \text{part fully coherent with } x_2 \\
& && \text{after } x_1 \text{ has been removed from } x_2 \text{ and } x_3 \\
&+ \left[1 - \gamma_{23\cdot1}^2(f)\right] S_{33\cdot1}(f) \qquad && \text{uncoherent with both } x_1 \text{ and } x_2 \qquad (11.20)
\end{aligned}
$$

This equation shows the role of the partial coherence function.

Note that by following the signal flow in Figure 11.5, one can easily verify that

$$H_1(f) = L_2(f) - L_1(f) L_3(f) \qquad (11.21)$$

and

$$H_2(f) = L_3(f) \qquad (11.22)$$

A multiple coherence function is defined in a manner similar to that of the ordinary coherence function. Recall that the ordinary coherence function for the system shown in Figure 11.3 can be written as $\gamma_{xy}^2(f) = (S_{yy}(f) - S_{nn}(f))/S_{yy}(f)$. Similarly, the multiple coherence function denoted $\gamma_{y:x}^2(f)$ is defined as

$$\gamma_{y:x}^2(f) = \frac{S_{yy}(f) - S_{nn}(f)}{S_{yy}(f)} \qquad (11.23)$$

That is the multiple coherence function $\gamma_{y:x}^2(f)$ is the fraction of output power accounted for via linear operations on the inputs; it is a measure of how well the inputs account for the measured response of the system. For the example shown above, it can be written as

$$\gamma_{y:x}^2(f) = \frac{S_{33}(f) - S_{33\cdot1,2}(f)}{S_{33}(f)} \qquad (11.24)$$

Note that the nearer $\gamma_{y:x}^2(f)$ is to unity, the more 'completely' does the linear model apply to the three components. Using Equations (11.12) and (11.18), the above equation can be written as

$$\gamma_{y:x}^2(f) = 1 - \left(1 - \gamma_{13}^2(f)\right)\left(1 - \gamma_{23\cdot1}^2(f)\right) . \tag{11.25}$$

which shows that it can be computed in terms of partial and ordinary coherence functions. See Sutton *et al.* (1994) for applications of the multiple coherence function.

## Computation of Residual Spectra and Interpretation in Terms of Gaussian Elimination

The above example illustrates the methodology. A more general computational formulation is given below which operates on the elements of the spectral density matrix of the measured signals. The residual spectral density function $S_{23\cdot1}(f)$ given in Equation (11.15) can be generalized as

$$S_{ij\cdot k}(f) = S_{ij}(f) - \frac{S_{ik}(f)S_{kj}(f)}{S_{kk}(f)} \tag{11.26}$$

This can be extended as

$$S_{ij\cdot k,l}(f) = S_{ij\cdot k}(f) - \frac{S_{il\cdot k}(f)S_{lj\cdot k}(f)}{S_{ll\cdot k}(f)} \tag{11.27}$$

We use the above expressions to 'condense' successive cross-spectral density matrices, e.g.

$$\begin{bmatrix} S_{11}(f) & S_{12}(f) & S_{13}(f) \\ S_{21}(f) & S_{22}(f) & S_{23}(f) \\ S_{31}(f) & S_{32}(f) & S_{33}(f) \end{bmatrix} \Rightarrow \begin{bmatrix} S_{22\cdot1}(f) & S_{23\cdot1}(f) \\ S_{32\cdot1}(f) & S_{33\cdot1}(f) \end{bmatrix} \Rightarrow \begin{bmatrix} S_{33\cdot1,2}(f) \end{bmatrix} \tag{11.28}$$

This can be extended to larger systems. This 'condensation' can be interpreted through Gaussian elimination (row manipulations) as follows (where $r_i$ is the $i$th row):

Step 1: $r_2 \rightarrow r_2 - r_1 \times \left(\dfrac{S_{21}(f)}{S_{11}(f)}\right); \quad r_3 \rightarrow r_3 - r_1 \times \left(\dfrac{S_{31}(f)}{S_{11}(f)}\right)$ gives

$$\begin{bmatrix} S_{11}(f) & S_{12}(f) & S_{13}(f) \\ S_{21}(f) & S_{22}(f) & S_{23}(f) \\ S_{31}(f) & S_{32}(f) & S_{33}(f) \end{bmatrix} \Rightarrow \begin{bmatrix} S_{11}(f) & S_{12}(f) & S_{13}(f) \\ 0 & S_{22\cdot1}(f) & S_{23\cdot1}(f) \\ 0 & S_{32\cdot1}(f) & S_{33\cdot1}(f) \end{bmatrix} \tag{11.29}$$

Step 2: $r_3 \rightarrow r_3 - r_2 \times \left(\dfrac{S_{32\cdot1}(f)}{S_{22\cdot1}(f)}\right)$ gives

$$\begin{bmatrix} S_{11}(f) & S_{12}(f) & S_{13}(f) \\ 0 & S_{22\cdot1}(f) & S_{23\cdot1}(f) \\ 0 & S_{32\cdot1}(f) & S_{33\cdot1}(f) \end{bmatrix} \Rightarrow \begin{bmatrix} S_{11}(f) & S_{12}(f) & S_{13}(f) \\ 0 & S_{22\cdot1}(f) & S_{23\cdot1}(f) \\ 0 & 0 & S_{33\cdot1,2}(f) \end{bmatrix} \tag{11.30}$$

i.e. the residual spectral density functions arise naturally.

Further interpretation can be obtained by starting from Figure 11.2 again. Since $X_3(f) = H_1(f)X_1(f) + H_2(f)X_2(f) + N(f)$, the cross-spectral density functions are written as

$$S_{13}(f) = H_1(f)S_{11}(f) + H_2(f)S_{12}(f)$$
$$S_{23}(f) = H_1(f)S_{21}(f) + H_2(f)S_{22}(f) \tag{11.31}$$

Solving for $H_1(f)$ and $H_2(f)$ using Gaussian elimination (eliminate the term $H_1(f)S_{21}(f)$ in the second Equation of (11.31)) gives

$$S_{11}(f)H_1(f) + S_{12}(f)H_2(f) = S_{13}(f)$$
$$\left(S_{22}(f) - \frac{S_{21}(f)S_{12}(f)}{S_{11}(f)}\right)H_2(f) = S_{23}(f) - \frac{S_{21}(f)S_{13}(f)}{S_{11}(f)} \tag{11.32}$$

Thus,

$$H_2(f) = \frac{S_{23\cdot1}(f)}{S_{22\cdot1}(f)} \quad \text{and} \quad H_1(f) = \frac{S_{13}(f)}{S_{11}(f)} - \frac{S_{12}(f)}{S_{11}(f)}\frac{S_{23\cdot1}(f)}{S_{22\cdot1}(f)}$$

## 11.3 PRINCIPAL COMPONENT ANALYSIS

Although residual spectral analysis is useful in source identification, condition monitoring, etc., the shortcoming of the method is that prior ranking of the input signals is often required (see Bendat and Piersol, 1980), i.e. a priori knowledge. Principal component analysis (PCA) is a general approach to explore correlation patterns (Otte et al., 1988).

Suppose we have three processes $x_1$, $x_2$, $x_3$. Then we start as before by forming the cross-spectral density matrix

$$\mathbf{S} = \begin{bmatrix} S_{11}(f) & S_{12}(f) & S_{13}(f) \\ S_{21}(f) & S_{22}(f) & S_{23}(f) \\ S_{31}(f) & S_{32}(f) & S_{33}(f) \end{bmatrix} \tag{11.33}$$

Note that this is a Hermitian matrix, i.e. $\mathbf{S} = \mathbf{S}^{*T} = \mathbf{S}^H$, where $\mathbf{S}^H$ is the conjugate transpose. If there is a linear relationship between the processes $x_i$, then the determinant of this matrix is zero (i.e. its rank is less than three). If there is no linear relationship then its rank is three.

Suppose the matrix is full rank (i.e. rank 3). Then eigenvalue (or singular value) decomposition gives

$$\mathbf{S} = \mathbf{U}\mathbf{\Lambda}\mathbf{U}^H \tag{11.34}$$

where $\mathbf{\Lambda}$ is a diagonal matrix that contains eigenvalues of $\mathbf{S}$, and $\mathbf{U}$ is a unitary matrix whose columns are the corresponding eigenvectors. We may describe the physical interpretation of this as follows. Suppose there exist three (fictitious) processes $z_1$, $z_2$, $z_3$ that are *mutually uncorrelated* and from which $x_1$, $x_2$, $x_3$ can be derived, i.e. for each frequency $f$,

$$\begin{bmatrix} X_1(f) \\ X_2(f) \\ X_3(f) \end{bmatrix} = \begin{bmatrix} m_{11}(f) & m_{12}(f) & m_{13}(f) \\ m_{21}(f) & m_{22}(f) & m_{23}(f) \\ m_{31}(f) & m_{32}(f) & m_{33}(f) \end{bmatrix} \begin{bmatrix} Z_1(f) \\ Z_2(f) \\ Z_3(f) \end{bmatrix} \tag{11.35}$$

i.e. $\mathbf{X}(f) = \mathbf{M}(f)\mathbf{Z}(f)$. Conceptually, this can be depicted as in Figure 11.6.

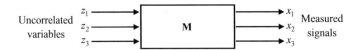

**Figure 11.6**  Virtual signals and measured signals

Then, forming the spectral density matrix gives

$$\boxed{S_{\mathbf{XX}}(f) = \mathbf{S} = \mathbf{M}^*(f)S_{\mathbf{ZZ}}(f)\mathbf{M}^T(f)}$$

(11.36)

Since the $z_i$ are mutually uncorrelated, $S_{\mathbf{ZZ}}(f)$ is a diagonal matrix. Thus, Equation (11.36) has the same form as Equation (11.34), i.e. $S_{\mathbf{ZZ}}(f) = \mathbf{\Lambda}$ and $\mathbf{M}^*(f) = \mathbf{U}$. So, the eigenvalues of $\mathbf{S}$ are the *power spectra* of these fictitious processes and their (relative) *magnitudes* serve to define the *principal components* referred to, i.e. $z_i$ are the principal components.

Note however, that, it is important not to think of these as physical entities, e.g. it is quite possible that more than three actual independent processes combine to make up $x_1$, $x_2$ and $x_3$. The fictitious processes $z_1$, $z_2$, $z_3$ are merely a convenient concept. These signals are called *virtual signals*. Note also that the power of these virtual signals is of course not the power of the measured signals. It is therefore interesting to establish to what degree each principal component contributes to the power of the measured signals. To see this, for example, consider $X_1(f)$ which can be written as (from Equation (11.35))

$$X_1(f) = m_{11}(f)Z_1(f) + m_{12}(f)Z_2(f) + m_{13}(f)Z_3(f)$$

(11.37)

Then, since the $z_i$ are uncorrelated the power spectral density function $S_{x_1 x_1}(f)$ can be written as

$$S_{x_1 x_1}(f) = |m_{11}(f)|^2 S_{z_1 z_1}(f) + |m_{12}(f)|^2 S_{z_2 z_2}(f) + |m_{13}(f)|^2 S_{z_3 z_3}(f)$$

(11.38)

and the power due to $z_1$ is $\gamma^2_{z_1 x_1}(f)S_{x_1 x_1}(f)$, where

$$\gamma^2_{z_1 x_1}(f) = \frac{\left|S_{z_1 x_1}(f)\right|^2}{S_{z_1 z_1}(f)S_{x_1 x_1}(f)}$$

(11.39)

This is a *virtual coherence function*. More generally, the virtual coherence function between the $i$th virtual input $z_i$ and the $j$th measured signal $x_j$ can be written as

$$\gamma^2_{z_i x_j}(f) = \frac{\left|S_{z_i x_j}(f)\right|^2}{S_{z_i z_i}(f)S_{x_j x_j}(f)}$$

(11.40)

Since the cross-spectral density function between $z_i$ and $x_j$ can be obtained by

$$\boxed{S_{z_i x_j}(f) = m_{ji}(f)S_{z_i z_i}(f)}$$

(11.41)

we see that the virtual coherence function: (i) can be computed from the eigenvalues and eigenvectors of $\mathbf{S}$; (ii) gives a measure of what proportion of $S_{x_j x_j}(f)$ comes from a particular component of $z_i$.

For the details of practical applications of principal component analysis, especially for noise source identification problems, see Otte *et al.* (1988).

## Relationship to the System Identification Methods[1]

It is interesting to relate principal component analysis (PCA) to the system identification methods we described in Chapter 9. Let $\mathbf{x}$ denote a column vector of observations (e.g. $x$ (input) and $y$ (output) in Figure 9.9 in Chapter 9) with correlation matrix

$$R_{xx} = E\left[\mathbf{x}\mathbf{x}^T\right] \tag{11.42}$$

Let $\mathbf{x}$ be derived from a set of uncorrelated processes $\mathbf{z}$ (through the transformation matrix $\mathbf{T}$) by

$$\mathbf{x} = \mathbf{T}\mathbf{z} \tag{11.43}$$

Then, the correlation matrix is

$$R_{xx} = \mathbf{T}E\left[\mathbf{z}\mathbf{z}^T\right]\mathbf{T}^T = \mathbf{T}R_{zz}\mathbf{T}^T \tag{11.44}$$

Since the elements of $\mathbf{z}$ are uncorrelated, $R_{zz} = \mathbf{\Lambda}$, where $\mathbf{\Lambda}$ is a diagonal matrix that contains eigenvalues of $R_{xx}$. So,

$$R_{xx} = \mathbf{T}\mathbf{\Lambda}\mathbf{T}^T \tag{11.45}$$

This is an eigendecomposition of the correlation matrix $R_{xx}$, and $\mathbf{T}$ is an orthogonal matrix whose columns are the corresponding eigenvectors, i.e.

$$\mathbf{T} = \begin{bmatrix} \mathbf{t}_1 & \mathbf{t}_2 \end{bmatrix} = \begin{bmatrix} t_{11} & t_{12} \\ t_{21} & t_{22} \end{bmatrix} \tag{11.46}$$

Let us apply this to the pair of variables ($x$ (input) and $y$ (output)) as in Figure 9.9 in Chapter 9. Assuming zero mean values, the correlation matrix is

$$R_{xx} = E\left[\begin{bmatrix} x \\ y \end{bmatrix} \begin{bmatrix} x & y \end{bmatrix}\right] = \begin{bmatrix} E[xx] & E[xy] \\ E[xy] & E[yy] \end{bmatrix} = \begin{bmatrix} \sigma_x^2 & \sigma_{xy} \\ \sigma_{xy} & \sigma_y^2 \end{bmatrix} \tag{11.47}$$

The eigenvalues are

$$\det(R_{xx} - \lambda \mathbf{I}) = \begin{vmatrix} \sigma_x^2 - \lambda & \sigma_{xy} \\ \sigma_{xy} & \sigma_y^2 - \lambda \end{vmatrix} = 0 \tag{11.48}$$

i.e.

$$\lambda_{1,2} = \frac{\sigma_x^2 + \sigma_y^2 \pm \sqrt{\left(\sigma_x^2 - \sigma_y^2\right)^2 + 4\sigma_{xy}^2}}{2} \tag{11.49}$$

---

[1] See Tan (2005).

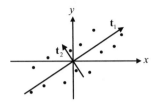

**Figure 11.7** Eigenvectors $\mathbf{t}_1$ and $\mathbf{t}_2$

The eigenvectors $\mathbf{t}_1$ and $\mathbf{t}_2$ corresponding to these eigenvalues are orthogonal and define a basis set of the data, as shown in Figure 11.7.

From the figure, the slope of the eigenvector corresponding to the largest eigenvalue is the PCA 'gain' relating $y$ to $x$. Using the eigenvectors $\mathbf{T} = [\,\mathbf{t}_1 \quad \mathbf{t}_2\,]$, Equation (11.43) can be expanded as

$$x = t_{11}z_1 + t_{12}z_2$$

$$y = t_{21}z_1 + t_{22}z_2 \tag{11.50}$$

We note that the first principal component $z_1$ is related to the largest eigenvalue, and the part due to $z_1$ is $t_{11}z_1$ for input $x$, and $t_{21}z_1$ for output $y$. The gain relating $y$ to $x$ (corresponding to the first principal component $z_1$) is then given by the ratio $t_{21}/t_{11}$. The ratio can be found from

$$(R_{\mathbf{xx}} - \lambda_1 \mathbf{I})\,\mathbf{t}_1 = 0 \tag{11.51}$$

and so

$$\frac{t_{21}}{t_{11}} = \frac{\sigma_y^2 - \sigma_x^2 + \sqrt{\left(\sigma_x^2 - \sigma_y^2\right)^2 + 4\sigma_{xy}^2}}{2\sigma_{xy}} \tag{11.52}$$

We see from this that Equation (11.52) is the total least squares gain ($a_T$, see Equation (9.59)). This equivalence follows from the fact that both the PCA approach and TLS minimize the power 'normal' to the 'principal' eigenvector.

# Appendix A

**Proof of** $\displaystyle \int_{-\infty}^{\infty} 2M\frac{\sin 2\pi a M}{2\pi a M}da = 1$

We first consider the contour integration of a function $F(z) = e^{jz}f(z) = e^{jz}/z$ around a closed contour in the $z$-plane as shown in Figure A.1, where $z = x + jy$.

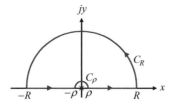

**Figure A.1**   A contour with a single pole at $z = 0$

Using Cauchy's residue theorem, the contour integral becomes

$$\oint \frac{e^{jz}}{z}dz = \int_{C_R} \frac{e^{jz}}{z}dz + \int_{-R}^{-\rho} \frac{e^{jx}}{x}dx + \int_{C_\rho} \frac{e^{jz}}{z}dz + \int_{\rho}^{R} \frac{e^{jx}}{x}dx = 0 \qquad \text{(A.1)}$$

From Jordan's lemma, the first integral on the right of the first equality is zero if $R \to \infty$, i.e. $\lim_{R\to\infty} \int_{C_R} e^{jz}f(z)dz = 0$. Letting $z = \rho e^{j\theta}$ and $dz = j\rho e^{j\theta}d\theta$, where $\theta$ varies from $\pi$ to $0$, the third integral can be written as

$$\int_{C_\rho} \frac{e^{jz}}{z}dz = j\int_{\pi}^{0} e^{j(\rho e^{j\theta})}d\theta \qquad \text{(A.2)}$$

*Fundamentals of Signal Processing for Sound and Vibration Engineers*
K. Shin and J. K. Hammond.   © 2008 John Wiley & Sons, Ltd

Taking the limit as $\rho \rightarrow 0$, this becomes

$$\lim_{\rho \rightarrow 0} \left\{ j \int_{\pi}^{0} e^{j(\rho e^{j\theta})} d\theta \right\} = j\theta|_{\pi}^{0} = -j\pi \tag{A.3}$$

Now, consider the second and fourth integral together:

$$\int_{-R}^{-\rho} \frac{e^{jx}}{x} dx + \int_{\rho}^{R} \frac{e^{jx}}{x} dx = \int_{-R}^{-\rho} \frac{\cos x + j \sin x}{x} dx + \int_{\rho}^{R} \frac{\cos x + j \sin x}{x} dx \tag{A.4}$$

Since $\cos(x)/x$ is odd, the cosine terms cancel in the resulting integration. Thus, Equation (A.4) becomes

$$\int_{-R}^{-\rho} \frac{e^{jx}}{x} dx + \int_{\rho}^{R} \frac{e^{jx}}{x} dx = 2j \int_{\rho}^{R} \frac{\sin x}{x} dx \tag{A.5}$$

Combining the above results, for $R \rightarrow \infty$ and $\rho \rightarrow 0$, Equation (A.1) reduces to

$$\lim_{\substack{\rho \rightarrow 0 \\ R \rightarrow \infty}} \left\{ 2j \int_{\rho}^{R} \frac{\sin x}{x} dx \right\} = 2j \int_{0}^{\infty} \frac{\sin x}{x} dx = j\pi \tag{A.6}$$

Thus, we have the following result:

$$\int_{0}^{\infty} \frac{\sin x}{x} dx = \frac{\pi}{2} \tag{A.7}$$

We now go back to our problem. We have written

$$\lim_{M \rightarrow \infty} 2M \frac{\sin 2\pi aM}{2\pi aM} = \delta(a)$$

in Chapter 3. In order to justify this, the integral of the function

$$f(a) = 2M \frac{\sin 2\pi aM}{2\pi aM}$$

must be unity. We verify this using the above result. Letting $x = 2\pi aM$ and $dx = 2\pi Mda$, we have

$$\int_{-\infty}^{\infty} f(a)da = \int_{-\infty}^{\infty} 2M \frac{\sin 2\pi aM}{2\pi aM} da = \int_{-\infty}^{\infty} 2M \frac{\sin x}{x} \frac{dx}{2\pi M} = \frac{1}{\pi} \int_{-\infty}^{\infty} \frac{\sin x}{x} dx \tag{A.8}$$

From Equation (A.7),

$$\int_{-\infty}^{\infty} \frac{\sin x}{x} dx = 2 \int_{0}^{\infty} \frac{\sin x}{x} dx = \pi$$

thus Equation (A.8) becomes

$$\int_{-\infty}^{\infty} 2M \frac{\sin 2\pi aM}{2\pi aM} da = 1 \tag{A.9}$$

This proves that the integral of the function in Figure A.2 (i.e. Figure 3.11) is unity.

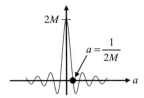

**Figure A.2**   Representation of the delta function using a sinc function

# Appendix B
## Proof of $|S_{xy}(f)|^2 \leq S_{xx}(f)S_{yy}(f)$

Suppose we have $Z_T(f)$ consisting of two quantities $X_T(f)$ and $Y_T(f)$ such that

$$Z_T(f) = u_1 X_T(f) + u_2 Y_T(f) \tag{B.1}$$

where $u_i$ are arbitrary complex constants. Then the power spectral density function $S_{zz}(f)$ can be written as

$$S_{zz}(f) = \lim_{T \to \infty} \frac{E[Z_T^*(f)Z_T(f)]}{T} = u_1^* S_{xx}(f)u_1 + u_2^* S_{yx}(f)u_1 + u_1^* S_{xy}(f)u_2 + u_2^* S_{yy}(f)u_2$$

$$= \begin{bmatrix} u_1^* & u_2^* \end{bmatrix} \begin{bmatrix} S_{xx}(f) & S_{xy}(f) \\ S_{yx}(f) & S_{yy}(f) \end{bmatrix} \begin{bmatrix} u_1 \\ u_2 \end{bmatrix}$$

$$= \mathbf{u}^H \mathbf{S} \mathbf{u} \tag{B.2}$$

where $\mathbf{S}$ is the cross-spectral density matrix.

$\mathbf{S}$ is a Hermitian matrix. Moreover, this cross-spectral density matrix is *positive semi-definite*, i.e. for all non-zero (complex) vectors $\mathbf{u}$, $\mathbf{u}^H \mathbf{S} \mathbf{u} \geq 0$ since the power spectral density function $S_{zz}(f)$ is non-negative for all frequencies $f$.

Since the matrix $\mathbf{S}$ is positive semi-definite, its determinant must be non-negative, i.e.

$$\begin{vmatrix} S_{xx}(f) & S_{xy}(f) \\ S_{yx}(f) & S_{yy}(f) \end{vmatrix} \geq 0 \tag{B.3}$$

or $S_{xx}(f)S_{yy}(f) - S_{xy}(f)S_{yx}(f) \geq 0$, i.e.

$$S_{xy}(f)S_{yx}(f) \leq S_{xx}(f)S_{yy}(f) \tag{B.4}$$

Since $S_{yx}(f) = S_{xy}^*(f)$, it follows that

$$|S_{xy}(f)|^2 \leq S_{xx}(f)S_{yy}(f) \tag{B.5}$$

Note that it can easily be verified that a multi-dimensional cross-spectral density matrix $\mathbf{S}$ (say, $n \times n$) is also a *positive semi-definite Hermitian* matrix.

---

*Fundamentals of Signal Processing for Sound and Vibration Engineers*
K. Shin and J. K. Hammond.    © 2008 John Wiley & Sons, Ltd

# Appendix C

## Wave Number Spectra and An Application

Rather than a 'time series' we may consider a function of 'space'. This might be demonstrated with height indications on a rough road, for example, as shown in Figure C.1.

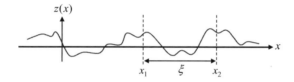

**Figure C.1**   Height profile of a rough road

If the process is 'spatially stationary' (homogeneous) we may characterize it by the autocorrelation function

$$R_{zz}(x_2 - x_1) = E[z(x_1)z(x_2)] \tag{C.1}$$

or $R_{zz}(\xi) = E[z(x_1)z(x_1 + \xi)]$, where $\xi = x_2 - x_1$ which is the spatial separation of the two points.

Now we shall consider a spectral analysis of the process. If the independent variable is time then we speak of $\omega$ (rad/s). Here we shall use $k$ (rad/m), and this is called the *wave number*. Note that $\omega = 2\pi/T$ shows how fast it oscillates (in radians) in a second, while $k = 2\pi/\lambda$ represents how many cycles (in radians) of the wave are in a metre, where $\lambda$ is the wavelength. Then, the wave number spectrum is defined as

$$S_{zz}(k) = \int_{-\infty}^{\infty} R_{zz}(\xi)e^{-jk\xi}\,d\xi \tag{C.2}$$

*Fundamentals of Signal Processing for Sound and Vibration Engineers*
K. Shin and J. K. Hammond.    © 2008 John Wiley & Sons, Ltd

Compare this expression with the usual spectral density function, $S_{xx}(\omega) = \int_{-\infty}^{\infty} R_{xx}(\tau)$ $e^{-j\omega\tau}d\tau$. We note that period $T = 2\pi/\omega$ is now replaced by wavelength $\lambda = 2\pi/k$. (See Newland (1984) for more details.)

### *Application*

Consider a vehicle moving over rough ground at speed $V$ as shown in Figure C.2. The equation of motion for this simple model is $m\ddot{y}(t) = -k[y(t) - z(t)] - c[\dot{y}(t) - \dot{z}(t)]$, so that

$$m\ddot{y}(t) + c\dot{y}(t) + ky(t) = c\dot{z}(t) + kz(t) \tag{C.3}$$

The problem we are concerned with is: given a specific road property $R_{zz}(\xi)$, calculate the value of the *variance* of $y(t)$ as the vehicle moves over the ground at constant speed $V$.

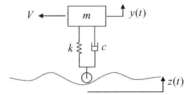

**Figure C.2**   A vehicle moving over rough ground at speed $V$

If we treat $z(t)$ as a stationary random variable, the variance of $y(t)$ can be written as (we assume $y(t)$ has a zero mean value)

$$E[y^2(t)] = \frac{1}{2\pi}\int_{-\infty}^{\infty} S_{yy}(\omega)d\omega = \frac{1}{2\pi}\int_{-\infty}^{\infty} |H(\omega)|^2 S_{zz}(\omega)d\omega \tag{C.4}$$

where the system frequency response function is

$$H(\omega) = \frac{k + jc\omega}{k - m\omega^2 + jc\omega}$$

Now, it remains to obtain $S_{zz}(\omega)$ while we only have $S_{zz}(k)$ at present, i.e. we must interpret a wave number spectrum as a frequency spectrum. We do this as follows. First, we convert a temporal autocorrelation to a spatial one as

$$R_{zz}(\tau) = E[z(t)z(t+\tau)] = E[z(x(t))z(x(t+\tau))] = E[z(x)z(x+V\tau)] = R_{zz}(V\tau) \tag{C.5}$$

and so

$$S_{zz}(\omega) = \int_{-\infty}^{\infty} R_{zz}(\tau)e^{-j\omega\tau}d\tau = \int_{-\infty}^{\infty} R_{zz}(V\tau)e^{-j\omega\tau}d\tau \tag{C.6}$$

Letting $V\tau = \xi$, this can be rewritten as

$$S_{zz}(\omega) = \frac{1}{V}\int_{-\infty}^{\infty} R_{zz}(\xi)e^{-j(\omega/V)\xi}d\xi = \frac{1}{V} S_{zz}(k)|_{k=\omega/V} \tag{C.7}$$

Thus, to obtain the frequency spectrum from the wave number spectrum we simply replace $k$ by $\omega/V$ and divide by $V$. Note that the speed can be expressed by

$$V = f\lambda = \frac{\omega}{2\pi}\lambda = \frac{\omega}{k}$$

For a simple example, if the road property is $R_{zz}(\xi) = e^{-0.2|\xi|}$ then the wave number spectrum and the power spectral density function are

$$S_{zz}(k) = \frac{0.4}{0.04 + k^2} \quad \text{and} \quad S_{zz}(\omega) = \frac{0.4V}{0.04V^2 + \omega^2}$$

respectively.

# Appendix D

## Some Comments on the Ordinary Coherence Function $\gamma_{xy}^2(f)$

### *The Use of the Ordinary Coherence Function*

If we wish to estimate the transfer function linking two signals from $S_{xy}(f) = H(f)S_{xx}(f)$, i.e. by forming the ratio $H(f) = S_{xy}(f)/S_{xx}(f)$, then we may compute the coherence function which is a direct measure of the 'validity' of this relationship. That is, if $\gamma_{xy}^2(f) \approx 1$ the transfer function $H(f)$ is well estimated; if $\gamma_{xy}^2(f)$ is low, the estimate of $H(f)$ is not trustworthy.

Also, this concept can be applied to multiple-source problems. For example, let $x(t)$ and $y(t)$ be the input and the output, and suppose there is another input $z(t)$ which we have not accounted for and it contributes to $y(t)$ as shown in Figure D.1.

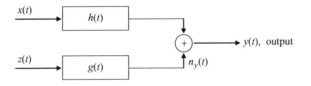

**Figure D.1** Multiple-source problems

If $z(t)$ is uncorrelated with $x(t)$, then its effect on the coherence function between $x(t)$ and $y(t)$ is the same as the measurement noise $n_y(t)$ as in Case (a), Section 9.2.

### *The Use of the Concept of Coherent Output Power*

Consider the experiment depicted in Figure D.2. A measurement $y_m(t)$ is made of sound from a plate being shaken with the addition of background noise $n(t)$ from a speaker, i.e.

$$y_m(t) = y(t) + n(t) \tag{D.1}$$

where $y(t)$ is the sound due to the plate and $n(t)$ is the noise due to the speaker.

*Fundamentals of Signal Processing for Sound and Vibration Engineers*
K. Shin and J. K. Hammond.    © 2008 John Wiley & Sons, Ltd

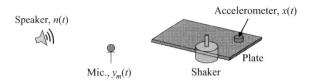

**Figure D.2**   Measurements of acoustic pressure resulting from a vibrating plate

From the coherence measurement $\gamma^2_{xy_m}(f)$ and the power spectral density $S_{y_m y_m}(f)$ we can calculate the power at the microphone due to the plate by

$$S_{yy}(f) = \gamma^2_{xy_m}(f)S_{y_m y_m}(f) \tag{D.2}$$

This is only satisfactory if $x(t)$ is a 'good' measurement of the basic source, i.e. the vibration signal $x(t)$ must be closely related to the radiated sound $y(t)$. An example where this might not be so is as shown in Figure D.3.

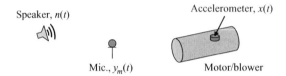

**Figure D.3**   Measurements of acoustic pressure due to a motor/blower

Now $x(t)$ will not be a good measurement of the primary noise source in general, i.e. the accelerometer will not measure the aerodynamic noise.

# Appendix E

## Least Squares Optimization: Complex-Valued Problem

Consider the least squares problem (Case 1 in Section 9.3) which finds the optimal parameter $a_1$ that fits the data such that $y = a_1 x$, where the objective function is given by

$$J_1 = \frac{1}{N} \sum_{i=1}^{N} (y_i - a_1 x_i)^2$$

If $x$ and $y$ are complex valued, then we may find an optimal complex parameter $a_1$, where the objective function is

$$J_1 = \frac{1}{N} \sum_{i=1}^{N} |y_i - a_1 x_i|^2 = \frac{1}{N} \sum_{i=1}^{N} (y_i^* - a_1^* x_i^*)(y_i - a_1 x_i) \qquad (\text{E.1})$$

Let $x_i = x_{i,\mathrm{R}} + j x_{i,\mathrm{I}}$, $y_i = y_{i,\mathrm{R}} + j y_{i,\mathrm{I}}$ and $a_1 = a_\mathrm{R} + j a_\mathrm{I}$. Then Equation (E.1) can be written as

$$J_1 = \frac{1}{N} \sum_{i=1}^{N} \left[ (y_{i,\mathrm{R}}^2 + y_{i,\mathrm{I}}^2) - 2a_\mathrm{R}(x_{i,\mathrm{R}} y_{i,\mathrm{R}} + x_{i,\mathrm{I}} y_{i,\mathrm{I}}) \right.$$
$$\left. + 2a_\mathrm{I}(x_{i,\mathrm{I}} y_{i,\mathrm{R}} - x_{i,\mathrm{R}} y_{i,\mathrm{I}}) + (a_\mathrm{R}^2 + a_\mathrm{I}^2)(x_{i,\mathrm{R}}^2 + x_{i,\mathrm{I}}^2) \right] \qquad (\text{E.2})$$

This is a real quantity. To minimize $J_1$ with respect to both $a_\mathrm{R}$ and $a_\mathrm{I}$, we solve the following equations:

$$\frac{\partial J_1}{\partial a_\mathrm{R}} = \frac{1}{N} \sum_{i=1}^{N} \left[ -2(x_{i,\mathrm{R}} y_{i,\mathrm{R}} + x_{i,\mathrm{I}} y_{i,\mathrm{I}}) + 2a_\mathrm{R}(x_{i,\mathrm{R}}^2 + x_{i,\mathrm{I}}^2) \right] = 0$$

$$\frac{\partial J_1}{\partial a_\mathrm{I}} = \frac{1}{N} \sum_{i=1}^{N} \left[ 2(x_{i,\mathrm{I}} y_{i,\mathrm{R}} - x_{i,\mathrm{R}} y_{i,\mathrm{I}}) + 2a_\mathrm{I}(x_{i,\mathrm{R}}^2 + x_{i,\mathrm{I}}^2) \right] = 0$$

$$(\text{E.3})$$

*Fundamentals of Signal Processing for Sound and Vibration Engineers*
K. Shin and J. K. Hammond.    © 2008 John Wiley & Sons, Ltd

Solving these equation gives

$$a_R = \frac{\sum\limits_{i=1}^{N} (x_{i,R} y_{i,R} + x_{i,I} y_{i,I})}{\sum\limits_{i=1}^{N} (x_{i,R}^2 + x_{i,I}^2)} \quad \text{and} \quad a_I = \frac{\sum\limits_{i=1}^{N} (x_{i,R} y_{i,I} - x_{i,I} y_{i,R})}{\sum\limits_{i=1}^{N} (x_{i,R}^2 + x_{i,I}^2)}$$

Thus the optimal complex parameter $a_1$ can be written as

$$a_1 = a_R + j a_I = \frac{\sum\limits_{i=1}^{N} \left[ (x_{i,R} y_{i,R} + x_{i,I} y_{i,I}) + j(x_{i,R} y_{i,I} - x_{i,I} y_{i,R}) \right]}{\sum\limits_{i=1}^{N} (x_{i,R}^2 + x_{i,I}^2)} = \frac{\sum\limits_{i=1}^{N} x_i^* y_i}{\sum\limits_{i=1}^{N} |x_i|^2} \tag{E.4}$$

Similarly, the complex form of $a_2$ and $a_T$ (Case 2 and Case 3 in Section 9.3) can be found as

$$a_2 = \frac{\sum\limits_{i=1}^{N} |y_i|^2}{\sum\limits_{i=1}^{N} y_i^* x_i} \tag{E.5}$$

$$a_T = \frac{\left( \sum\limits_{i=1}^{N} |y_i|^2 - \sum\limits_{i=1}^{N} |x_i|^2 \right) + \sqrt{\left( \sum\limits_{i=1}^{N} |x_i|^2 - \sum\limits_{i=1}^{N} |y_i|^2 \right)^2 + 4 \left| \sum\limits_{i=1}^{N} x_i^* y_i \right|^2}}{2 \sum\limits_{i=1}^{N} y_i^* x_i} \tag{E.6}$$

Note the location of conjugates in the above equations, and compare with the frequency response function estimators $H_1(f)$, $H_2(f)$ and $H_T(f)$ given in Section 9.3.

# Appendix F
## Proof of $H_W(f) \rightarrow H_1(f)$ as $\kappa(f) \rightarrow \infty$

We start from Equation (9.67), i.e.

$$H_W(f) = \frac{\tilde{S}_{y_m y_m}(f) - \kappa(f)\tilde{S}_{x_m x_m}(f) + \sqrt{\left[\tilde{S}_{x_m x_m}(f)\kappa(f) - \tilde{S}_{y_m y_m}(f)\right]^2 + 4\left|\tilde{S}_{x_m y_m}(f)\right|^2 \kappa(f)}}{2\tilde{S}_{y_m x_m}(f)} \qquad \text{(F.1)}$$

Let $\kappa(f) = 1/\varepsilon$; then the right hand side of the equation can be written as

$$\frac{f(\varepsilon)}{g(\varepsilon)} = \frac{\tilde{S}_{y_m y_m}(f)\varepsilon - \tilde{S}_{x_m x_m}(f) + \sqrt{\tilde{S}^2_{x_m x_m}(f) - 2\tilde{S}_{x_m x_m}(f)\tilde{S}_{y_m y_m}(f)\varepsilon + \tilde{S}^2_{y_m y_m}(f)\varepsilon^2 + 4\left|\tilde{S}_{x_m y_m}(f)\right|^2 \varepsilon}}{2\tilde{S}_{y_m x_m}(f)\varepsilon}$$

$$\qquad \text{(F.2)}$$

Now, taking the limit $\varepsilon \rightarrow 0$ (instead of $\kappa \rightarrow \infty$) and applying L'Hôpital's rule, i.e.

$$\lim_{\varepsilon \to 0} \frac{f(\varepsilon)}{g(\varepsilon)} = \lim_{\varepsilon \to 0} \frac{f'(\varepsilon)}{g'(\varepsilon)}$$

we obtain

$$\lim_{\varepsilon \to 0} \frac{f'(\varepsilon)}{g'(\varepsilon)} = \frac{\tilde{S}_{y_m y_m}(f) + \frac{1}{2}\left(\tilde{S}^2_{x_m x_m}(f)\right)^{-1/2}\left(-2\tilde{S}_{x_m x_m}(f)\tilde{S}_{y_m y_m}(f) + 4\left|\tilde{S}_{x_m y_m}(f)\right|^2\right)}{2\tilde{S}_{y_m x_m}(f)}$$

$$= \frac{\tilde{S}_{y_m y_m}(f) - \tilde{S}_{y_m y_m}(f) + 2\left(\tilde{S}_{x_m x_m}(f)\right)^{-1}\left|\tilde{S}_{x_m y_m}(f)\right|^2}{2\tilde{S}_{y_m x_m}(f)} = \frac{\left(\tilde{S}_{x_m x_m}(f)\right)^{-1}\left|\tilde{S}_{x_m y_m}(f)\right|^2}{\tilde{S}_{y_m x_m}(f)}$$

$$= \frac{\tilde{S}^*_{x_m y_m}(f)\tilde{S}_{x_m y_m}(f)}{\tilde{S}_{x_m x_m}(f)\tilde{S}_{y_m x_m}(f)} = \frac{\tilde{S}_{y_m x_m}(f)\tilde{S}_{x_m y_m}(f)}{\tilde{S}_{x_m x_m}(f)\tilde{S}_{y_m x_m}(f)}$$

$$= \frac{\tilde{S}_{x_m y_m}(f)}{\tilde{S}_{x_m x_m}(f)} = H_1(f) \qquad \text{(F.3)}$$

This proves the result.

*Fundamentals of Signal Processing for Sound and Vibration Engineers*
K. Shin and J. K. Hammond.    © 2008 John Wiley & Sons, Ltd

# Appendix G

## Justification of the Joint Gaussianity of $X(f)$

If two random variables $X$ and $Y$ are jointly Gaussian, then the individual distribution remains Gaussian under the coordinate rotation. This may be seen from Figure G.1. For example, if $X'$ and $Y'$ are obtained by

$$\begin{bmatrix} X' \\ Y' \end{bmatrix} = \begin{bmatrix} \cos\phi & -\sin\phi \\ \sin\phi & \cos\phi \end{bmatrix} \begin{bmatrix} X \\ Y \end{bmatrix} \tag{G.1}$$

then they are still normally distributed. For a complex variable, e.g. $Z = X + jY$, the equivalent rotation is $e^{j\phi}Z$. If two random variables are Gaussian (individually) but not jointly Gaussian, then this property does not hold. An example of this is illustrated in Figure G.2.

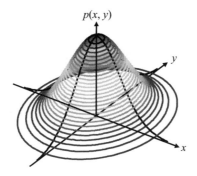

$p(x, y)$

**Figure G.1**  Two random variables are jointly normally distributed

Now, consider a Gaussian process $x(t)$. The Fourier transform of $x(t)$ can be written as

$$X(f) = X_c(f) + jX_s(f) \tag{G.2}$$

*Fundamentals of Signal Processing for Sound and Vibration Engineers*
K. Shin and J. K. Hammond.    © 2008 John Wiley & Sons, Ltd

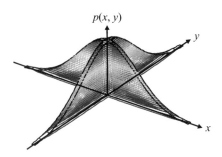

**Figure G.2**   Each random variable is normally distributed, but not jointly

where $X_c(f)$ and $X_s(f)$ are Gaussian since $x(t)$ is Gaussian. If these are *jointly* Gaussian, they must remain Gaussian under the coordinate rotation (for any rotation angle $\phi$). For example, if $e^{j\phi}X(f) = X'(f) = X'_c(f) + jX'_s(f)$ then $X'_c(f)$ and $X'_s(f)$ must be Gaussian, where $X'_c(f) = X_c(f)\cos\phi - X_s(f)\sin\phi$ and $X'_s(f) = X_c(f)\sin\phi + X_s(f)\cos\phi$.

For a particular frequency $f$, let $\phi = -2\pi f t_0$. Then $e^{-j2\pi f t_0}X(f)$ is a pure delay, i.e. $x(t - t_0)$ in the time domain for that frequency component. If we assume that $x(t)$ is a stationary Gaussian process, then $x(t - t_0)$ is also Gaussian, so both $X'_c(f)$ and $X'_s(f)$ remain Gaussian under the coordinate rotation. This justifies that $X_c(f)$ and $X_s(f)$ are *jointly normally distributed*.

# Appendix H

## Some Comments on Digital Filtering

We shall briefly introduce some terminology and methods of digital filtering that may be useful. There are many good texts on this subject: for example, Childers and Durling (1975), Oppenheim and Schafer (1975), Oppenheim *et al.* (1999) and Rabiner and Gold (1975). Also, sound and vibration engineers may find some useful concepts in White and Hammond (2004) together with some other advanced topics in signal processing.

The reason for including this subject is because we have used some digital filtering techniques through various MATLAB examples, and also introduced some basic concepts in Chapter 6 when we discussed a digital LTI system, i.e. the input–output relationship for a digital system that can be expressed by

$$y(n) = -\sum_{k=1}^{N} a_k y(n-k) + \sum_{r=0}^{M} b_r x(n-r) \qquad (\text{H.1})$$

where $x(n)$ denotes an input sequence and $y(n)$ the output sequence. This difference equation is the general form of a digital filter which can easily be programmed to produce an output sequence for a given input sequence. The $z$-transform may be used to solve this equation and to find the transfer function which is given by

$$H(z) = \frac{Y(z)}{X(z)} = \frac{\displaystyle\sum_{r=0}^{M} b_r z^{-r}}{1 + \displaystyle\sum_{k=1}^{N} a_k z^{-k}} \qquad (\text{H.2})$$

By appropriate choice of the coefficients $a_k$ and $b_r$ and the orders $N$ and $M$, the characteristics of $H(z)$ can be adjusted to some desired form. Note that, since we are using a finite word length in the computation, the coefficients cannot be represented exactly. This will introduce some arithmetic round-off error.

*Fundamentals of Signal Processing for Sound and Vibration Engineers*
K. Shin and J. K. Hammond.    © 2008 John Wiley & Sons, Ltd

In the above equations, if at least one of coefficients $a_k$ is not zero the filter is said to be recursive, while it is non-recursive if all the coefficients $a_k$ are zero. If the filter has a finite memory then it is called an FIR (Finite Impulse Response) filter, i.e. the impulse response sequence has a finite length. Conversely, an IIR (Infinite Impulse Response) filter has an infinite memory. Note that the terms 'recursive' and 'non-recursive' do not refer to whether the memory is finite or infinite, but describe how the filter is realized. However, in general, the usual implementation is that FIR filters are non-recursive and IIR filters recursive.

There are many methods of designing both types of filters. A popular procedure for designing IIR digital filters is the discretization of some well-known analogue filters. One of the methods of discretization is the 'impulse-invariant' method that creates a filter such that its impulse response sequence matches the impulse response function of the corresponding analogue filter (see Figure 5.6 for mapping from the $s$-plane to $z$-plane). It is simple and easy to understand, but high-pass and band-stop filters cannot be designed by this method. It also suffers from aliasing problems. Another discretization method, probably more widely used, is the 'bilinear mapping' method, which avoids aliasing. However, it introduces some frequency distortion (more distortion towards high frequencies) which must be compensated for (the technique for the compensation is called 'prewarping').

FIR filters are often preferably used since they are always stable and have linear phase characteristics (i.e. no phase distortion). The main disadvantage compared with IIR filters is that the number of filter coefficients must be large enough to achieve adequate cut-off characteristics. There are three basic methods to design FIR filters: the window method, the frequency sampling method and the optimal filter design method. The window method designs a digital filter in the form of a Fourier series which is then truncated. The truncation introduces distortion in the frequency domain which can be reduced by modifying the Fourier coefficients using windowing techniques. The frequency sampling method specifies the filter in terms of $H(k)$, where $H(k)$ is DFT$[h(n)]$. This method is particularly attractive when designing narrow-band frequency-selective filters. The principle of optimal filter design is to minimize the mean square error between the desired filter characteristic and the transfer function of the filter.

Finally, we note that IIR filters introduce phase distortion. This is an inevitable consequence of their structure. However, if the measured data can be stored, then 'zero-phase' filtering can be achieved by using the concept of 'reverse time'. This is done by filtering the data 'forward' and then 'backward' with the same filter as shown in Figure H.1.

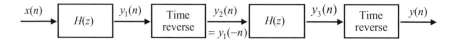

**Figure H.1**   Zero-phase digital filtering

The basic point of this scheme is that the reverse time processing of data 'undoes' the delays of forward time processing. This zero-phase filtering is a simple and effective procedure, though there is one thing to note: namely, the 'starting transients' at each end of the data.

# References

Ables, J. G., 'Maximum entropy spectral analysis', *Astronomy and Astrophysics Supplement Series*, Vol. 15, pp. 383–393, 1974.

Allemang, R. J. and Brown, D. L., 'Experimental modal analysis', Chapter 21, in *Harris' Shock and Vibration Handbook*, Fifth Edition, ed. Harris, C. M. and Piersol, A. G., McGraw-Hill, 2002.

Bencroft, J. C., 'Introduction to matched filters', *CREWES Research Report*, Vol. 14, pp. 1–14, 2002.

Bendat, J. S., 'Solutions for the multiple input/output problem', *Journal of Sound and Vibration*, Vol. 44, No. 3, pp. 311–325, 1976a.

Bendat, J. S., 'System identification from multiple input/output data', *Journal of Sound and Vibration*, Vol. 49, No. 3, pp. 293–208, 1976b.

Bendat, J. S., 'Statistical errors in measurement of coherence functions and input/output quantities', *Journal of Sound and Vibration*, Vol. 59, No. 3, pp. 405–421, 1978.

Bendat, J. S. and Piersol, A. G., *Engineering Applications of Correlation and Spectral Analysis*, John Wiley & Sons, Inc., 1980.

Bendat, J. S. and Piersol, A. G., *Random Data: Analysis and Measurement Procedures*, Third Edition, a Wiley-Interscience, 2000.

Bingham, C., Godfrey, M. and Tukey, J. W., 'Modern techniques of power spectrum estimation', *IEEE Transactions on Audio and Electroacoustics*, Vol. AU-15, No. 2, pp. 56–66, 1967.

Bogert, B. P., Healey, M. J. R. and Tukey, J. W., 'The quefrency analysis of time series for echoes: cepstrum, pseudo-autocovariance, cross-cepstrum and saphe cracking', *Proceedings of the Symposium on Time Series Analysis*, ed. Rosenblatt, Murray, pp. 209–243, John Wiley and Sons, Inc., 1963.

Brigham, E. O., *The Fast Fourier Transform and Its Applications*, Prentice Hall, 1988.

Carter, G. C., Knapp, C. H. and Nuttall, A. H., 'Estimation of the magnitude-squared coherence function via overlapped fast Fourier transform processing', *IEEE Transactions on Audio and Electroacoustics*, Vol. AU-21, No. 4, pp. 337–344, 1973.

Childers, D. and Durling, A., *Digital Filtering and Signal Processing*, West Publishing, 1975.

Cohen, L., 'Time-frequency distributions – a review', *Proceedings of the IEEE*, Vol. 77, No. 7, pp. 941–981, 1989.

Cooley, J. W. and Tukey J. W., 'An algorithm for the machine computation of complex Fourier series', *Mathematics of Computation*, Vol. 19, pp. 297–301, 1965.

Davies, P., 'A recursive approach to Prony parameter estimation', *Journal of Sound and Vibration*, Vol. 89, No. 4, pp. 571–583, 1983.

de Prony, B. G. R., 'Essai éxperimental et analytique: sur les lois de la dilatabilité de fluides élastique et sur celles de la force expansive de la vapeur de l'alkool, à différentes temperatures', *Journal de l'École Polytechnique*, Vol. 1, cahier 22, pp. 24–76, 1795.

Duhamel, P. and Vetterli, M., 'Fast Fourier transforms: a tutorial review and a state of the art', *Signal Processing*, Vol. 19, pp. 259–299, 1990.

Fahy, F. and Walker, J., *Fundamentals of Noise and Vibration*, Spon Press, 1998.

Gao, Y., Brennan, M. J. and Joseph, P. H., 'A comparison of time delay estimator for the detection of leak noise signal in plastic water distribution pipes', *Journal of Sound and Vibration*, Vol. 292, pp. 552–570, 2006.

Hammond, J. K. and White, P. R., 'The analysis of non-stationary signals using time-frequency methods', *Journal of Sound and Vibration*, Vol. 190, No. 3, pp. 419–447, 1996.

Harris, F. J., 'On the use of windows for harmonic analysis with the discrete Fourier transform', *Proceedings of the IEEE*, Vol. 66, No. 1, pp. 51–83, 1978.

Hsu, Hwei P., *Fourier Analysis*, Revised Edition, Simon & Schuster, 1970.

Jenkins, G. M. and Watts, D. G., *Spectral Analysis and its Applications*, Holden-Day, 1968.

Kay, S. M. and Marple Jr, S. L., 'Spectrum analysis – a modern perspective', *Proceedings of the IEEE*, Vol. 69, No. 11, pp. 1380–1419, 1981.

Lee, Y.-S., 'Active control of smart structures using distributed piezoelectric Transducers', PhD Thesis, Institute of Sound and Vibration Research, University of Southampton, 2000.

Leuridan, J., De Vis, D., Van der Auweraer, H. and Lembregts, F., 'A comparison of some frequency response function measurement techniques', *Proceedings of the 4th International Modal Analysis Conference IMAC, Los Angeles, CA*, pp. 908–918, 1986.

Levi, E. C., 'Complex-curve fitting', *IRE Transactions on Automatic Control*, Vol. AC/4, pp. 37–43, 1959.

Marple Jr, S. L., *Digital Spectral Analysis with Applications*, Prentice Hall, 1987.

Newland, D. E., *An Introduction to Random Vibrations and Spectral Analysis*, Longman Scientific & Technical, 1984.

Oppenheim, A. V. and Schafer, R. W., *Digital Signal Processing*, Prentice Hall International, 1975.

Oppenheim, A. V., Willsky, A. S. and Hamid Nawab, S., *Signals & Systems*, Second Edition, Prentice Hall International, 1997.

Oppenheim, A. V., Schafer, R. W. and Buck, J. R., *Discrete-Time Signal Processing*, Second Edition, Prentice Hall International, 1999.

Otnes, R. K. and Enochson, L., *Applied Time Series Analysis Vol. 1. Basic Techniques*, John Wiley and Sons, Inc., 1978.

Otte, D., Fyfe, K., Sas, P. and Leuridan, J., 'Use of principal component analysis for dominant noise source identification', *Proceedings of the Institution of Mechanical Engineers, International Conference: Advances in the Control and Refinement of Vehicle Noise*, C21/88, pp. 123–132, 1988.

Papoulis, A., *Signal Analysis*, McGraw-Hill, 1977.

Papoulis, A., *Probability, Random Variables, and Stochastic Processes*, McGraw-Hill, 1991.

Priestley, M. B., *Spectral Analysis and Time Series*, Academic Press, 1981.

Proakis, J. G. and Manolakis, D. G., *Introduction to Digital Signal Processing*, Macmillan, 1988.

Rabiner, L. R. and Gold, B., *Theory and Applications of Digital Signal Processing*, Prentice Hall, 1975.

Randall, R. B., *Frequency Analysis*, Third Edition, Bruel and Kjaer, 1987.

Schmidt, H., 'Resolution bias errors in spectral density, frequency response and coherence function measurement, III: application to second-order systems (white noise excitation)', *Journal of Sound and Vibration*, Vol. 101, No. 3, pp. 377–404, 1985a.

Schmidt, H., 'Resolution bias errors in spectral density, frequency response and coherence function measurement, IV: time delay bias errors', *Journal of Sound and Vibration*, Vol. 101, No. 3, pp. 405–412, 1985b.

Smith, J. O., *Mathematics of the Discrete Fourier Transform (DFT)*, http://ccrma.stanford.edu/~jos/mdft/, 2003.

Spitznogle, F. R. and Quazi, A. H., 'Representation and analysis of time-limited signals using a complex exponential algorithm', *Journal of the Acoustical Society of America*, Vol. 47, No. 5(1), pp. 1150–1155, 1970.

Sutton, T. J., Elliot, S. J., McDonald, A. M. and Saunders, T. J., 'Active control of road noise inside vehicles', *Noise Control Engineering Journal*, Vol. 42, No. 4, pp. 137–147, 1994.

Tan, M. H., 'Principal component analysis for signal-based system identification', PhD Thesis, Institute of Sound and Vibration Research, University of Southampton, 2005.

Welch, P. D., 'The use of fast Fourier transform for the estimation of power spectra: a method based on time averaging over short, modified periodograms', *IEEE Transactions on Audio and Electroacoustics*, Vol. AU-15, No. 2, pp. 70–73, 1967.

Wellstead, P. E., 'Non-parametric methods of system identification', *Automatica*, Vol. 17, pp. 55–69, 1981.

White, P. R. and Hammond, J. K., 'Signal processing techniques', Chapter 1, in *Advanced Applications in Acoustics, Noise and Vibration*, ed. Fahy, F. J. and Walker, J. G., Spon Press, 2004.

White, P. R., Tan, M. H. and Hammond, J. K., 'Analysis of the maximum likelihood, total least squares and principal component approaches for frequency response function estimation', *Journal of Sound and Vibration*, Vol. 290, pp. 676–689, 2006.

Wicks, A. L. and Vold, H., 'The $H_s$ frequency response function estimator', *Proceedings of the 4th International Modal Analysis Conference, IMAC, Los Angeles, CA*, pp. 897–899, 1986.

Zadeh, L. A. and Desoer, C. A., *Linear System Theory: The State Space Approach*, McGraw-Hill, 1963.

# Index

*Fundamentals of Signal Processing for Sound and Vibration Engineers*
K. Shin and J. K. Hammond.   © 2008 John Wiley & Sons, Ltd

Printed and bound by CPI Group (UK) Ltd, Croydon, CR0 4YY

17/04/2025

14658865-0001